THE GREAT ICE AGE

The Great Ice Age is also our age. The four and a half billion year evolution of the Earth and its biota culminated in the appearance of a life form capable of studying and manipulating the planetary systems. With human consciousness focusing as never before on the world around us, we are now beginning to realise the enormity of the global experiment we are conducting as the 'conscious forcing function' in the evolution of the Earth System.

The Great Ice Age documents and explains the natural climatic and palaeoecologic changes that have occurred during the past 2.6 million years, outlining the emergence and global impact of our species during this period. Exploring a wide range of records of climate change, the authors demonstrate the interconnectivity of the components of the Earth's climate system, show how the evidence for such change is obtained, and explain some of the problems in collecting and dating proxy climate data.

One of the most dramatic aspects of humanity's rise is that it coincided with the beginnings of major environmental changes and a mass extinction that has the pace, and maybe magnitude, of those in the far-off past that stemmed from climatic, geological and occasionally extraterrestrial events. This book reveals that anthropogenic effects on the world are not merely modern matters but date back perhaps a million years or more.

R.C.L. Wilson is Professor of Earth Sciences and **S.A. Drury** is a Lecturer in Earth Sciences, both at the Open University, Milton Keynes; **J.L. Chapman** is Honorary Research Associate of the Sedgwick Museum, Department of Earth Sciences, Cambridge University, UK.

Published in association with the Open University.

THE GREAT ICE AGE

Climate Change and Life

R.C.L. Wilson, S.A. Drury and J.L. Chapman

London and New York

First published 2000
by Routledge
11 New Fetter Lane, London EC4P 4EE

Simultaneously published in the USA and Canada
by Routledge
29 West 35th Street, New York, NY 10001

Routledge is an imprint of the Taylor & Francis Group

Typeset in Garamond by J&L Composition Ltd, Filey, North Yorkshire
Printed and Bound in Great Britain by St Edmundsbury Press,
Bury St Edmunds, Suffolk

British Library Cataloguing in Publication Data
A catalogue record for this book is available from the British Library

Library of Congress Cataloging in Publication Data
Wilson, R.C.L.
The great ice age: climate change & life / R.C.L. Wilson, S.A.
Drury, and J.L. Chapman
p.   cm.
Includes bibliographical references.
1. Climatic changes. 2. Glacial epoch. 3. Paleoecology—
Quaternary. 4. Nature—Effect of human beings on. I. Drury, S.
A. (Stephen A.), 1946–   . II. Chapman, J.L. (Jenny L.)
III. Title.
QC981.8.C5W55   1999
551.6′09′01—dc21   99–18443

ISBN 0–415–19841–0 (hbk)
ISBN 0–415–19842–9 (pbk)

CONTENTS

FIGURES

BOXES

TABLES

PREFACE

This book documents the dramatic climatic changes that occurred during the last 2.6 million years of the Earth's history, and the ways in which these affected life and the evolution of humanity.

The Great Ice Age is also our age. The four and a half billion year evolution of the Earth and its biota culminated with the appearance of a life form capable of studying and manipulating environmental systems. That we are now a significant factor that contributes to global change – the 'conscious forcing function' in the evolution of the Earth System – hardly needs stating when the effects of human activities are so obvious in the world around us. We are beginning to realise the enormity of the global experiment we are conducting. This book is not an attempt to describe the impact our industrialised society has on planetary systems. Our purpose is to document and explain natural climatic and ecological changes that occurred during the past 2.6 million years, and to outline the emergence and early global impact of our species during this period.

The embryonic phase of this book began in the womb of an Open University Course Team preparing a new course entitled *Earth and Life* presented for the first time in 1997. The course examines the co-evolution of our planet and life upon it. This text was to be the last of five books, but by the time we had delivered our overlength second draft, realisation that the entire course was too long had dawned, and so 'The Great Ice Age' was unfortunately jettisoned. By then we had learnt so much (none of us are Quaternary climate change specialists) that we decided to complete the project.

We have two overarching aims. The first is to relate the history of climate change in the context of the interconnectivity between the different components of the Earth's climate system in the style and at the level adopted by the new *Earth and Life* course. This is addressed in Part 1. Our text is written on the assumption that its readers have a reasonable understanding of the workings of the Earth's climate system, including atmospheric and oceanic circulation and the greenhouse effect. We have, however, provided some revision material in Chapter 2. The second aim is to complement the geoscience-based explanation of climate change and Earth systems with a discussion of how life on land responded to climate change. This includes the evolution of humans, and the ways in which we have modified our planet in such a dramatic way in an exceedingly short period of geological time. This is the rationale behind Part 2, which provides an overview of human evolution, though not providing the same level of evidence and argument included in Part 1.

Following the style of the *Earth and Life* course, we have deliberately avoided citing in the text all the sources of information and ideas we have summarised. At the end of each chapter suggestions for further reading are given, and the list of figure sources at the end of the book provides more bibliographic information.

We are indebted to many people for help they have given during the gestation of our text but stress that responsibility for errors and omissions rests with the authors. Special thanks are due to the *Earth and Life* course team, especially Angela

Colling, Peter Francis, Charles Turner and John Wright, for their comments on various drafts. We thank Angela Colling for letting us include part of the course text entitled *The Dynamic Earth* as the first part of Chapter 2. The constructive comments of our external reviewers Dr Mark Maslin (Part 1) and Professor Bernard Wood (Part 2) are much appreciated as are those provided by Dr Cynthia Burek from her perspective as an Open University Associate Lecturer. Comments by Professor David Bowen on a very early draft ensured that the interconnectivity between components of the Earth's climate system are addressed more clearly. Thanks are due also to Giles Clark of the OU's Publishing Division for finding us an enthusiastic publisher, and for Sarah Lloyd of Routledge for taking the plunge and agreeing to publish a book written by non-specialists. Last but not least, we thank Joan Gomes, Jann Matela, Jo Morris and Denise Swann for their patience and diligence in providing secretarial support.

Chris Wilson
Steve Drury
Jenny Chapman
December 1998

PART 1

CYCLES OF CLIMATIC CHANGE: EVIDENCE AND EXPLANATIONS

The Ice Age is not a single age at all, but a series of cold periods separated by times when the climate was as warm, or even much warmer than it is today. Glacial periods alternated with interglacial: but 'Hot and Cold Ages' does not have the same drama about it as 'Ice Age'. It is known that even upon the greater cycles there were smaller cold–warm climatic fluctuations, described as stadials and interstadials. The historian focuses his microscope on ever finer detail, as more precise methods of analysis are invested; he discovers more, and racks the focus up yet again, which suggests further questions and so on.

It is not so long ago that the influence of ice upon our scenery was denied by many reputable observers: other explanations were in currency for the scratches on Scottish boulders, or what we would now recognise as fossil moraines, or scraped and deepened valleys. The clues that allowed recent description of the sequence of events through the Pleistocene were derived from deep-sea cores of sediment. These were obtained many miles away from the direct influence of ice, from oozes underlying quiet ocean wastes. This is where change could be cheated. A steady and unbroken rain of shells of tiny foraminiferans falling to the sea floor were different from those in the warm phases; there are also changes to the isotopes in the crystals making up the shells. I grew up with five cycles of glaciation named in the textbooks, all of which were named after tributaries of the Danube . . . but thanks to these new oceanic data the number of cycles is now known to have been many more. Every year, it seems that more wiggles are added to the temperature curve for the last two million years as the investigator focuses more and more clearly upon the past.

Richard Forte. 1993. The Hidden Landscape, Jonathan Cape, London.

Signposting

Much of Part 1 is a detective story, discussing the clues – proxy climate indicators – that palaeoclimatologists use to reconstruct past changes and how they attempt to explain them. Chapter 1 contrasts the Earth today with what parts of it were like when ice sheets last reached their maximum extent 18 000 years ago, and introduces some examples of how past climates can be interpreted from geological clues. Chapter 2 gives a general overview of the Earth's climate system, and the problems faced by the climate detective in collecting evidence and ascertaining the timing of changes. As the most spectacular manifestation of global climate change was the growth and demise of ice sheets, Chapter

3 explains how they accumulate and move and introduces the astronomical pacemaker that controlled the timing of glaciations and deglaciations. Chapters 4, 5 and 6 review the array of evidence from oceanic and terrestrial realms that show that climate change occurred over a range of time scales from millions of years, through tens of thousands of years, to more rapid changes that occurred over centuries or decades. From time to time, we take short excursions into possible explanations, but as with all good detective stories the attempt to solve the crime is left to the end. Here, in Chapter 7, you will find a number of explanations for climate change, none of which adequately explains *all* the evidence documented in earlier chapters. The challenge of formulating an all embracing model that explains all the clues that have been amassed drives palaeoclimatologists to seek more knowledge of the past. Their work also illuminates the possible courses of future climate change, for if models can explain the past, our confidence in their ability to make predictions about future climate trends will be increased. This is the only way we can evaluate the possible consequences of the global environmental experiment that humanity is conducting.

Note: The abbreviations 'Ma' and 'ka' are used in this book to refer to the dates of past events. Ma: million years ago; ka: thousand years ago.

1

THE GREAT ICE AGE

INTRODUCTION

This book examines the Earth's history over a very short period of time – geologically speaking that is. It focuses on the climatic history since 2.6 million years ago during which time large ice sheets repeatedly advanced and retreated across significant areas of the Northern Hemisphere. Parts of Antarctica had been ice-covered for much longer (since almost 50 million years ago), but the establishment of Northern Hemisphere ice sheets heralded a major change in the Earth's climate system. The Great Ice Age was not the first time that the Earth has been covered by extensive ice sheets: they also developed several times during the Pre-Cambrian, and during the Ordovician and Permian-Carboniferous.

We have used the nineteenth century concept of 'The Great Ice Age' in the title of this book to describe the period during which major glaciers and ice sheets regularly advanced and retreated across the Northern Hemisphere. This 2.6 million year time interval spans the Pleistocene and Holocene Epochs of the Quaternary Period (Figure 1.1).

If we compress Earth history into one year, with the planet forming at one minute past midnight on 1 January, then The Great Ice Age began at about 7 pm on 31 December. Why is this short period of geological time so important? Because studying it helps us understand our past and our future. It is a unique period of time in the Earth's history for a number of reasons:

- it is the only time when both poles were/are covered by ice sheets

- it is characterised by repeated, and relatively regular, patterns of climate change over time scales ranging from hundreds of thousands (10^5) to less than 100 (10^2) years;
- these changes can be studied to a much greater degree of resolution than for any other period of geological time;
- humans evolved during this period of climatic change and occupied every continent, and virtually every type of environment, be it hot or cold, or wet or dry;
- much of our natural heritage of landforms and wildlife is a relic of the last glacial period that ended some 20 thousand years ago;
- the Great Ice Age is not over yet, so understanding the past may help us predict future climatic and ecological changes.

The last point about predictions is worth exploring a little further in the context of forecasting the weather and longer term climatic trends. Daily changes in the weather are a familiar fact of life. Likewise, we expect seasonal changes in temperature and precipitation. We have made some progress in predicting the weather, but even with the aid of super computers, our ability to make accurate forecasts only extends to a week or two ahead. This is because weather systems exhibit chaotic behaviour. In contrast, seasonal changes are triggered by latitudinal variations in solar insolation caused by the fact that the Earth's axis of rotation is inclined with respect to its orbital plane around the sun. This is a very simple model in which there is a broadly linear relationship between

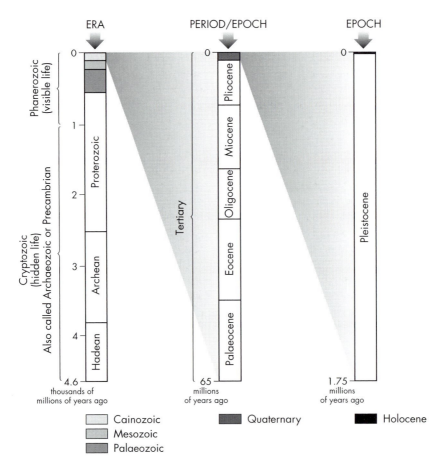

Figure 1.1 Some of the major divisions of geological time. The 'Great Ice Age' spans the Pleistocene and Holocene epochs; the Holocene began only 10 000 years ago.

radiation received by each of the Earth's hemispheres and seasonal changes in temperature.

Climate is 'averaged weather' (usually defined over a period of 30 years) – the average (as determined from historical meteorological measurements) temperature, pressure, wind speed and variability, precipitation and evaporation, plus seasonal variations over a given area of the Earth's surface (Box 1.1).

Box 1.1 Climate and Life

On a totally water-covered Earth, there would be a simple latitudinal distribution of climate belts (Figure 1.2). A hot equatorial zone with heavy rainfall would straddle the equator, and be flanked by subtropical highs – areas with high evaporation and low rainfall. At higher latitudes in both hemi-

spheres heavy precipitation would occur where warm tropical air masses and cold polar ones would meet along the polar front along which low pressure zones would form. Low temperatures and precipitation, and high pressure, would characterise the polar regions. The presence of large continental areas modifies this simple pattern. This is not only because land areas have different thermal properties than oceans, but also because they significantly modify the direction of movement of atmospheric and oceanic currents.

None the less elements of the four basic climate zones – tropical, arid, temperate and polar – can be seen in any climate map of the world. More than four climate zones, however, are required to depict the full range of climate variability over the continents. The problem of depicting on maps a combination of variables such as temperature, precipitation, evaporation, wind, and their seasonal variation, is usually overcome by incorporating, to varying degrees, the distribution of biomes. Biomes are assemblages of similar ecosystems, such as tropical rain forest, desert, or temperate forest. Their distribution reflects the strong relationship between climate and life: similar

environments, today and in the past, lead to the evolution of plants and animals that have similar morphologies and functions. Figure 1.3 shows the relationship between two key climatic variables (temperature and precipitation) and different biomes.

Even maps that show tropical, dry, warm temperate, cool temperate and cold climate zones have to resort to some degree to using biome distributions in order to subdivide some major zones (e.g. tropical climate is subdivided into equatorial forest, monsoon forest and savannah zones). Figure 1.4 shows the potential global distribution of biomes today: it ignores the effects of human activities (see Figure 1.21), such as agriculture and forest clearance. Elements of the simple latitudinal distribution of climatic belts can be seen on Figure 1.4, as can the effects of oceanic circulation patterns on climate.

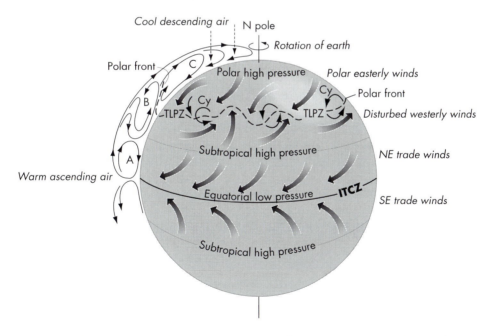

Figure 1.2 Idealised planetary atmospheric circulation on a totally water covered Earth, showing three main cells (A, B, C) in the Northern Hemisphere (the same pattern is mirrored in the Southern Hemisphere). Latitudinal variation in solar heating causes convection currents which are deflected longitudinally by the effect of the Earth's rotation (known as the Coriolis force). *Key*: TLPZ: temperate low pressure zone; Cy: cyclone; ITCZ: inter tropical convergence zone.

☐ Why do different types of forest grow at the same latitudes on either side of the Atlantic (e.g. in the UK and Newfoundland)?

■ NW Europe is warmed by the North Atlantic Drift (a warm surface current flowing to the north-east across the northern Atlantic), so that the distribution of climatic zones and biomes is shifted significantly northwards.

Take a moment to consider, using Figures 1.3 and 1.4, what would happen to the distribution of biomes if the intensity of the North Atlantic Drift diminished, and what would happen if global mean temperatures dropped by 5°C, or rose by 1°C. As we shall see, all these changes have occurred in the past, and significantly altered the distribution of life on the Earth.

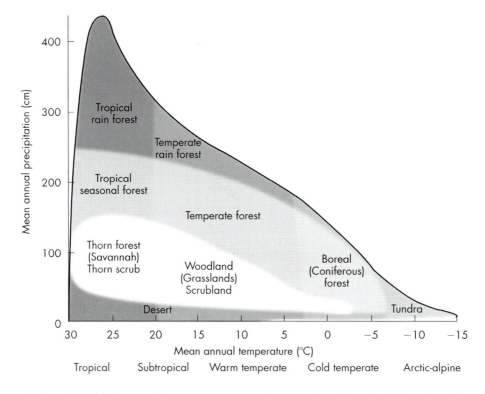

Figure 1.3 The relationship between biomes, temperature and precipitation. Note that the boundaries between different biome fields are approximate. The white area includes a wide range of environments in which either grassland or one of the biomes dominated by woody plants may form the dominant vegetation in different areas.

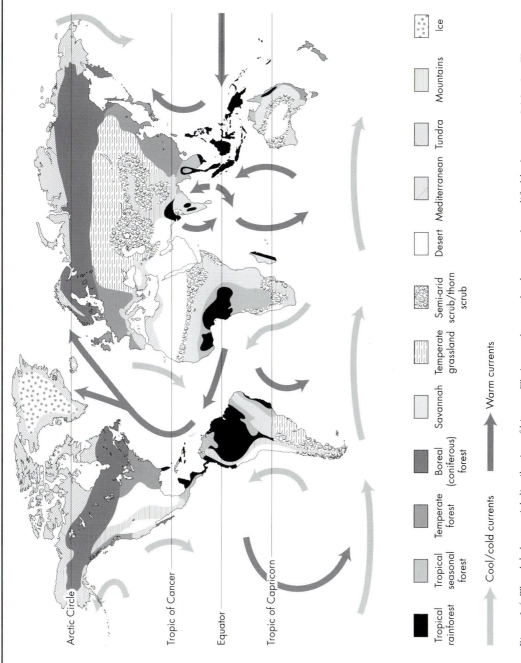

Key (left to right):

Tropical rainforest | Tropical seasonal forest | Temperate forest | Boreal (coniferous) forest | Savannah | Temperate grassland | Semi-arid scrub/thorn scrub

Desert | Mediterranean | Tundra | Mountains | Ice

→ Cool/cold currents   → Warm currents

Arctic Circle
Tropic of Cancer
Equator
Tropic of Capricorn

Figure 1.4 The global potential distribution of biomes. Their actual extent has been greatly modified by human activities (see Figure 1.21).

Climate variability describes differences between averages of the same kind, such as the differences between two summers in the same region. Climate change involves longer timescales – longer than that used to define the climate of an area. Palaeoclimatology is the study of major climate changes over long periods of time, based on a relatively limited data set. The reverse is true for modern climatology, which has plenty of data to draw on but only over very short historical periods.

The results originating from the models used to produce weather forecasts can be tested against what actually happens a few days or weeks later, but climate prediction involves timescales of tens to thousands of years. So how can climate models be tested over very long timescales, other than by leaving future generations to evaluate their accuracy? The answer is by determining whether they can success-fully explain what happened in the past (i.e. be used to 'retrodict' rather than predict the future). In other words, if climate models can simulate today's conditions, and those in the past reconstructed from geological evidence, then our confidence in their predictive capability will be increased.

We should not delude ourselves that studies of the geological record of the past 2.6 million years were motivated by the need to make predictions about the future. They were driven by simple curiosity – a desire to understand the past. Only recently have we realised that this 'pure science' approach enables us to construct models that may have a bearing on our future.

The rest of this chapter provides a broad-brush picture of the awesome extent of changes in climate that have beset the planet since 2.6 Ma. We focus mainly on the contrast between two climatic extremes: the

Figure 1.5(a) Ice-smoothed and striated rock surface and perched erratic block exposed after the Morterasch glacier in SE Switzerland retreated (ice covered this area late in the nineteenth century).

coldest conditions during the maximum advance of ice sheets, and the warm interglacial conditions that we enjoy today. The principal purpose of the rest of this chapter is scene setting, although some key concepts and examples of fossil climatic indicators are introduced.

CLIMATIC DETECTIVE WORK

Reconstructing past climates and a timetable of climatic change is akin to detective work. It requires a painstaking search for clues and the ability to piece them together in a logical and consistent way. Fortunately the Earth's climate system in the past did not commit the perfect crime, for it left many clues behind. Some of these are obvious, and examples of them are described below, whereas others are only revealed by sophisticated analytical techniques, some of which are described later in this book.

Most people living in North West and Central Europe and northern North America are never far from evidence indicating that rapid climatic changes have occurred during the past few hundred thousand years. In New York's Central Park, smooth rock surfaces traversed by parallel striations are testament to the former presence of a thick ice sheet. Similar features can be observed in many parts of North America and NW Europe. The base of this ice sheet was armoured with rock fragments so that as the ice moved, the underlying rock was ground to a smooth surface – a process akin to using sandpaper to smooth wood. As larger rock fragments were dragged across the rock surface, they cut the parallel striations (Figure 1.5).

Many highland regions in the world bear the scars

Figure 1.5(b) Ice-smoothed and striated rock surface overlain by boulder clay (or till) showing wide range of sizes of pebbles 'floating' in clay. These features were produced as an ice sheet advanced and retreated during the last glaciation over Red Wharf Bay, Anglesey (North Wales). The pen indicates the trend of the ice flow.

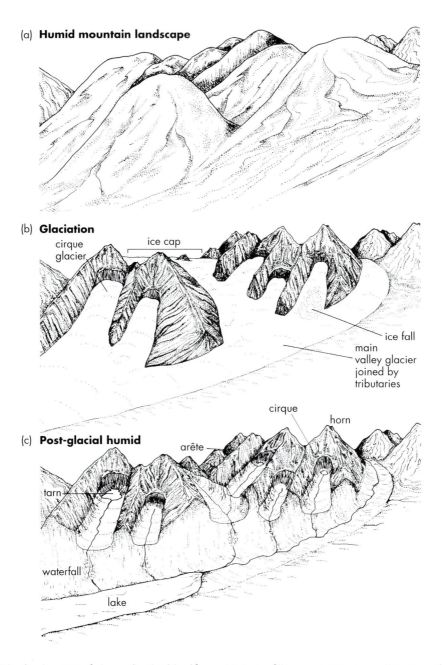

Figure 1.6 The development of glaciated upland landforms. (a) As ice fills pre-existing river valleys, it modifies their shape by erosion. Rock debris incorporated into the ice makes it extremely abrasive, like the sand on sandpaper. (b) As the ice flows down the valleys, it changes their former V-shaped cross section profiles to U-shapes, and produces characteristic landforms shown in the middle sketch. (c) When the ice melts, a new landscape emerges, carrying the unmistakable features of glacial erosion, such as U-shaped valleys, bowl-shaped depressions known as cirques, corries (in Scotland) or cwms (in Wales), and hanging valleys.

of erosion by ice though today they are ice free. This changed the cross-sectional V-shaped profile of river valleys to the characteristic U-shapes cut by glaciers (Figure 1.6).

A variety of features can be used to determine ice flow trends and directions. These include the orientation of drumlins (Figure 1.7) which are elongated dome-shaped hills sculptured beneath moving ice in rock or glacial debris.

Although the most spectacular evidence for glaciation occurs in mountainous areas, the history of repeated glaciation can best be worked out in lowland areas where sediments deposited directly from ice, or from meltwater, occur (Box 1.2). Mapping the distribution of sediments and topographic features formed beneath and in front of ice as it melts enables the limits of former ice advances to be mapped.

So far we have focused on the clues left behind by advancing and retreating glaciers and ice sheets, but records of warmer conditions may also be preserved, a notable example of which was discovered beneath the Trafalgar Square area of London. In 1957, excavations for the foundations of Uganda House uncovered fossil

bones that included the extinct straight tusked elephant, two extinct species of rhinoceros, hippopotamus and cave lion (Figure 1.9). This assemblage of large mammals might suggest a past sub-tropical climate, but the fossil beetles and spores and pollen of plants found with them indicated that average summer temperatures were only 2–3°C higher than those experienced in London today. The fossils and sediments in which they were found were deposited during the last interglacial episode that occurred between approximately 130 ka and 117 ka. So why were such animals present in what is now the London area during the last interglacial yet are absent there today? The reason is that during the present interglacial human hunting and agricultural activities prevented their return – it is not because the present interglacial is significantly cooler than previous ones.

In many parts of the world, landscapes, fossils and sediments tell the same story of repeated advances and retreats of ice during successive cold and warm intervals. Each advance and retreat was accompanied by corresponding falls and rises of global sea-level as water was alternatively withdrawn from and released to the oceans. The last rise in sea level since 18 ka

Figure 1.7(a)

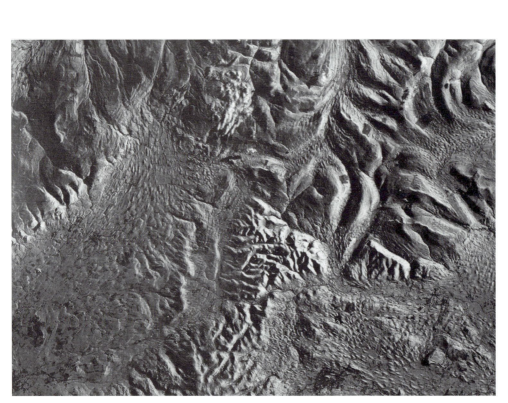

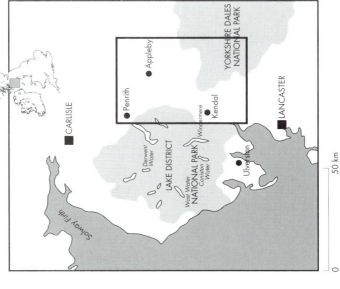

Figure 1.7 Topographic features oriented parallel to former ice flow directions. Ice flow trends in northern England. (a) A swarm of drumlins at Ribblehead, Yorkshire. Their elongate shape indicates the trend of the flow of the ice beneath which they formed. The steeper ends face downstream indicating flow from left to right (i.e., west to east). (b) Satellite view of the English Lake District and adjacent areas of the Pennines. The alignment of the long axes of drumlins enables ice flow lines to be interpreted (e.g. NNE–SSW in SE corner, and swinging around from E–W to NE–SE in NW corner).

Box 1.2 Evidence of Past Ice Limits and Movement

Moving ice transports anything from fine, clay-sized particles to large boulders. Some particles are plucked from underlying bedrock at the base of the ice and others may fall from the sides of mountains remaining above it. Moving ice armoured with rock debris is a very powerful agent of erosion, leaving behind characteristic highland landforms when it melts (Figure 1.6).

Deposition takes place where debris is released from the ice, either at its contact with the underlying rocks, or where it terminates. Sediment deposited directly from ice is called boulder clay or till (Figure 1.5(b)).

Figure 1.8 shows the features of an ice sheet, and the depositional features remaining after it melted. Boulder clay may form sheets that blanket pre-existing landscapes, but in some areas it is moulded beneath moving ice into clusters of small, low elliptical mounds called drumlins (Fig-

ure 1.7(a)). Piles of till (Figure 1.5(b)) accumulate at the edges of ice sheets and glaciers to form terminal and lateral moraines. These can be used to map the maximum advance of past ice sheets but determining their age is problematic, because they usually contain no fossils or other material that can be dated. Sometimes their age can be inferred by dating water lain sediments occurring above and below them. Sediments of glacial origin that are deposited from glacial meltwater streams are known as glacio-fluvial sediments. They consist of a mixture of sand and gravel. Sediments deposited in lakes associated with other glacial features (i.e. ice dammed lakes, and in kettle holes which are hollows formed when the last remnants of ice encased in morainic material melted) often include muds within which spores and pollen are preserved, which can be used to determine the age of the deposits (see Chapter 8).

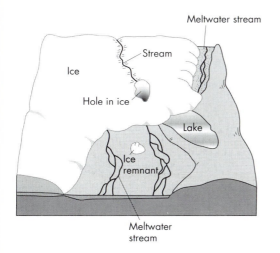

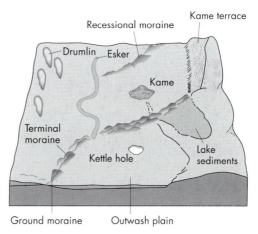

Figure 1.8 Sketches of an ice sheet before and after it melted, showing the depositional features that may form beneath and at the edges of it.

Figure 1.9 Trafalgar Square between about 130 ka–120 ka during the period that preceded the last glacial advance. The present day buildings are shown in the background.

drowned continental shelf areas around the world, resulting in the separation of the British Isles from the rest of Europe, and Alaska from Russia. Changes of sea-level that are world-wide in extent are referred to as eustatic, with ice-related ones being described as glacioeustatic.

The last ice retreat has left behind a legacy of crustal uplift and subsidence that is still continuing.

Box 1.3 Isostasy and Ice

In contrast to Earth movements caused by mountain building, vertical movements caused by loading or unloading the Earth's crust are termed isostatic uplift or subsidence. The growth of a large volcano, or a large ice-cap (Figure 1.10) causes the underlying crust to sink. This subsidence continues until a mass of mantle material equivalent to the load on the crust has been displaced. The displacement of the mantle material results in a crustal bulge forming around the loaded area.

The rate at which the crust and mantle respond to loading is much slower than the build-up or melting of ice caps. This is why areas that were buried beneath several kilometres of ice 18 ka are still rising today, thousands of years after the ice sheet melted away (100 m of ice loading depresses continental crust by 27 m). This time-lag between the disappearance of ice sheets and crustal unloading (and vice versa) is an important feature of the Earth's climate system. The rise of sea-level consequent on the melting of ice caps loads crust beneath the oceans and under shallow seas at the edges of continents.

□ Will a 100 m water loading of the crust depress it more or less than the same thickness of ice?

■ More, because water is denser than ice. In fact, a 100 m rise or fall in sea-level results in 30 m of isostatic compensation. This means that although the amount of water trapped in ice caps during the last glaciation was equivalent to removing a 165 m layer of water from the oceans, the net drop in sea-level around the world was only 115 m. This was because the water loading caused the crust beneath the oceans to drop by ~50 m.

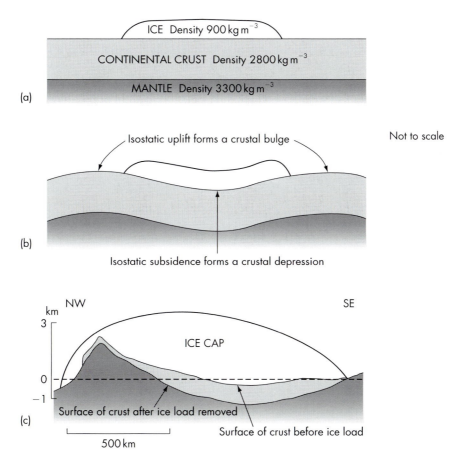

Figure 1.10 Ice and isostasy. Because the Earth's crust 'floats' on denser mantle (a), ice loading causes a crustal depression which displaces the mantle, forming a crustal bulge beyond the ice (b). (c) A cross section across the present day Greenland ice sheet, showing the crustal depression beneath.

This is because the accumulation of thick ice sheets depresses the Earth's crust beneath.

□ So what happens when the ice load is removed by melting?

■ Isostatic compensation results in crustal uplift (see Box 1.3).

As we shall see later, the slow response of the crust to ice-loading and unloading is an important factor that models of the global climate system must take into account. Scandinavia and Canada have only been uplifted by half the amount needed to compensate for the loading of the ice sheets that once covered them – they still have several hundred metres to go over the next few thousand years.

As can be seen in Figure 1.11(a), isostatic compensation for the removal of ice over northern Britain continues today, with the maximum rate of uplift occurring in the Highlands of Scotland – where the ice sheet was thickest. This has resulted in former beach deposits being left high and dry as raised beaches. Subsidence in southern Britain results from the collapse of a crustal bulge that developed around the area depressed by ice-loading. The legacy of crustal subsidence left by former thick ice-sheets affects many parts of the world (Figure 1.11(b)), caus-ing a relative sea-level rise that threatens many pop-ulous coastal areas.

We have sampled some of the lines of evidence that indicate the nature and extent of past changes in climate during the Great Ice Age. We now need to take a global perspective, and examine just how different planetary environments were at the end of the last glacial period compared to today's interglacial conditions.

A GLOBAL PERSPECTIVE

If we travelled back in time and toured the Earth as it was at the last glacial maximum 18 thousand years ago, or during many similar previous maxima, we would hardly recognise our planet: it would appear to be an alien world (Figure 1.12(b)).

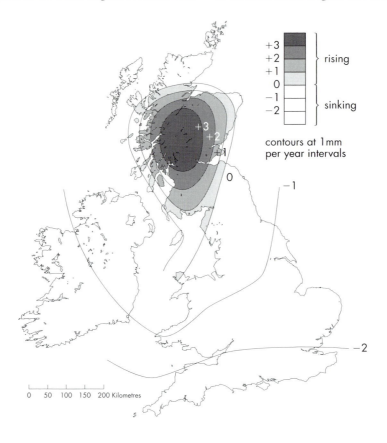

Figure 1.11(a)

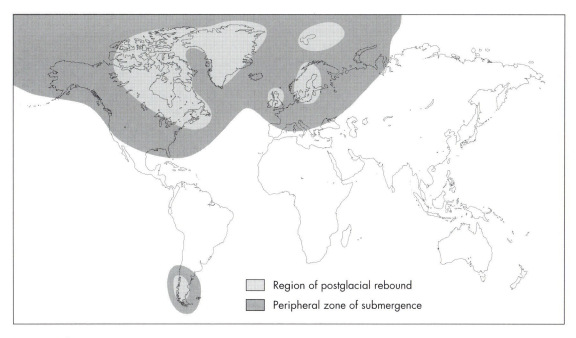

Region of postglacial rebound

Peripheral zone of submergence

Figure 1.11(b)

Figure 1.11 Ice loading and crustal movements. (a) Rates of crustal uplift and subsidence around the British Isles today, due to release of ice-loading, and the collapse of the crustal bulge around the loaded area. (b) The global distribution of areas of uplift and subsidence due to the former presence of thick ice sheets. The emergent areas lay beneath the thickest parts of former ice sheets.

Many parts of the Northern Hemisphere that are densely populated today were frozen wastes: glaciers reached New York and Berlin, and the ground beneath London and Paris was permanently frozen. Ice sheets over 3 km thick covered much of North America and Scandinavia. Winds blowing down from the ice sheets reached speeds of 300 km per hour. Much of Western Europe was largely treeless with a tundra-like landscape. Much of the northern part of the North Atlantic Ocean was frozen although the Norwegian Sea remained seasonally ice free. Sea ice extended down to the latitude of the British Isles, and floating ice and large icebergs, even during the summer months, would have been common between Ireland and Newfoundland (Box 1.4), and would occasionally have been seen off the Portuguese coast.

There are two major conclusions that can be drawn from data shown in Box 1.4. The first is that at 18 ka the oceanic polar front (i.e., the boundary between cold polar surface waters and warmer sub-tropical waters) had moved down to a roughly west–east position, stretching from Portugal across to Newfoundland (Figure 1.16, see p. 23). Studies of younger sediments have enabled the position of the oceanic polar front to be mapped at successive younger time intervals.

The second conclusion is that warm-water currents of the Gulf Stream and North Atlantic Drift were not operating to warm the north-east Atlantic as they do today. Consequently, this area shows a very strong temperature difference between glacial and inter-glacial conditions compared with other areas. This is shown very clearly on maps showing temperature differences between 18 ka and today (Figure 1.12). This shows how inaccurate it is to speak of a general shift of so many degrees in temperature between glacial and interglacial conditions. Temperature changes were modified everywhere by local circumstances, particularly by changes in ocean circulation.

The global average change for surface-water temperatures was in the order of 2–4°C, but, in some areas, such as the north-eastern Atlantic, the downward shift in temperature may have been as much as 10°C. In parts of the tropics CLIMAP data (see caption for Figure 1.15 on p. 22) suggest that it was almost as warm during the ice age maxima as it is today but more recently results indicate that this is a conservative view of temperature changes in such areas, particularly as there is evidence that some high tropical mountains were glaciated. It is clear that the temperature differences in the North Atlantic between today and 18 ka are unusually large, suggesting that changes in circulation patterns here may be linked to the growth of Northern Hemisphere ice sheets.

It was possible to walk from England to France 18 thousand years ago, apart from needing a boat to cross rivers that flowed across what is now the English Channel. Global sea-levels were over 100 m lower than today, for huge amounts of water were frozen into ice sheets. Alaska and Siberia were united, as were Australia and Papua New Guinea. Only a few hundred kilometres of ocean separated Australia and Asia – a crossing achieved by humans at 40 ka (or possibly even earlier) when sea-levels were only a little higher than they were at 18 ka.

Had air travel existed at 18 ka it would have been a

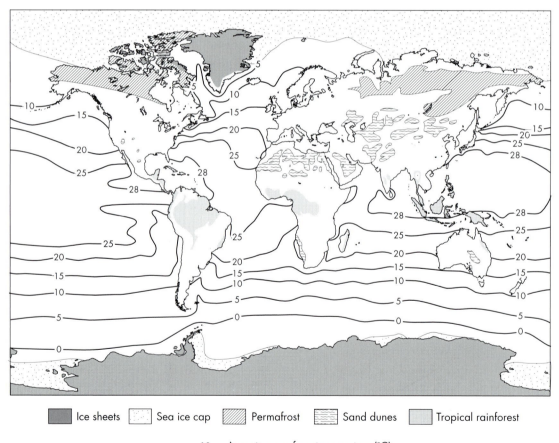

Ice sheets   Sea ice cap   Permafrost   Sand dunes   Tropical rainforest

—10— August sea-surface temperature (°C)

Figure 1.12(a)

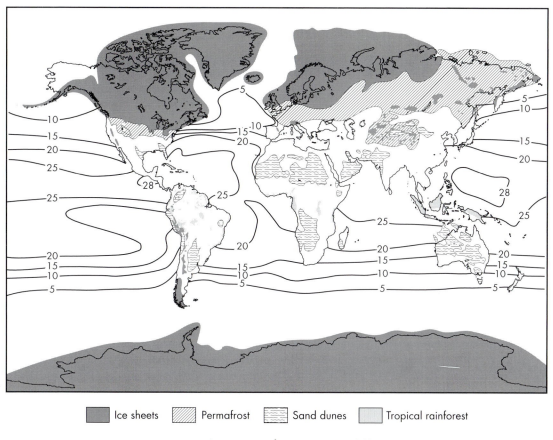

Figure 1.12(b)

Figure 1.12 The Earth today (a) and at 18 ka (b). On land the distribution of ice sheets, sea ice, permanently frozen ground (permafrost), active desert sand dunes, and tropical rainforests is shown. Compare with Figure 1.2. In the oceans, the summer (August) surface temperatures today and at 18 ka is plotted (i.e. see Box 1.4 for an explanation of how past sea-surface temperatures are determined). Note that the contraction of equatorial rainforests into 'refugia' shown in (b) is highly controversial.

much riskier mode of travel than it is today. This is because the glacial world was much windier and dustier. Atmospheric dust, whipped up from sediments deposited at the margins of ice sheets and from deserts, would have quickly etched windows and damaged engines. Uncomfortable journeys due to air turbulence would have been more common. Once at your holiday destination in the Mediterranean or the Canary Islands, your stay on the beach would have been frequently interrupted by dust storms, and you would have had to brace yourself to swim in seas several degrees colder than their equivalents today. Perhaps the tropical latitudes would have offered more amenable havens for time travellers – or our beleaguered ancestors? Only in some places, for desert areas were much more extensive than those of today, and much more sand and dust was being blown around them, and into regions beyond them. But they would have been even more spectacular places to visit, with sand dunes being much larger. This was because wind

Box 1.4 Foraminifera and Past Sea-surface Temperatures

Foraminifera ('forams' for short) are microscopic single celled organisms that secrete shells composed of calcium carbonate. Different species live at or near the sea surface (i.e. planktonic forams) or on the sea bottom (i.e. benthonic forams) even in the deepest oceans.

The planktonic species *Neogloboquadrina pachyderma* lives in high latitude oceans, and occurs in two varieties that coil in different directions. Today, the variety that coils to the right lives in ice-free water, whereas the left-coiling form is able to live in waters in which sea ice is common. In fact, in the Greenland and Labrador seas, the left-coiling variety forms almost 100 per cent of the total population of planktonic forams. These little forams are very effective palaeoenvironmental indicators. Their distribution in cores of ocean floor sediments shows that in the Northern Atlantic Ocean there were periods when the sea surface was ice free, and others when it was largely ice covered (Figure 1.13 (b)).

Foraminiferal studies not only enable past ice cover conditions of oceans to be determined, but enable sea-surface temperature maps to be constructed. This is because studies of assemblages of species of modern forams show that their distribution in the world's oceans is strongly temperature dependent. For example, of the sixteen most common North Atlantic planktonic foraminifera, one is largely limited to high latitudes, five flourish in middle latitudes, five in both middle and low latitudes, and a further five are predominantly low latitude. This enables assemblages of foraminifera to be identified that characterise polar, subpolar, subtropical and tropical waters today, and in cores of ocean floor sediments. Studies of many cores from the Atlantic enable a north–south profile to be drawn summarising change through time in the assemblages of planktonic foraminifera (Figure 1.14). This shows that whereas little or no change in temperature occurred south of Spain, high latitudes were

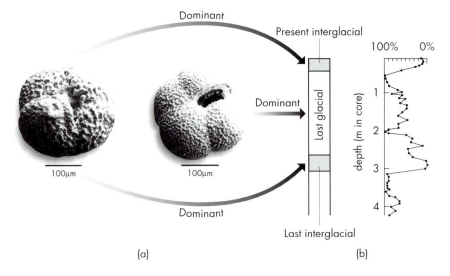

Figure 1.13 The coiling habit of the foraminifera *Neogloboquadrina pachyderma* and sea-surface temperatures. (a) Electron microscope pictures of the two forms. (b) Diagram summarising the intervals in a core from the Norwegian Sea in which the left and right-coiling forms dominate. (c) The frequency of the left-coiling form as a percentage of total planktonic forams in a core from the Mid-North Atlantic.

affected by incursions of polar waters during the last (75–15 thousand years ago) and penultimate (>125 thousand years ago) glacial periods.

The data on the modern composition of planktonic faunal and floral assemblages and the physical environments in which they live have been analysed by multivariate statistical techniques and used to build a mathematical model of this part of the marine ecosystem. A feature of such a model is that, given faunal assemblage data from a present day sea floor site, it should be able to predict the likely temperature and salinity conditions there; these predictions can be checked against actual observations in order to check the accuracy of the model. It is a logical step to analyse data on fossil assemblages in older horizons in the cores in the same way in order to determine past sea-surface temperatures.

□   What critical assumption is made in doing this?

■   The assumption is that the estimates used to model a modern ecosystem also apply to an ecosystem in, say, glacial times. This approach is often referred to as the principle of uniformitarianism (i.e. the present is the key to the past) and is commonly applied to the interpretation of past environments.

This method of using a mathematical model of one situation to interpret data from another is known as the application of transfer functions. Transfer functions are increasingly important in the field of palaeoecology, but their operation and evaluation have to be left to competent mathematicians!

By applying transfer functions to the planktonic assemblage data it is possible to estimate, by studying the abundance of species that flourish seasonally, the seasonal temperatures of the waters in which they were living. These estimates can be mapped as surface-water isotherms for February and August, 18 ka (Figure 15 (a) and (b)). Compare the temperature distribution then with present-day August conditions (Figure 15(c)).

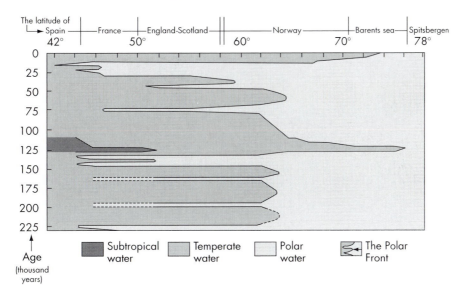

Figure 1.14 The changing distribution of planktonic foraminiferal assemblages in the North Atlantic during the past 225 thousand years. During cold periods, latitudinal surface water temperature gradients were enhanced in mid-latitudes.

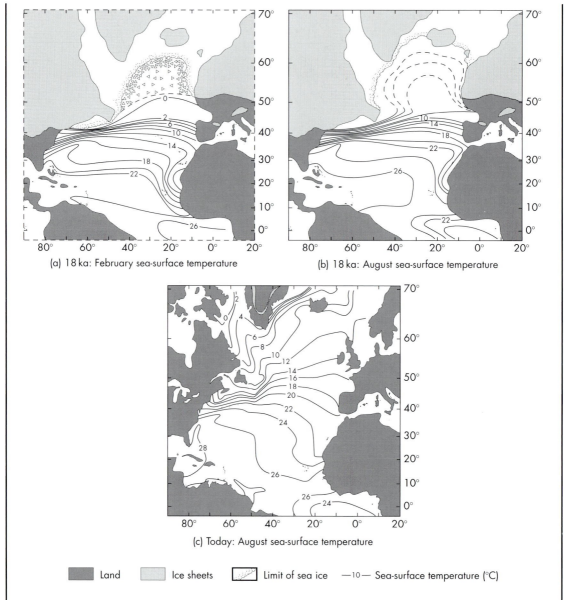

(a) 18 ka: February sea-surface temperature

(b) 18 ka: August sea-surface temperature

(c) Today: August sea-surface temperature

Land   Ice sheets   Limit of sea ice   —10— Sea-surface temperature (°C)

Figure 1.15 Prevailing sea-surface temperatures in the North Atlantic in February and August at 18 ka (a and b), and today in August (c). These maps were published by the CLIMAP (Climate: Long-range: Investigation Mapping and Prediction) Project in 1981. More recent work suggests sea ice retreated in the Norwegian Sea allowing warmer waters to reach this area during summers.

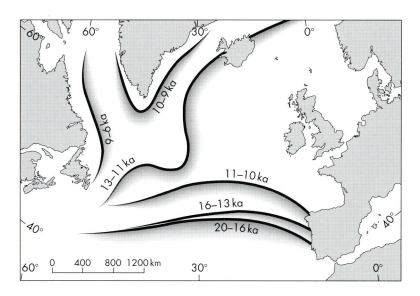

Figure 1.16 Changes in the position of the oceanic polar front (the boundary between relatively cold and warm surface waters) between 20 ka and 9 ka.

speeds were much higher for longer periods of time. Areas of active desert dunes today represent the relatively minor reworking of much larger dunes formed some 18 thousand years ago (Figure 1.17). This is because today wind speeds only reach velocities high enough to transport sand for a few days during a year, whereas during the last glacial maximum speeds were probably maintained for months at a time.

Perhaps what is even more surprising about the glacial world at 18 ka is that areas of tropical rain forests may have contracted to relatively small isolated patches dotted across South America, Central Africa and the Far East. The evidence for this distribution – which is still very controversial – is described in Chapter 8.

So far, our discussion of climate change has only compared today with the last glacial maximum at 18 ka. But are conditions we enjoy today typical of interglacial conditions? Not entirely, because there is a variety of evidence that suggests that global mean temperatures were about 1°C warmer at 4–8 ka. This temperature rise was sufficient to increase evaporation rates, and intensify monsoonal rains in tropical areas (Box 1.5). This resulted in most present day desert areas receiving significantly more rainfall, so that per-

manent lakes were developed, and large mammals were able to feed on plants growing on sand dunes formed during the last glacial maximum (Box 1.5). This warmer period between 8–4 ka is termed the Holocene thermal maximum. Understanding the climate during this time, and the accompanying global and regional shifts in the distribution of plants and animals, helps to predict what the consequences might be of global warming caused by the increase in atmospheric CO_2 due to burning fossil fuels.

COLDER AND DRIER, WARMER AND WETTER

We have now reviewed some of the evidence from low latitude areas that shows that significant changes in climate occurred in tandem with the advance and retreat of polar ice sheets. The expansion of northern and southern ice sheets was not just accompanied by the narrowing of climatic belts between them. Glacial and interglacial stages were characterised by markedly different planetary conditions. The planet was colder and drier during the former, and warmer and wetter during the latter.

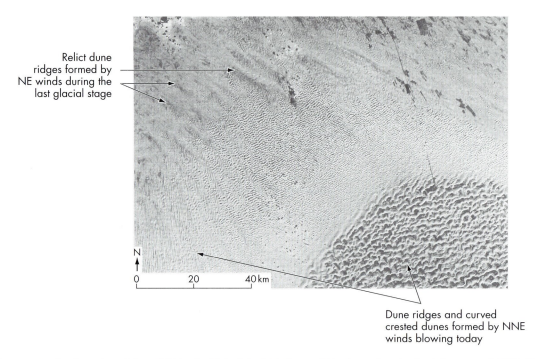

Relict dune
ridges formed by
NE winds during the
last glacial stage

N

0   20   40 km

Dune ridges and curved
crested dunes formed by NNE
winds blowing today

Figure 1.17 Satellite view of dune field in the United Arab Emirates. Older larger longitudinal dunes trend NW–SE, parallel to the dominant wind direction during the last glacial. They are being eroded and replaced by smaller transverse (trending WSW–ENE) and longitudinal dunes (trending NNW–SSE) formed by the present day NNW wind direction blowing from the Arabian Gulf.

Box 1.5 Greening the Sahara

Between about 9 ka and 6 ka, much of what is now extremely hot almost lifeless Saharan desert was covered by grass and open woodland. Antelopes, rhinos, giraffes, elephants and lions were at least as abundant as they are now in game parks in East Africa. They could drink from permanent freshwater lakes in which crocodiles lurked and hippos wallowed. Our ancestors thrived here too, hunting and fishing in a region that is now only visited by nomads and explorers (Figure 1.18). Wetter conditions were also characteristic of earlier interglacial episodes.

The map and description of a humid Sahara are based on fossil evidence, most of which is obtained from fossil lake sediments. You may think that long lived lakes equate with reasonably

high rainfall, but this is only partly true. The fossil lakes in the Sahara are almost all situated above permeable strata. If rain falls, and does not evaporate, it will immediately drain away into the permeable rock, unless the precipitation is so heavy that temporary drainage by surface run-off in streams and rivers occurs. Lakes will only form if the water-table (the top of the zone within the underlying rock that is saturated with groundwater) rises above the ground level within a saucer shaped topographic basin. The level of the water-table is partially controlled by the amount of rainfall, but also by the amount of evaporation. So the occurrence of fossil lakes in the Sahara indicates higher rainfall *and* lower evaporation rates compared to those experienced at the present time.

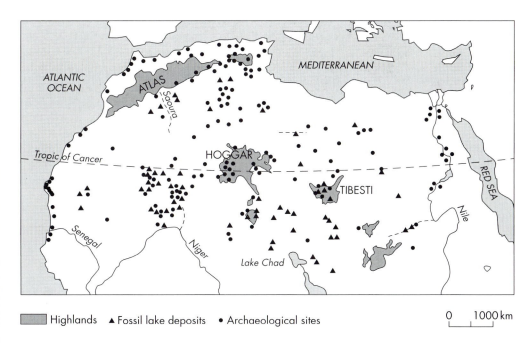

Figure 1.18 The occurrence of fossil lake deposits and archaeological sites indicating human settlement in the Sahara, dated between 10 ka and 2 ka years in age.

Higher water-table levels enable plant roots to reach water more easily, either by reaching the water-table itself, or the damp rock or soil immediately above it.

Fossil lake sediments found in present day desert regions may contain a variety of sediment types: windblown sand, carbonate and sulphate minerals produced during evaporation, fossil plant material and fish, freshwater shells, and diatomite. The occurrences of fossils and diatomite are particularly important, as they enable the former presence of freshwater to be identified. Diatomite is a sediment composed of the siliceous shells of diatoms (a type of planktonic unicellular algae), one group of which lives in freshwater, and another in salty or brackish water. The two types are morphologically distinct, and only the freshwater variety occurs in Saharan fossil lake sediments.

☐ Why should a colder world be drier than a warmer one?

■ Because lower temperatures reduce the amount of evaporation from the oceans, resulting in lower rainfall over land areas.

An important cause of climatic change at low latitudes is change in the strength of monsoonal winds and the rainfall that accompanies them. Monsoon winds result from the seasonal control that large mid-latitude air masses exert on atmospheric circulation so that the idealised system of wind belts across the planet (Figure 1.2) is significantly modified. During the winter, cold denser air develops over large land-masses resulting in an outward flow, resulting in clear weather that persists for months. In summer, as the land area heats up, the situation is reversed (Figure 1.19). A hot, low pressure air mass develops, and winds blow inwards from surrounding oceans, bringing with them moist air and rain. In Asia, the monsoon is accentuated by the high Tibetan plateau, and so the summer rainfall, brought by southerly winds from the Indian Ocean is much heavier than it is in areas further west. Africa is subjected to monsoonal winds too. During the summer, warm moist winds blow from the southwest from the Atlantic, and southeast from the Indian Ocean, usually bringing much needed rainfall to the Sahara, the Horn of Africa and Arabian Peninsula.

The summer monsoon must have been much stronger 6 thousand years ago, bringing more rainfall to the Saharan region. It was very weak at 18 ka ago, so that then this region was even drier than it is today. Figure 1.20 shows the changing limits of monsoonal rainfall at 6 ka and 18 ka compared with that of today. The southern limit of active sand desert today coincides approximately with the 100 mm annual rainfall contour: this marks the northern limit to today's summer monsoonal rainfall. At the last glacial maximum, it was located 300–400 km to the south, and during the Holocene climatic maximum (6–9 ka) it was almost 1000 km north of its present position. During the Holocene maximum, rainfall may have exceeded the present day annual amounts

over the Sahara by between 100 and 400 mm, with temperatures being up to 2°C higher. Increased summer cloudiness would have reduced evaporation rates so that aquifers were charged to higher levels, resulting in lake formation. Temperatures may have been as much as 8°C lower, at 18 ka, but annual rainfall was much less (perhaps 500 mm less), particularly on the southern borders of the Sahara (the Sahel).

As we shall see in succeeding chapters, there have been 49 cold–warm oscillations since 2.6 Ma, and for the last one million years these have been of the same magnitude as that described for the last glacial–interglacial cycle. These relatively short-lived climatic extremes, however, were not typical of conditions during the last one million years. For example, in Europe and North America periglacial conditions prevailed for about three-quarters of the time.

Climatic change caused major extinctions, particularly at the onset of major Northern Hemisphere glaciations at 2.6 Ma. Life on the planet responded by migrating across continents or retreating to highland refuges. In contrast humans were able to adapt to climatic change and to colonise most of the Earth by the end of the last glacial. Not long after the distribution

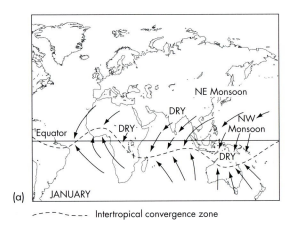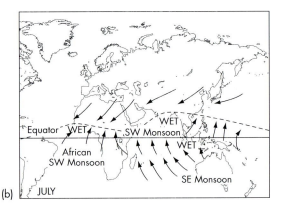

- - - - - - - - - Intertropical convergence zone

Figure 1.19 The seasonal changes in monsoonal winds. In January (a), Eurasia is very cold, resulting in cold dense dry air flowing southwards to the warmer oceans. During the summer (b), the continental area is very hot, causing air to rise, drawing northwards cooler moist air from over the oceans to the south. This results in heavy rainfall over Asia, and to a lesser extent, Africa. The pattern of northeasterly (winter) and southeasterly (summer) monsoonal winds is the result of the Coriolis force modifying the outward and inward flow direct from continental and oceanic areas respectively.

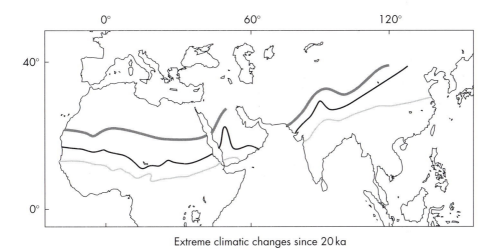

Extreme climatic changes since 20 ka

Monsoon precipitation range: ── Present ── Last glacial maximum ▰▰ Holocene optimum

Figure 1.20 Changes in the limits of monsoonal rains during the Holocene climatic optimum and the Last Glacial Maximum.

of biomes had adjusted to the last global warming into the present interglacial, agricultural practices began to be developed in the Middle East at about 11 ka. During a single interglacial period, our species freed itself from the self-regulation that had limited the size of populations of all species for hundreds of millions of years, so that today there are very few areas of the world where human activity does not dominate the environment (Figure 1.21).

SUMMARY

1 In this book, 'The Great Ice Age' is used to describe the last 2.6 million years, during which time major glaciers and ice sheets regularly advanced and retreated in the Northern Hemisphere.

2 The Great Ice Age is a unique period of time in Earth history because it is characterised by repeated episodes of major climate change that can be studied to a much greater degree of resolution than in any other period of geological time.

3 Studying the Great Ice Age helps us understand our past and contemplate our future because (i) humans evolved during this time, and colonised every continent, and came to dominate planetary ecology, and (ii) the glacial–interglacial cycles will continue. Understanding the past, therefore, may help us predict future climatic and ecological changes.

4 Climate is 'averaged weather', that is the average temperature, pressure, wind speed and variability, precipitation, evaporation, plus seasonal variations of these parameters, over a given area of the Earth's surface.

5 Biomes are distinctive types of plant assemblages, the distribution of which is largely controlled by climate. Climate maps of the world use biome distributions to varying degrees to overcome the problem of depicting on a single map the variables listed in 4 above.

6 Global mean temperatures during successive glacial maxima over the past 800 thousand years were about 5°C lower than today's value (15°C). Temperatures during interglacial stages (including the present one) were 1–3°C higher than today.

7 During glacial stages, ice sheets advanced to cover large areas of the Northern Hemisphere,

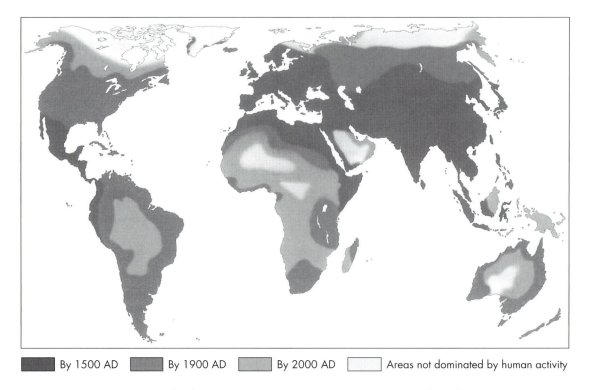

By 1500 AD   By 1900 AD   By 2000 AD   Areas not dominated by human activity

Figure 1.21 Human dominance of global environments. The map shows how recently and rapidly human activities have controlled the ecology of most of the land area of the Earth.

reaching New York, Berlin and London. They have left characteristic landscapes in both highland (e.g., U-shaped valleys) and lowland (e.g., drumlins) areas, and a legacy of crustal uplift or subsidence due to the loading effect of thick ice. Water locked in continental ice sheets resulted in lower global sea-levels.

8 Palaeoecological studies of planktonic foraminifera enable past sea-surface temperatures to be determined at the last glacial maximum. The results show that the largest temperature differences between sea-surface temperatures today and those at 18 ka occur in the North Atlantic. Here, the position of the oceanic polar front (where colder polar waters meet warmer subtropical waters) has been traced at different time intervals. It extended between Northern Spain and Newfoundland until 13 ka.

9 The glacial Earth was not only colder, but drier and windier, and hence dustier. Lower temperatures reduced evaporation rates over the oceans, and the larger thermal gradient between the Equator and the poles resulted in wind speeds much higher than those of today. Desert areas were more extensive, and sand dunes within them larger than those forming today.

10 Between 8 and 4 ka, an increase in global mean temperature of about 1°C increased evaporation rates. The resultant intensification of monsoonal rains almost eliminated sub-tropical deserts.

FURTHER READING

Anderson, B.G., and Borns, H.W. 1994. *The Ice Age World.* Scandinavian University Press, Oslo. This expensive book contains some excellent colour illustrations of glacial processes and features and of the glacial/interglacial history of the Earth (including palaeogeographic maps) over the past few million years.

Lowe, J.J., and Walker, M.J.C. 1997. *Reconstructing Quaternary Environments* (2nd edition). Longman, London. A good follow up to *The Great Ice Age*, with chapters reviewing the geomorphological, lithological and biological records of environmental change, and relevant dating methods and stratigraphy.

Van Andel, T.H. 1994. *New Views on an Old Planet: a History of Global Change* (2nd edition). Cambridge University Press, Cambridge. A wonderfully readable book that documents global changes that have occurred on Earth since its formation. It introduces the key concepts discussed in Part 1 of *The Great* *Ice Age*, but does not treat most of them at the same level of detail.

Williams, M.A.J., Dunkerley, D.L., De Deckker, P., Kershaw, A.P., and Stokes, P. 1993. *Quaternary Environments*. Arnold, London. The content of this book overlaps that of *The Great Ice Age*, but its approach is completely different. As its title suggests, its focus is on environments rather than reviewing different timescales of climate change and focusing on the interlinked nature of components of the climate system as attempted in Part 1 of *The Great Ice Age*. *Quaternary Environments* also covers many of the themes examined in Part 2 of *The Great Ice Age*.

See also references to figure sources for this chapter (pp. 253–4).

2

UNDERSTANDING PRESENT AND PAST CLIMATES

INTRODUCTION

'Climate' is a term used across a range of spatial scales. It can be used to describe the 'average weather' for a given region over decades or more, or be used to refer to conditions across the entire planet. The nature of the Earth's climate at both the regional and global scale is the result of an *interconnected* system driven by the uneven heating of the planet by solar radiation. This system transfers heat from low to high latitudes, and involves all of the Earth's spheres: the atmosphere, hydrosphere (mainly the oceans), cryosphere, biosphere, and lithosphere (Figure 2.1). It tends to keep the Earth's radiation budget in balance globally so that incoming radiation is equalled by outgoing radiation. This balance was not maintained

during periods when the Earth cooled and warmed into and out of glacial episodes.

From the descriptions given in the previous chapter, it is clear that the Earth's climate system must have behaved differently during the last glacial episode (and for that matter during all previous glacials) to how it behaves during the interglacial conditions we enjoy today. The fact that both interglacial and glacial conditions persisted for thousands of years suggests not only that the climate system can exist in at least two equilibrium states, but also that one or more factors may have caused it to change states. That it did not experience runaway cooling or warming indicates that feedback mechanisms must operate to limit the amount of change.

Although this book is written on the assumption

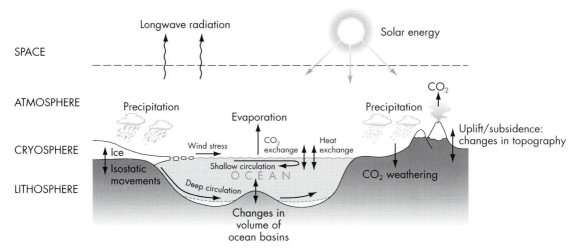

Figure 2.1 Diagram showing the major components of the Earth's climate system, and the links between them.

that its readers already have a sound understanding of the workings of the climate system today, the first section of this chapter gives a brief review of the key components of the system that will figure in later chapters. We then examine how forcing factors may change the climate and how feedback mechanisms may both reinforce and limit the effects of such factors. The spatial and time scales over which climatic changes and forcing factors operate are then reviewed. Finally, we discuss the kinds of evidence available to study climates of the past.

MOVING HEAT AROUND

The average amount of solar energy received at the top of the atmosphere is ~343 W m^{-2}: this is the effective solar flux. Only a proportion of this energy is retained by the Earth because some of it is either reflected or radiated back into space. It is important to remember that the process of radiation is completely different from reflection, in which energy is not absorbed and re-radiated, but 'bounced-off' with its frequency and wavelength unaffected (as with light reflected from a mirror). Absorption of shorter wavelength energy by gases (particularly water vapour, carbon dioxide (CO_2) methane (CH_4) and nitrous oxides) in the atmosphere and the Earth's surface and its re-radiation as longer wavelength radiation is a key factor in the greenhouse effect. This results from the absorption of outgoing longwave radiation and the subsequent re-radiation, much of it back towards the Earth's surface.

The amount of solar radiation that reaches the Earth's surface depends on (i) the angle at which the sun's rays intercept the Earth's surface (they are spread over a larger area as the angle becomes more oblique at higher latitudes), and (ii) the fact that rays travel through greater thicknesses of atmosphere as the angle becomes more oblique. For these reasons, surfaces at high latitude areas receive less solar energy per unit area than those at low latitudes (Figure 2.2, curve A). Not all the effective solar flux is available to heat the atmosphere, oceans and land surface because about 30 per cent of it is reflected back into space,

mainly from the tops of clouds but also from land and sea (Figure 2.2, curves B and C). In other words, the albedo of the Earth as a whole – the percentage of incoming solar radiation that is reflected from it – is about 30 per cent. Thus on average, the Earth's surface (surface of ocean and continents) receives ~240 W m^{-2}.

☐ Examine Figure 2.2: why does the amount of solar radiation reflected by the Earth's surface increase towards the poles.

■ Ice has a very high albedo (up to 90 per cent), and so its presence at high latitudes accounts for much of the increase in the amount of solar radiation reflected at high latitudes. In addition, more solar energy is reflected by ocean surfaces at high latitudes due to the low angle of incidence of the sun's rays.

The amount of solar energy reflected by clouds diminishes at high latitudes because in these regions the atmosphere has a much lower water content (because it is colder) than it does at lower latitudes and so on average is less cloudy.

In Figure 2.2, curves D and E show that there is a net loss at high latitudes, and a net gain at low latitudes. The areas of loss and gain between the two curves are equal as would be expected if the Earth's radiation budget is in balance. This meridional difference in loss and gain of radiation drives the atmospheric and oceanic circulatory systems.

The atmosphere is responsible for transporting about one and a half times as much heat to the poles as the oceans, although the role of the latter becomes dominant in the tropics (Figure 2.3). Atmospheric and oceanic circulation are modified by the effect of the Earth's rotation (the Coriolis force) (Figure 1.2), the configuration of oceans and continents, and their topographic relief. Over long timescales crustal movements may modify both atmospheric and oceanic circulation patterns, resulting in regional or even global changes in climate.

The most obvious manifestation of atmosphere–ocean coupling is the way in which winds drive surface currents. The oceans, however, can drive atmospheric

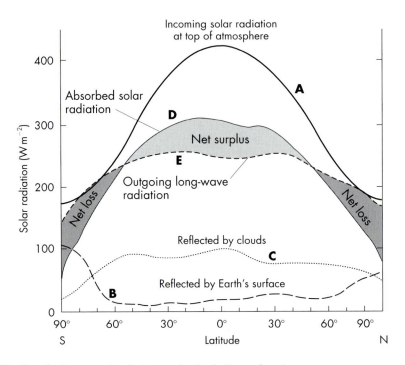

Figure 2.2 The fate of solar energy impinging on the Earth. For explanation, see text.

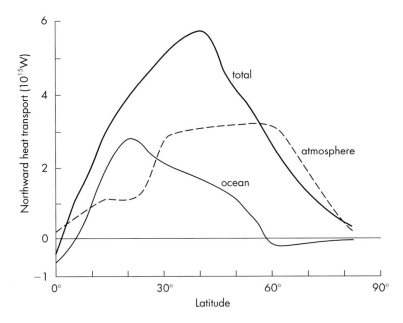

Figure 2.3 Estimates of the contributions to poleward heat transport by the oceans and the atmosphere in the Northern Hemisphere.

circulation. For example cyclones (hurricanes and typhoons) are only generated above ocean waters with a surface temperature in excess of 28°C: once formed, they transport large amounts of heat away from low latitudes. Cyclones and other weather systems (e.g. mid latitude depressions) operate over very short timescales (a few days) but other larger scale atmosphere–ocean interactions occur over longer periods of time – from a few years to hundreds of years.

El Niño events were seldom out of the news in the 1990s (Box 2.1). They occur in the tropical Pacific every three to six years, but their effects are felt in many parts of the world. This is because changes in the atmospheric pressure field over the Pacific cannot occur independently of other pressure systems. The strength of Asian monsoons is affected by El Niño events, drastically reducing rainfall over tropical rainforests which may result in the uncontrollable spread of fires which in normal times would be doused by heavy rains. In contrast, the Pacific coasts of the Americas experience heavy rainfall in seasons that are normally relatively dry.

Similar oscillations between pressure systems occur in the Atlantic and Indian Oceans. For example the North Atlantic Oscillation (NAO) is a change in the pressure system between Iceland (65°N) and the Azores (40°N). This results in alternating severe and mild winters in Labrador, Greenland and NW Europe.

Another type of atmosphere–ocean interaction involves deep circulation in the oceans on a global scale and over long periods of time (hundreds to thousands of years). Such deep circulation has a profound influence on global climates, not only because it transports heat, but also because, as we shall see in the next section and in Chapter 5 (Box 5.2), it affects the CO_2 content of the atmosphere.

Deep circulation is caused by density changes in oceanic water resulting from changes in temperature and salinity, and so is referred to as thermohaline circulation. Cold winds may cool surface water so that it becomes denser than the water beneath it. The salinity of seawater can be changed by adding freshwater (as rain or snow, or from melting ice), or removing freshwater by evaporation or by freezing to ice (sea ice is composed of freshwater, so its formation results in an increase in salinity of the adjacent seawater). The densification of seawater by cooling and freezing to ice contributes to the formation of deep thermohaline circulation (Box 2.2).

A highly schematic representation of the three key components to the global system of thermohaline circulation is shown in Figure 2.6(a). They are:

- The Atlantic Conveyor;
- The Southern Ocean Raceway; and
- The Pacific and Indian Ocean Anticonveyor.

Cold saline southward flowing Arctic bottom currents are augmented in the North Atlantic by surface water cooled by winter winds to form North Atlantic Deep Water (NADW). The water sinking from the surface is delivered by the Gulf Stream, and so is initially quite warm (12–15°C); evaporation and cooling increase its density causing it to sink. It is estimated that in winter about 1300 km^3 of water per day is cooled down to 1–4°C and sinks from the surface of the North Atlantic.

□ If the warm surface water cools by 11°C, how much heat is released every day into the atmosphere? Assume that:

- the specific heat of water is $4.18 \times 10^3 \, \text{J kg}^{-1} \, °\text{C}^{-1}$.
- 1 m^3 of water has a mass of 10^3 kg.

■ The total amount of heat released per day is given by:
specific heat $\times$ mass of water $\times$ temperature drop
$= 4.18 \times 10^3 \, \text{J kg}^{-1} \, °\text{C}^{-1})$
$\times (1300 \times 10^{12} \, \text{kg}) \times 11°\text{C}.$
$= 59\,774 \times 10^{15} \approx 60\,000 \times 10^{15} \, \text{J or } 6 \times 10^{19} \, \text{J}.$

This is equivalent to the heat output of about half a million large power stations.

This enormous amount of heat is four orders of magnitude more than that supplied by solar radiation at ~55°N in winter – no wonder NW Europe has

Box 2.1 El Niño Events

El Niño means the Christ Child, and was originally the name given by fishermen to a warm current, associated with good fish catches, that flows south along the coast of South America around Christmas time. However, the name has come to be associated with a change in the Pacific equatorial current system, first noticeable around the turn of the year, that brings only disaster.

El Niño events occur every three to six years or so, and generally last about a year. The South-East Trade winds become weaker than usual and bursts of westerly winds occur in the western equatorial Pacific. The collapse of the South-East Trades allows a pool of warm water usually situated in the western equatorial Pacific to spread into the central and eastern Pacific, and then polewards along the coasts of North and South America. Regions of vigorous convection and heavy rain, normally over Indonesia, move into the Central Pacific as the ITCZ (Inter Tropical Convergence Zone: see Figures 1.2 and 1.19) moves southwards (Figure 2.4).

Figure 2.4 El Niño events. (a) Conditions in the Pacific during a normal year. In the upper plan-view diagram, the darkest area has the highest sea-surface temperatures; the lighter tone represents high biological productivity due to upwelling. As shown in the cross-section along the Equator, the South-East Trades drive surface waters westward, with the result that the sea-surface slopes up towards the west and the thermocline slopes downwards. The shallowness of the thermocline (the boundary between warm surface water and deeper cooler water) in the eastern tropical Pacific allows water beneath the thermocline to be mixed up to the surface, with the result that there is high biological productivity here. The vertical scale is greatly exaggerated. (b) During an El Niño event, the weakness of the South-East Trade winds, and westerlies in the western Pacific, cause the slopes of the sea-surface and thermocline to collapse, allowing warm water to flow eastwards along the Equator. Any upwelling now occurs from within the surface layer. All El Niño events are slightly different, but the features shown here seem to occur in most events.

El Niño's notoriety is due to its associated severe events that result in the loss of human life and livestock, and the damage to natural ecosystems caused by it. Cyclones occur much further west than usual; regions that are usually dry experience torrential rain which in turn causes mudslides and trigger epidemics; other areas, usually with plentiful rainfall, may be stricken with drought. One important consequence of El Niño is that nutrient-rich water is no longer brought to the surface, or 'upwelled', along the coast of Peru and Chile, or in the vicinity of the Equator in the

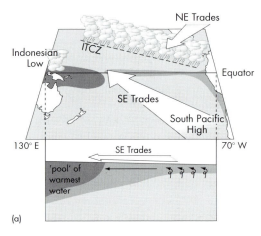

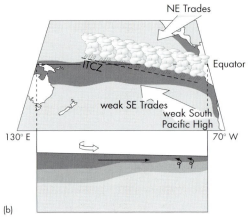

eastern Pacific. Normally, these upwelling regions sustain high primary productivity and hence productive fisheries. During El Niño events, the fisheries crash.

El Niño events are sometimes referred to as **El Niño–Southern Oscillation (ENSO)** events. They occur when the pressure difference between the high pressure region centred in the south-eastern Pacific and the low pressure region usually over Indonesia is at a minimum. This pressure difference continually rises and falls on a time-scale of several years (a phenomenon known as the Southern Oscillation; see Figure 2.5), so it could well be argued that El Niño events are not perturbations of climate at all, but a natural though relatively uncommon state of the atmosphere–ocean system. Indeed, many climatologists now see the climate system of the tropical Pacific as oscillating between extreme states, and have named the periods of particularly large pressure differences and very strong South-East Trade winds 'La Niña' (meaning 'the Girl', as opposed to 'the Boy', the literal translation of El Niño.)

El Niño events are a clear example of the atmosphere and ocean acting as one system: it is not even possible to say whether an El Niño event is initiated in the ocean or in the atmosphere, so closely are the two coupled together. The vigorous

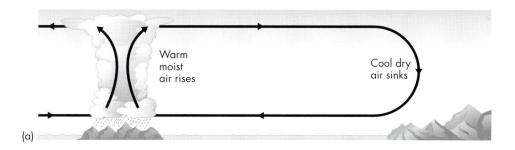

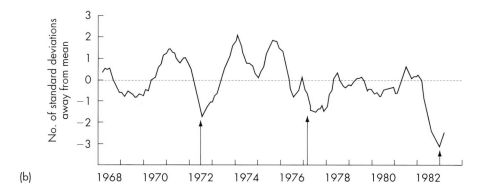

Figure 2.5 El Niño events. (a) Schematic diagram to show the atmospheric circulation between the Indonesian Low and the South Pacific High. The greater the surface pressure difference between the two, the stronger the South-East Trades. (b) The variation of the Southern Oscillation Index between 1967 and 1982. The two meteorological stations used here are Tahiti for the South Pacific High and Darwin, Australia, for the Indonesian Low, and the Index is expressed in terms of the departure from the normal difference in sea-level atmospheric pressure between the two stations divided by the standard deviation for the appropriate month. Negative values mean pressure differences less than usual. The arrows indicate the start of El Niño events in recent times.

convection in the atmosphere in the central and eastern Pacific – the cyclones and the heavy precipitation – are all a consequence of the eastward movement of the 'pool' of warm water usually in the western Pacific, but the eastwards flow of warm water is itself a result of changes occurring in the atmosphere.

mild winters compared to Newfoundland and Labrador!

The lower limb of the Atlantic conveyor joins the Southern Ocean Raceway which circulates around Antarctica. The Raceway is also supplied by dense waters spilling off the continental shelves of Antarctica. These dense waters are formed both by the seasonal formation of sea ice that holds the temperature of surface waters near to freezing point, and the release of salt-rich water resulting from the formation of ice. The blend of dense deep water produced by the giant mixer of the Southern Ocean Raceway flows northwards into the Pacific and Indian Oceans as bottom hugging currents. The northward flow direction of deep denser water is the opposite of that which occurs in the Atlantic, so the term 'anti-conveyor' is used.

The global circulating system illustrated in Figure 2.6(a) results in about one thousandth of the total volume of ocean water being formed as new deep water each year. It takes about one thousand years for a parcel of water that sinks in the North Atlantic to be delivered to the Northern Pacific.

Without the densification processes described earlier, the delivery of freshwater by atmospheric circulation transporting water vapour from low to high latitudes to polar regions would tend to shut down deep water circulation.

FORCING FUNCTIONS AND FEEDBACKS

Any factor that may cause changes in the Earth's climate system is described as a forcing function. The term originates from computer modelling studies of the climate, particularly those that simulate climate change, and 'function' here means the mathematical expression of the 'forcing' or driving mechanism. In general, you can think of a forcing function as something that causes a system to respond in a way other than it would do if left alone (i.e. something forcing it to change). Climate modellers describe the size of the response of the climate system to a given magnitude as the climate's sensitivity. An analogy can be drawn with human diet. If we maintain a steady

Box 2.2 The Global Oceanic Deep Thermohaline Circulation System

The Atlantic loses water at a rate that is slightly more than the flow of the Amazon River. Were it not for the influx of less saline Pacific water, the salinity of the Atlantic would rise by one part per thousand every thousand years. You may be wondering why the overall circulatory systems in the Atlantic and Pacific are different. There are two reasons that relate to the distribution of land and mountains around the two oceans. First, the Gulf of Mexico acts like a simmering saucepan making warm saline water that flows north-eastwards into the Atlantic to form the Gulf Stream, which becomes North Atlantic Drift as it loses its heat through evaporation, becomes cooler and denser, finally contributing to the formation of NADW. Second, the distribution of mountains around the Atlantic and their relationship to the direction of the prevailing winds means that the Atlantic loses more water via evaporation that it gains by precipitation, run off from rivers and melting ice. This results in the Atlantic being more saline than the Pacific, so that when its surface waters cool to 2°C or 3°C, they are almost as dense as water beneath the sea ice ring-ing Antarctica. So does this mean that the formation of NADW can be turned off by a decrease in the salinity of surface waters? We will return to this question in Chapters 6 and 7.

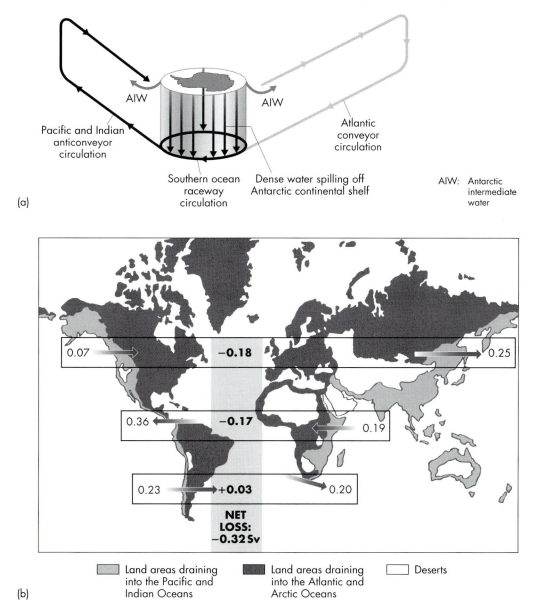

Figure 2.6 The global oceanic deep thermohaline circulation system. (a) Sketch showing the three key components of the system; for explanation see main text. (b) The hydrologic budget for the Atlantic, showing why it is more saline than the Pacific and Indian Oceans. More water vapour is carried away from the Atlantic by winds than is brought in them. The transport units shown are Sverdrups (Sv); one Sverdrup $= 10^6 \, \text{m}^3 \, \text{sec}^{-1}$ (more than the flow rate of three Amazon rivers).

regime of exercise and eating, our weight stabilises at a certain figure, characteristic of our own metabolism. If we change our eating pattern, and consistently eat 10 per cent more, our weight will increase until it stabilises at a different, higher level. This will almost certainly not be 10 per cent more than our initial body weight, nor will different individuals show the same change: we each have a different sensitivity to the same forcing. Unfortunately, the Earth's sensitivity to climate is hard to measure, and hard to predict. Theoretical modelling suggests that an increase in the average amount of incoming solar radiation of 4.0–4.5 W m^{-2} could cause a change in temperature of anything from 1.5 to 5.5°C, which means that the climate's sensitivity is of the order of 0.5–1.0°C for W m^{-2} of radioactive forcing (say 0.75°C per W m^{-2} of forcing).

□   So how much would the amount of incoming solar radiation have to be reduced to cause global cooling of 5°C?

■   Using an average of 0.75°C/W m^{-2} gives ~7 W m^{-2} (i.e. 5/0.75).

As the average amount of solar radiation reaching the Earth's surface is ~240 W m^{-2}, this reduction is about 3 per cent. This is far greater than estimates of fluctuations in the total solar flux (<0.2 per cent) and so other forcing functions must have operated to change the Earth's climate system from glacial to interglacial states.

Figure 2.7 is based on Figure 2.1, showing some of the differences between the Earth's climate system during interglacial and glacial intervals.

□   What forcing functions do you think might have operated to cause global cooling during glacial episodes?

■   The amount of solar radiation warming the *surface* of the Earth could have been reduced in three ways: first, a reduction in the solar radiation reaching the top of the atmosphere (i.e. changing the solar flux), or second by an increase in the Earth's albedo; third, a reduction in the proportion of greenhouse gases in the atmosphere would have

reduced the amount of longwave energy retained in the atmosphere.

□   Would the Earth's overall albedo have been lower or higher during glacial intervals?

■   It would have been significantly higher, as a much greater area was covered by ice (Figure 1.12(b)).

Not only would ice covered areas have increased the Earth's albedo, but an increase in land area due to sea-level fall, a reduction in the extent of forests, and an expansion of tundra and desert areas, would all cause more short wave radiation to be reflected directly back into space. However, as we saw in Chapter 1, a colder Earth is also a drier Earth, so cloud cover may have been reduced and so have partly compensated for the changes that increased the Earth's albedo. Increases or decreases in the proportion of greenhouse gases in the atmosphere may raise or lower global mean surface temperatures. As explained earlier, this is because these gases absorb outgoing long-wave radiation and re-radiate energy, much of it back to the Earth's surface.

It has been estimated that the climate system's sensitivity to atmospheric CO_2 is about 1.4°C for each 100 ppm change in CO_2 concentration.

□   So could CO_2 changes be solely responsible for global cooling of 5°C?

■   Clearly they could not, as this would require – if the sensitivity value is correct – the virtual elimination of atmospheric CO_2.

The discussion of how increasing the Earth's albedo or decreasing the proportion of CO_2 in its atmosphere might cause global cooling begs the question of what might have triggered such changes. It also ignored the interconnectivity of components in the climate system. Perhaps a small change in radiative forcing might have caused ice sheets to expand sufficiently to change slightly the Earth's albedo: this would have reduced global temperatures a fraction more, causing more ice sheet growth, and so on. This amplification of a small forcing factor is an example of a feedback process.

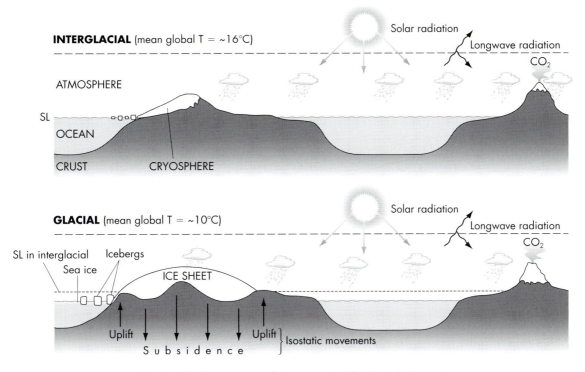

Figure 2.7 A comparison of the Earth's climate system during interglacial and glacial intervals.

□ Is the example described above a negative or positive feedback effect?

■ It is positive, as it would amplify changes in the climate system. A negative feedback acts to counter (reduce) change within a system, such as the reduction in cloud cover mentioned above that characterised a drier Earth during glacial periods.

The role of water vapour in the atmosphere is another example of positive feedback. As temperature increases, more water evaporates, increasing the concentration of water vapour in the atmosphere. Water vapour is a greenhouse gas, so the temperature rises further, causing more water to evaporate and so on. Clearly, positive feedback cannot continue to cause global cooling or warming indefinitely. If it did, the Earth would freeze or boil over! There is a limit to how extensive ice sheets can become, or to the water vapour content of the atmosphere.

There also appear to be limits (in the absence of human activities such as the burning of fossil fuels, cement making and forest clearance) to the CO_2 content of the atmosphere. Analyses of cores taken from polar ice caps (Chapters 5 and 6) show that the proportion of CO_2 in the atmosphere fluctuated by 30 per cent between the last two glacial maxima (200 ppm) and the last and present interglacial (~280 ppm) before humans began burning fossil fuels. A variety of feedback mechanisms may have contributed to maintaining atmospheric CO_2 levels within these limits. They are described later, and include rock weathering, changes in the amount of plant life (biomass, particularly on land), and burial of organic matter (particularly beneath shelf seas and the deep oceans). Most important of all, however, are changes in the amount of CO_2 dissolved in the oceans, for they contain over 98 per cent of the gas present in the ocean-atmosphere system. Changes in the CO_2

content of the oceans occur over timescales of 500–1000 years, whereas the other processes listed would take longer to alter atmosphere CO_2 concentrations.

Changes in oceanic temperature and deep circulation result in changes in atmospheric CO_2 content. The solubility of any gas increases as ocean water cools, so positive feedback would operate as the Earth cooled into a glacial period, as more CO_2 was dissolved. If the surface of the oceans cooled by 5°C, about 50 ppm of CO_2 could have been removed from the atmosphere. Higher wind speeds during glacials would not only have enhanced the exchange of CO_2 from the air into the oceans due to the increased roughness of surface waters, but also have caused increased upwelling of deeper waters carrying nutrients to the surface. The consequent enhancement of planktonic productivity and the resultant rain of organic debris falling to the ocean floors would remove more carbon from the ocean-atmosphere system.

Many attempts have been made to model the Earth's climate system in order to try to determine the contributions of various forcing functions to global cooling. The outcome of one example is shown in Figure 2.8: it shows that lowered sea surface temperatures (which caused a reduction of cloud cover and the proportion of water vapour in the atmosphere) may have been responsible for about 45 per cent of the global cooling 18 thousand years ago,

with ice-albedo effects (caused by land and sea ice) contributing about 33 per cent.

Feedback processes operate on different time scales, depending on the rate of response to change that is possible within components of the climate system. Continental areas heat up faster than oceans, and the circulation of the atmosphere is much more rapid than that in the oceans. Therefore we should expect there to be a time lag between the onset, or change in magnitude, of a forcing factor, and the response to it by different parts of the climate system.

☐ Can you think of an example of a time lag related to changing seasonal temperatures?

■ The top of the atmosphere above us (in the Northern Hemisphere) receives the maximum amount of solar energy on June 21/22, but the hottest month of the year is July. Likewise, sea-surface temperatures around the UK usually reach their maximum temperatures in August or early September.

The existence of such time lags in the Earth's climate system, and the plethora of possible feedback mechanism make it very difficult to estimate its sensitivity to forcing factors. This is why the estimate of the climate's sensitivity to solar forcing ranges between 0.5°C to 1.0°C per $W\,m^{-2}$ of changes in incoming solar radiation. This highlights the need,

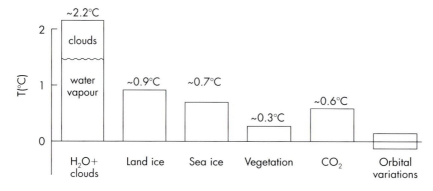

Figure 2.8 Estimates of the contributions to the difference between global mean temperatures today and at 18 ka obtained by a global climate model. Note that this model assumed a lower atmospheric CO_2 sensitivity than that quoted in the text, and did not include the effects of atmospheric dust or changes in deep ocean circulation.

when reading later chapters, to keep in mind the factors that may contribute to climate change, and the time scales over which they operate (Box 2.3). The variations in past temperatures, ice volumes, atmospheric composition, planetary windiness, vegetation, etc. may not be synchronous with changes in possible forcing factors, or with each other. It is important to

remember the point about the interconnectiveness of the components of the climate system. We will examine components of the climate system in a simplified way in later chapters: but in the real world, changes in one part of the system may trigger responses by other components that may reinforce or reduce global changes in climate.

Box 2.3 The Earth's Climate and Forcing Functions in Time and Space

As you read later chapters in Part 1 of this book that document climate changes that occurred over different time scales, you will need to keep in mind possible forcing functions that might have caused or contributed to them. When considering possible causes of change, it is important to match the spatial and time scales over which they occur with forcing functions that operate over similar ranges of scale.

Table 2.1 summarises the factors that may contribute to climate changes, and indicates the timescales over which they operate. Figure 2.9 adds the spatial dimension to this information by plotting the time scales and areal extent of different climate related phenomena on logarithmic scales. Both the table and figure include phenomena that are explained in later chapters of the book. The bars along the base of Figure 2.9 show

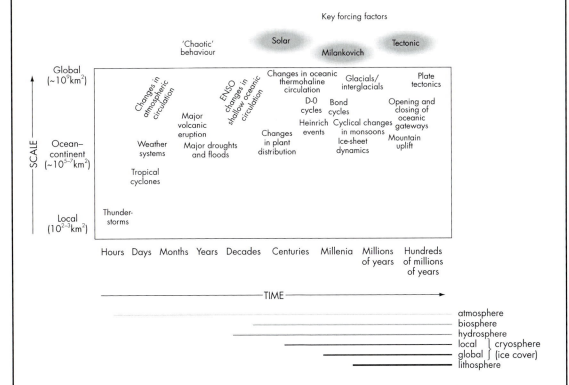

Figure 2.9 The spatial and time dimensions of the Earth's climate system plotted on logarithmic scales. The possible role of a variety of forcing functions is also shown. D-O: Dansgaard-Oeschger cycles.

the time scales over which different 'spheres' of the Earth (listed in Table 2.1) operate. This illustrates an important point. This is that longer term changes in climate probably involve progressively more components of the Earth system.

Table 2.1 Factors that may contribute to climate change (see also Figures 2.1 and 2.9) and the time scales over which they operate

Major 'sphere' in which factors occur	Individual factors	Time scales over which factors operate	Sections in this book where phenomenon is discussed

Astronomical	Changing solar flux:
	  long term
	  short term
	Changes in the Earth's orbit and orientation of its axis of rotation (astronomical, or Milankovich cycles)
	  eccentricity
	  obliquity
	  precession
	  harmonics of the above
		10^9 years
		10–200 years
		
		10^5 years
		4×10^4 years
		2×10^4 years
		Few thousand years
			Box 3.4, pages 62–5

Lithosphere	Continental drift/plate tectonics
	Uplift/mountain building (modifies atmospheric circulation and weathering rates affecting atmospheric CO_2 levels)
	
	Isostatic response to loading by:
	  mountain building
	  ice sheet growth/melting
	Ocean ridge growth:
	  global sea-level change
	Volcanism (atmospheric CO_2, aerosols)
		10^{6-8} years
		
		10^{6-8} years
		
		10^{5-8} years
		10^{4-5} years
		
		10^{6-8} years
		<1 year–10^8 years
			pages 142–4
			
			Box 5.2, pages 108–10, pages 145–6
			
			Box 1.3, pages 14–15
			
			page 140
			Box 7.1, page 142

Cryosphere	Ice sheet growth and melting	10^{2-5} years	Box 1.3, pages 14–15
Ocean	Global sea-level changes due to land ice formation and melting Deep water circulation Surface water circulation Upwelling CO_2 content	10^{2-5} years ~10^3 years 10^{1-2} years 1–10 years ~10^2 years	Box 1.3, pages 14–15 Box 2.2, pages 36–7
Atmosphere	Composition of greenhouse gases Long term Glacial/interglacial Circulation	10^{7-8} years 10^{3-5} years <1–10 years	Box 5.2, pages 108–10 Figure 5.17, page 107
Biosphere	Forest growth, swamp and peat bog formation	10^2 years	
Human activities	Fossil fuel burning, land use	10^{1-2} years	

THE EVIDENCE

Proxy Data

By now you may be wondering how it is possible to determine changes in global mean temperatures before reliable observations were made and recorded by humans. Historical records concerning changing agriculture, such as the cultivation of vines in Britain during the Medieval climatic optimum (in the twelfth and thirteenth centuries), or the direct effects of cooling causing extensive freezing of lakes and rivers during the Little Ice Age (between about 1500 and 1850 AD) can be used to infer global warming and cooling trends during the past one thousand years or so. This approach uses indirect information or proxy data, much as detective work pieces together the details of a crime. The word picture of a glacial Earth painted in Chapter 1 is not based on fantasy, but on the geological record. This enables today's world to be compared with the glacial world at 18 ka (Figure 1.12), and the pattern of global warming and cooling to be traced back several million years as will be shown in later chapters.

The limits of past ice sheets can be mapped by tracing the glacial debris (Box 1.2) left behind as they began to retreat from their maximum extent. Fossil sand dunes can be mapped beyond the limits of present day deserts. Lake sediments show that a few thousand years ago, the Sahara supported abundant life (Box 1.5). Sea surface temperatures can be determined by both palaeontological (Box 1.4) and chemical analyses (Chapter 4) of ocean sediments, and land temperatures can be estimated by studying spores, pollen and beetles (Chapter 8), and by analysing cores drilled into ice sheets (Chapters 5 and 6). These are all examples of proxy-data that enable past climate changes to be determined.

It is important to keep in mind that climate proxy indicators do not yield data comparable to that collected by weather stations today. Most palaeoclimatic indicators record a response to climate change, such as a change in vegetation or global sea-levels, from which temperature changes can be inferred. Even in cases where isotopic techniques enable the ambient temperature in which an organism grew to be determined, the results do not have the same resolution as a thermometer reading - either in terms of temperature, or just as important, in terms of *time*.

The Importance of Time

Just as weather stations record temperatures at particular times, we need to be able to determine the age of climatic proxy indicators. Again, the analogy with criminal detective work is appropriate: without knowing the timing of a crime, it would be difficult to prove who the perpetrator was. Without a reliable time framework, it is not possible to compare palaeoclimatic data from different parts of the world in order to describe what the planet was like 5 thousand, 20 thousand or 100 thousand years ago. Neither would it be possible to evaluate the possible links between forcing factors and the changes detected from the proxy data.

In this book, and in other literature discussing relatively recent climate change, you will find that past events and intervals of time are referred to in two ways, using either names (Figure 1.1), or ages in thousands of years before the present (ka). This is because the repeated alternation of past glacial and interglacial episodes was discovered before methods were available to determine their ages quantitatively. A further complication is that different names for glacial/interglacial stages are used in different parts of the world. We have avoided using many of these terms, preferring to give quantitative age estimates wherever possible.

The Problem of Resolution

Palaeoclimatologists have to live with the fact that the resolution of their data diminishes the further back they go in time: this applies not only to accurately determining environmental changes (such as determining past temperatures) but their timing as well. Figure 2.10 illustrates this problem. Over the past 150 years or so, historical records of direct temperature readings around the world have enabled a record of global temperature changes to be determined to a resolution of one year. The resolution for the past thousand years drops to around the decadal level, and over the past 15 thousand it is even less (though as we will see in Chapter 6, there are some high resolution records, to the decadal level or better, that are available for this long time span).

Even using historical records of temperature measurements to determine mean annual global temperatures is fraught with a variety of difficulties. Prior to the nineteenth century, reliable records are only available from Western Europe. The study that resulted in the earlier part of Figure 2.10(a) identified 3000 sets of records around the world that were potentially suitable to contribute to the global data base, but just over 1200 of them were eliminated after closer inspection revealed inconsistencies, such as changes in location of the recording station (e.g. from a valley to a hill) or the influence of localised warming due to

urbanisation. The historical study did not face the problem of dating the available temperature readings, but careful selection was needed to obtain reliable temperature readings. This enabled the authors of the study to produce a plot of variations in temperature with time: this is an example of what is known as a time series (in scientific terminology).

Time Series

The construction of time series, extending over periods ranging from thousands to several millions of years, is one of the main goals of researchers studying the Great Ice Age. They face three major problems: dating the record, finding enough sample points that contain material that can be dated, and coping with limits to the resolution of the data that can be obtained.

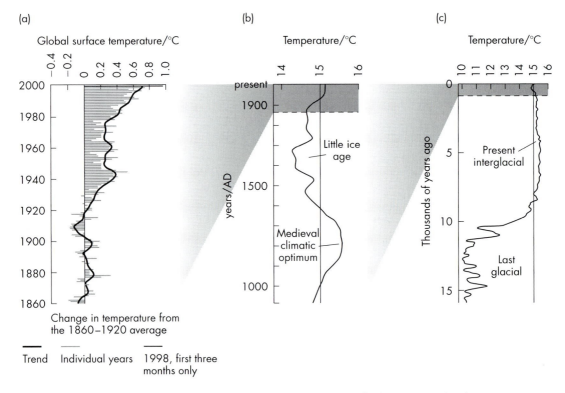

Figure 2.10 A summary of changes in global mean temperatures (a) since 1860 showing annual and 10 year average temperatures; (b) for the past one thousand years; (c) for the past fifteen thousand years.

The first problem concerns dating the geological record: this encompasses two difficulties. The first of these is that proxy climate data is obtained from a great variety of materials in the geological record. The most common ones are listed in Figure 2.11. As can be seen from this figure, there is no single dating method that can be used to determine the age of all these materials. This problem is compounded by the fact that different methods are applicable across different age ranges. The resolution of each method is different, and the method by which the 'clock' is set to zero and starts ticking in each material is not the same. There is no need to remember all the details of Figure 2.11, but you should carry from it the message that dating even the most recent part of the geological record is not a simple matter.

The second difficulty concerning the dating of palaeoclimatic records is that usually a limited number of points in the time series can be dated. This may be because only a few levels are amenable to dating, such as volcanic ash layers within a succession of deep sea sediments, or because of limitations imposed by the investigator's time and budget. This means that the ages between the sample points have to be estimated by interpolation, which assumes that the rate of deposition of the succession remained constant.

Not only do palaeoclimatologists face the problem of dating their climatic proxy data, they have a second, sampling problem, of a much greater magnitude than faced the authors of Figure 2.10(a). Carefully chosen cores of deep ocean sediment or ice have the potential to reveal a continuous time series. But there is a limit, due to constraints of technique, time and money, to the number of samples that can be obtained and analysed.

The third problem involves the inherent resolution

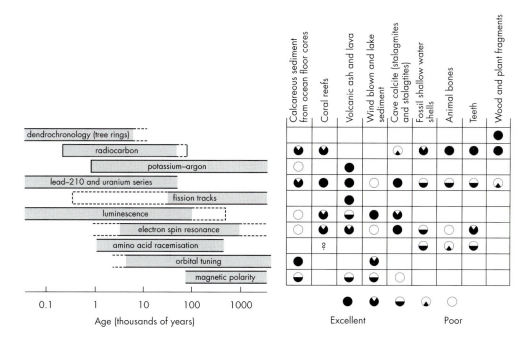

Figure 2.11 Age dating techniques: is there a 'best buy'? The details of the dating methods shown are not explained in this book. The purpose of the illustration is to show their applicability to different materials, their reliability, and the age ranges that they cover. Full circles indicate methods that are very reliable for the material shown. Less than full circles indicate less reliability, and/or doubts about the applicability of the method to the material indicated. In fact there is no 'best buy' method that can be used to date all the materials shown across ages from one hundred to millions of years.

that is possible from the material being investigated. As we shall see, many records yield data that can only be resolved to a hundred years or even a thousand years, but there are some that permit annual records to be obtained.

UNDERSTANDING CLIMATE CHANGE

Returning to the analogy between palaeoclimatology and criminal investigation alluded to earlier, just as a detective collects evidence to reconstruct a crime, determine a motive and discover the culprit, scientists use their understanding of the Earth's climate system to interpret the proxy data they collect. The way the climate system may have been changed can be demonstrated by correlating observed effects with postulated causes. Usually this only achieves a qualitative understanding, which does not yield many insights concerning changes in the interconnectivity between different components of the climate system.

Palaeoclimatologists make extensive use of computer climate models both to aid their interpretation of proxy climate data, and to explore the possible effects of changes in different climatic forcing agents. Two types are used: simulation and sensitivity models.

The more ambitious simulation models attempt to simulate past climates. The models require that quantitative definitions of various components of the climate system are fed into the programme, such as solar insolation, the extent of ice sheets, deep and surface oceanic temperatures, land albedo, cloud cover, CO_2 and water vapour content of the atmosphere. These general circulation models (GCMs) can only be run on very large supercomputers, and until recently, most of them could only treat the oceans as a 'wet carpet', and did not simulate shallow and deep circulatory patterns. Producing models of regional and global oceanic circulation is also a complex task but significant progress is being made. The reliability of the output of simulation models can be assessed by comparing their outputs to present day temperature and wind patterns. They can be used to produce low resolution simulations of past climates, say during the Cretaceous period. Once we have models that satisfactorily simulate observations of present and past observations of the behaviour of the atmosphere, oceans and cryosphere, the next step will be to couple them together to refine simulations of possible past climatic conditions, and to predict more confidently future climatic changes.

Despite the development of ever more sophisticated climate computer models the problem that still has to be faced is that apart from at the last glacial maximum 18 thousand years ago, there is not enough high resolution palaeoclimate data available from numerous locations around the world on which to base computer based simulations of past global climate systems. For this reason, many climate modelling studies use simplified models of the climate system to conduct sensitivity experiments (such as the one used to generate Figure 2.8) rather than trying to reproduce every detail of a past global climatic episode. In other words, models are used to explore the sensitivity of the climate system to changes in specific forcing factors, such as changes in the proportion of greenhouse gases in the atmosphere, or the distribution of land and sea around the Earth.

The next chapter explores the behaviour of the Earth's cryosphere and introduces the Milankovich theory of ice ages which suggests that astronomical forcing factors are the pacemakers of past glacial and interglacial events. The following three chapters examine the evidence of climatic change recorded in different realms. This will give you insights into the problems of obtaining proxy climate data, and the difficulties of dating the changes inferred. Such knowledge is essential in order to evaluate the possible feedback mechanisms that amplified the Milankovich pacemaker over timescales of tens of thousands of years (Chapter 4), and drove significant climatic changes at the millennial to centennial scale (Chapter 5).

SUMMARY

1 The Earth's climate at both the global and regional scales is the result of an interconnected system

involving the atmosphere, oceans, cryosphere (ice sheets, glaciers and sea-ice), biosphere and crust. The climate system is driven by the uneven heating of the planet, which results in heat being transferred from low to high latitudes by the coupled atmosphere–ocean system.

2 Forcing functions, such as changes in the proportion of atmospheric CO_2, or astronomical effects, may cause the way the climate system operates to change. The size of this response is known as the climate's sensitivity, and it results from feedback processes that amplify the effects of the initial forcing mechanism.

3 El-Niño–Southern Oscillation (ENSO) events are a good example of atmosphere – ocean coupling within the climate system. They occur when the intensity of the low pressure area over Indonesia is at a minimum, resulting in a reduction of the strength of Asian monsoons and their associated rainfall, and an increase in rainfall along the Pacific coasts of the Americas. When the pressure difference is high, the resultant 'La Niña' brings very strong SE trade winds, heavy rains occur in Asia but the Pacific coast of the Americas experience reduced rainfall.

4 The large scale density driven circulation of the oceans is known as thermohaline circulation. It plays a significant role in redistributing heat around the Earth, and is particularly important today in the North Atlantic, keeping NW Europe much warmer than areas at similar latitudes on the east coast of North America. Densification of ocean surface waters occurs due to temperature decreases and/or an increase in salinity caused either by evaporation or the removal of freshwater during the formation of sea ice. Evaporation and warming of waters in the Gulf of Mexico feeds the Gulf Stream, the subsequent cooling and further evaporation of which forms North Atlantic Deep Water (NADW). There are three key components to the global thermohaline circulatory system: the Atlantic Conveyor, the Southern Ocean Raceway and the Pacific and Indian Ocean Anti-conveyor (Figure 2.6(a)).

5 Feedback processes operate at different timescales, dependent on the rate of response to change that is possible within the components of the climate system involved. For example, the response time of deep thermohaline ocean circulation is slower than that of surface currents ($\sim10^3$ years compared to 1–10 years) whereas that of the atmosphere is very short – days to a few years (Table 2.1).

6 Interpretations of the nature of past climates are based on climate-dependent signatures left in the geological record. These are known as climate proxies, and include the record left in deep ocean sediments, ice caps, fossil flora and fauna contained in sediments, etc.

7 In order to reconstruct climates at different times in the past, and to determine rates of climate change, the age of proxy data must be determined. Dating is complicated by the fact that there is no common age determination method for all materials that yield proxy climate data, and no single method is able to provide dates back to 2.6 million years ago.

8 The resolution of time series of climatic data generally diminishes with increasing age of the source material, but there are notable exceptions.

9 Two types of computer modelling studies are used to explore how the Earth's climate system responds to different forcing functions. Simulation models try to match the behaviour of the system with high resolution palaeoclimate data, such as for the end of the last glacial stage 18 thousand years ago. Sensitivity experiments based on observed climate changes, or postulated forcing factors, are used to explore how individual forcing factors may have changed past climate systems.

FURTHER READING

Barry, R.G., and Chorley, R.J. 1998. *Atmosphere, Weather and Climate* (7th edition). Routledge, London. An advanced text providing an excellent follow up to Chapter 2 and *The Dynamic Earth* (see below).

Colling, A., Dise, N., Francis, P., Harris, N., and Wilson, C. 1997. *The Dynamic Earth*. The Open University, Milton Keynes. This Open University text is part of a second level science course entitled *Earth and Life* that examines the co-evolution of our planet and life upon it. *The Dynamic Earth* provides a detailed explanation of all key aspects of the Earth's climate system, parts of which are reproduced in Chapter 2 of this book.

Duff, P.McL.D. (ed.). 1993. *Holmes' Principles of Physical Geology* (4th edition). Chapman and Hall, London. See Chapter 21 on 'Ice Ages and Climate Change' and Chapter 14 on 'Dating and Pages of Earth History'.

See also references to figure sources for this chapter (p. 254).

3

UNDERSTANDING THE CRYOSPHERE

INTRODUCTION

In Chapter 1 we saw how changes in the extent and volume of land ice was a dominant feature of past climate changes at high latitudes. It probably played a crucial role in causing global changes and so it is important to understand the mechanisms that control the growth and demise of the Earth's covering of ice – the cryosphere (from the Greek word for frost: kruos).

In this chapter, we explore how the cryosphere interacts with other components of the 'Earth System' (Figure 3.1). In particular, we will focus on glaciers and explore how:

- they gain and lose ice;
- between glacier formation and climate are linked;
- they move;
- their growth is limited;
- past climatic changes may have periodically changed and triggered periods of glacier growth and melting.

The cryosphere consists of glaciers, sea ice, and permanently frozen ground (Box 3.1). Today, about 14 per cent of land areas are covered by ice, or underlain by frozen ground, and approximately 4 per cent of the oceans are covered by sea ice. During glacial

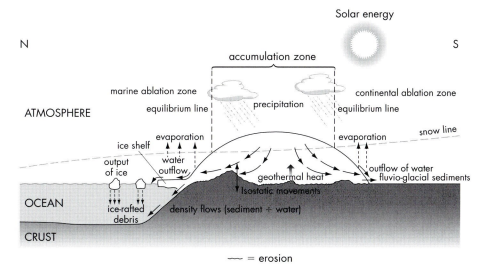

Figure 3.1 A sketch showing the features of ice caps discussed in Chapter 3 and the principal linkages between them and other components of the earth system.

maxima, these proportions increased to 25 per cent and 6 per cent respectively.

HOW ICE SHEETS WORK

Accumulation and Ablation

Snow and ice will only accumulate where one winter's snowfall is not totally melted away during the following summer. The limit of the area of accumulation on a glacier can be clearly seen during the summer: it is a bright white area contrasting with the darker colour of older glacier ice exposed in downstream areas (Figure 3.2). The white area is composed of firn: this material is formed as an intermediate step in the transformation of low density snow (0.1 kg m^3) to higher density ice (>0.8 kg m^3). It is produced when snow crystals lose their complex shape by melting and refreezing to produce a granu-lar, porous accumulation of ice crystals which later compact to become ice.

Glaciers flow under the influence of gravity. The ice moves from a zone of net accumulation to a zone of net ablation (Box 3.2) where ice is removed by melting, or if it flows into the sea, by melting and by being broken into icebergs by wave and tidal forces (Figure 3.3). The boundary between the zones of accumulation and ablation is the equilibrium line — this is the limit of the firn shown on Figure 3.2.

If a glacier neither increases or decreases in size, net accumulation must be balanced by ice flow into the zone of ablation. An increase in the volume of a glacier occurs when net accumulation exceeds ablation, and vice versa.

Glaciation and Climate

Glaciers and ice sheets can only be initiated in areas where winter snowfall is not completely melted away

---

Box 3.1 The Cryosphere

Glaciers, permafrost and sea-ice are the main components of the global mass of ice – the cryosphere.

Glaciers are masses of moving ice. They are divided into two groups:

- dome-shaped masses of ice that flow radially away from the highest part of the dome, and which largely mantle the underlying surface of the crust;
- ice flow in channels, constrained by flanking mountains.

Dome-shaped glaciers are subdivided on the basis of their size. Ice sheets are continent-wide in extent. Only two exist today, in Antarctica (27 × 10^6 km^3) and Greenland (3.6 × 10^6 km^3). During the last glacial maximum (and during several previous maxima), ice sheets also occurred over north-west Europe and North America (with a combined volume of 37 to 42 × 10^6 km^3). Ice caps are small ice domes with a volume of only a few thousand cubic kilometres (e.g. Vatnajökoll in Iceland). Smaller ice caps (with volumes measured in tens or a few hundreds of cubic kilometres) occur on the summits of flat topped mountains.

Glaciers flowing in channels include cirque and valley glaciers (Figure 1.6) occupying amphitheatre-like depressions on mountain sides. These may occur as isolated patches of ice, or feed valley glaciers, which may also be fed by summit ice caps .

Perennially frozen ground, or permafrost, occurs where mean annual temperatures are less than −2°C. Freezing may occur down to several hundred metres, but during the summer, the top one or two metres often melts.

Figure 3.2 Aerial view of a valley glacier showing the firn limit, which approximates to the equilibrium line. The white area consists of firn that has survived through the summer, and so marks the zone of accumulation. The darker area consists of exposed glacier ice and is an area where ablation exceeds accumulation.

during the summer. For continued growth, accumulation must exceed ablation over long periods of time. Precipitation and temperature, therefore, are crucial factors in the initiation and growth of ice sheets. The importance of these factors on glacier formation may be illustrated by considering three widely separated areas: Antarctica, Siberia and Scandinavia. Some glaciated regions, such as the central wastes of Antarctica, have extremely low levels of precipitation, because they are far from the influence of the

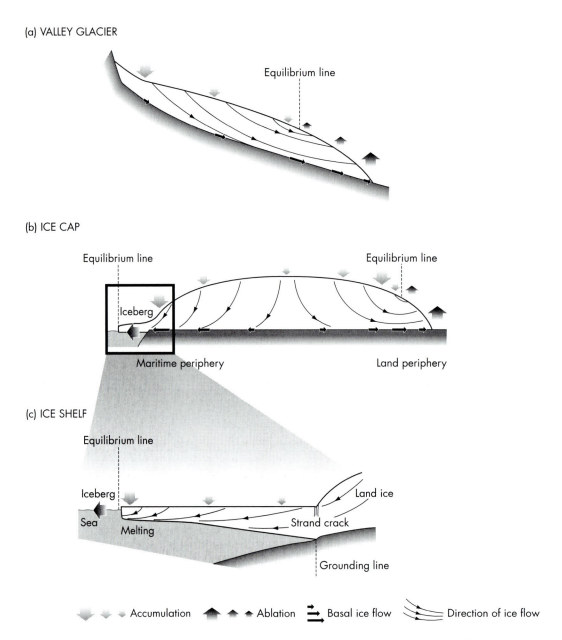

Figure 3.3 Idealised cross sections through (a) a valley glacier (b) an ice sheet, and (c) one margin of a continental ice sheet and its associated ice shelf, showing zones of accumulation and ablation separated by equilibrium lines. Ice flow and basal ice flow are explained in Box 3.3.

Box 3.2 A Glacial Mass Budget

Figure 3.4 is an idealised cross section across an ice sheet, showing seasonal variations in accumulation and ablation at three localities. Near the summit, no melting occurs, and so the cumulative plot of accumulation shows a continuous rise, steepest in the winter and early spring. The middle plot at a lower altitude shows that some melting occurs in the summer, but this does not remove all the winter snowfall. The third plot in the ablation zone shows a higher winter accumulation of snow, but that melting starts in the spring and continues into the autumn removing all this precipitation, and melting some of the ice that has flowed down from the accumulation zone.

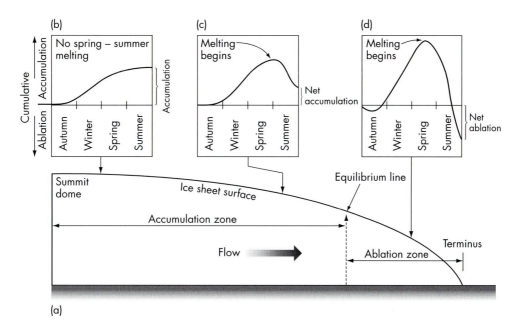

Figure 3.4 Idealised section through an ice sheet (a) showing its cumulative mass budget at three locations (b–d). At (b), the temperature never rises above freezing, so there is no ablation. At (c), there is some ablation during the summer, but not enough to melt all the snow that fell during the preceding winter and spring. However, at (d), late spring and summer ablation exceeds earlier accumulation.

open oceans. All the precipitation that does fall is in the form of snow, and, because the rate of ablation in summer is almost negligible, glacier ice can form. However in parts of the Arctic and much of mainland Siberia, not only is annual precipitation low (because they are remote from the open oceans and moisture bearing winds) but evaporation by strong dry winds may actually give a net precipitation deficit which prevents the formation of glaciers despite the very low mean annual temperatures. In more maritime areas, where the climate is affected by the proximity of major oceans with comparatively warm waters offshore, as in Scandinavia, very heavy winter snowfalls in the zone of accumulation may permit extensive glacier growth, even though mild summer conditions favour rapid ablation in the lower parts of the glaciers. So it is clear that proximity to the open ocean is an important factor that affects the likelihood of glaciation.

Glacier formation is today closely associated with mountain areas. This is the case in Antarctica and Greenland, where the ice sheets are partly fringed by mountain massifs. The critical altitude above which glaciers can be initiated is called the glaciation limit. Usually its altitude is assessed by plotting the levels at which small glaciers form in each mountain area, but this can be taken as approximately within 200 metres of the lowest firn limits and the regional snowline (the lowest limit of permanent snowbanks).

The altitude of the glaciation limit in the Scandinavia region is shown in Figure 3.5.

□ Why does the attitude of the glaciation limit increase inland?
■ Because precipitation is highest near the coast, and decreases inland.

Mountain ranges have a strong effect on precipitation patterns, particularly where they lie across the paths of moisture-bearing winds. Air masses are forced to rise to pass over the mountains, and in

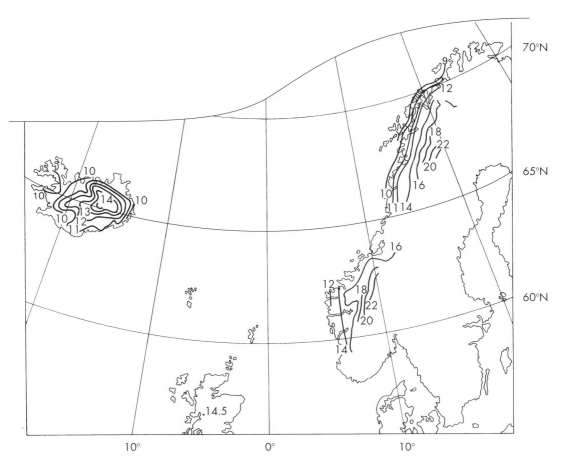

Figure 3.5 Contours of the altitude in hundreds of metres of the actual or potential altitude at which glaciers may be initiated (i.e. the glacial limit) at the present time around the North Atlantic. Note how the altitude diminishes towards coastlines, and higher latitudes.

doing so they cool and drop part of their moisture load as rain or snow onto the windward slopes of the mountain ranges. Given the prevailing circulation patterns in the Northern Hemisphere, this generally means the west facing slopes. In Figure 3.5 you can see how the greatly increased levels of precipitation on the west side of the Scandinavian mountain chain duly lower the glaciation limit. Increased precipitation in the area in recent years has caused glaciers to advance by up to 15 cm per day, in contrast to significant retreat in many other parts of the world.

☐ Does the altitude of the glaciation limit remain the same from south to north along the Norwegian coast?

■ No it does not: it is 1200 m in the south, and 1000 m in the north.

☐ Why is this?

■ Because mean annual temperatures decrease northwards.

So although altitude is an important factor controlling the formation of glaciers, the link between latitude and temperature, and the link between precipitation amounts and proximity to oceans are both very important. This results in the relationship between annual precipitation and the altitude of regional snowlines/glaciation limits shown in Figure 3.6. It is probable that the North Atlantic region has

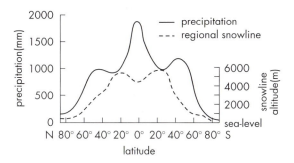

Figure 3.6 Curves showing the latitudinal variation of the snowline altitude and average precipitation. Note how high precipitation depresses snowlines in the cool temperate latitudes (40–60°) and near the equator.

a critical combination of latitude, altitude and moisture supply that leads to glaciation.

How Glaciers Move

As we shall see in Chapter 6, there is evidence to suggest that during the last glacial stage, Northern Hemisphere ice sheets discharged huge armadas of icebergs into the North Atlantic for short (<1000 years) periods of time. This indicates that flow rates within ice sheets may fluctuate significantly. So how do ice sheets move? Flow occurs within the ice due to internal deformation. This movement may be augmented by sliding of ice over bedrock, or by deformation of water saturated sediments beneath the ice (Box 3.3). The extent to which these processes operate depends on the temperature of the glacial ice.

Basal sliding and movement along water saturated sediments only occurs at the bases of warm ice glaciers. Beneath a cold ice glacier, the ice is frozen to the bedrock, and waterlogged sediments are frozen and so cannot deform. Warm ice glaciers tend to occur in mountainous areas, and cold ice glaciers in polar areas. However, both kinds of ice may exist in polar and non polar regions, and even in different parts of the same glacier. At first sight, this may seem surprising, but there are significant variations in temperature within ice sheets.

Figure 3.8 shows temperature profiles for three glaciers. The alpine glacier consists of warm ice, and so flow occurs via internal flow and basal sliding: water flows from beneath the ice at the glacier's snout. Temperature increases with depth in the Antarctica and Greenland ice sheets.

☐ Why should this be so?

■ The increase in temperature with depth occurs because there is a heat source at their bases: geothermal energy.

In addition, heat may be introduced at the base of glaciers by friction as it moves over bedrock. Heat is only transferred slowly upwards through the ice, because it is a good insulator. The combination of

Box 3.3 Glacier Movement

Most glaciers flow at velocities ranging from 3–300 m per year, but flows between 1–2 km per year may occur on steep slopes or where accumulation exceeds ablation to a large degree. Large outflows from the ice sheets in Greenland and Antarctica known as ice streams can flow at rates in excess of 10 km per year.

Three processes contribute towards glacier movement: internal flow, basal sliding across bedrock, and by deformation of underlying water-saturated sediments (Figure 3.7(a–c)). Internal flow occurs only in cold glaciers, whereas the other two modes require the base of the glacier to be above the freezing point of water: beneath

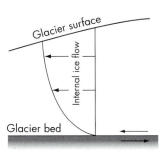

(a) Velocity profile for a glacier resting on a frozen rock or sediment surface

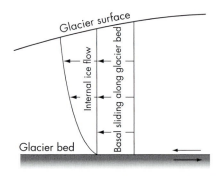

(b) Velocity profile for a glacier melting basally and resting on bedrock

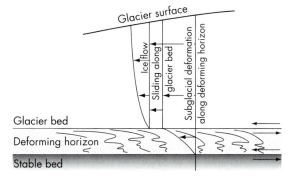

(c) Velocity profile for a glacier melting basally and resting on soft deformable sediment

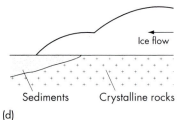

(d)

Figure 3.7 Glacier movement. (a) Idealised cross sectional velocity profile of a cold glacier, the base of which is frozen to the underlying bedrock. Flows occurs entirely by internal deformation of the ice. (b) Velocity profile of a warm glacier, where a thin film of water occurs at the ice–rock interface. Flow takes place by basal sliding and internal deformation. (c) Velocity profile of a warm glacier, where a slurry of water saturated sediments occurs beneath the ice. Flow takes place by deformation of the sediments, basal sliding and internal deformation. Movement due to internal deformation may dominate, as the sediment slurry is much weaker than the ice. (d) Sketch showing the change in surface profile of a warm glacier where it flows from hard bedrock to water saturated sediments.

thick ice sheets increased pressure lowers this temperature to −1.7°C. In some cases where water rich slurries of sediment lie beneath the ice (Figure 3.7(c)), movement may not be primarily due to internal ice flow because the slurry is much weaker than the ice. This results in more rapid flow compared to that which may occur over hard bedrock, and so the surface slope of the glacier is reduced (Figure 3.7(d)). Such a situation probably occurred beneath both the Scandinavian and North American ice sheets as they flowed outward from older, harder igneous and metamorphic rocks onto softer younger sediments. As we shall see in a later Chapter (Box 6.1) this relationship between ice and two contrasting underlying rock types can lead to instability and collapse of ice sheets.

geothermal heating and the fact that the melting point of ice decreases slightly with increasing pressure, has resulted in the formation of an extensive sub-glacial 'lake' beneath the Russian Vostok research site in Antarctica (for location, see Figure 5.12(a)).

The continued effects of geothermal heating and the flow of colder ice from summit regions results in the bases of glaciers changing laterally from frozen to melting conditions. In the example of the Greenland ice sheet shown in Figure 3.9, the temperature of surface ice decreases with increasing altitude towards the summit. As the colder, near summit ice flows outwards, it results in *decreasing* temperatures with depth until the warming effect of geothermal heat reverses the cooling trend. The warming results in part of the base of the ice melting, and meltwater being forced under pressure towards the ice-margin to produce a zone of partial melting. Note that beneath the thin margin of the ice sheet, the base is frozen. This is because it is no longer insulated by a thick ice cover from cold atmospheric temperatures.

☐ What might happen at the ice margin if the climate warmed?
■ Higher temperatures would lead to higher ablation rates, so that the glacier would retreat towards the melting zone (providing there was no compensatory increase in accumulation).

This retreat could lead to the collapse of the ice margin, causing a very rapid retreat and the release of icebergs where the ice reaches the sea. Rapid glacial retreat, if it involved the melting of large volumes of ice, will result in a global sea-level rise. This would further destabilise ice sheet margins resulting in a positive feedback process that would accelerate their break up, and possibly lead to the rapid collapse of entire ice sheets.

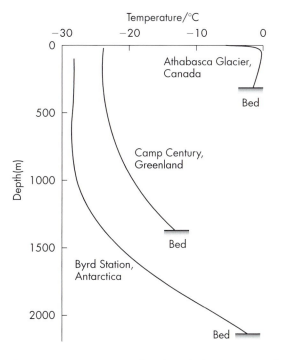

Figure 3.8 Temperature profiles through the Antarctic and Greenland ice sheets, and a mountain glacier. The slight decrease in temperature in the upper part of the Athabasca Glacier is caused by outward flow of colder ice. The temperature gradient in the Antarctica and Greenland ice sheets is caused largely by the input of geothermal heat.

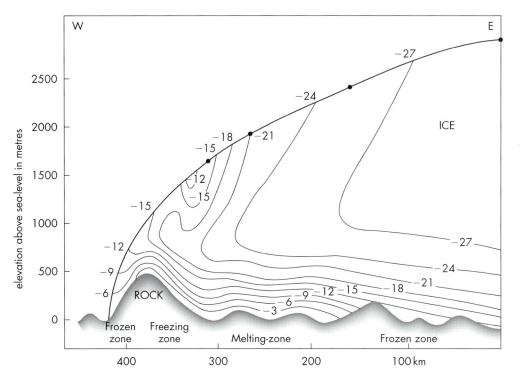

Figure 3.9 Temperature structure and cross section of the Greenland ice sheet at 70°N, showing the conditions at the base of the ice.

Limits to Ice Sheet Expansion

We have seen how altitude and latitude are crucial controls on glacier formation because they affect temperature and precipitation which in turn affect accumulation and ablation rates. Other, but not entirely unrelated factors may limit the extent to which ice sheets can expand from their sites of initiation.

Expansion of the Antarctic ice sheet to lower latitudes did not occur because the land ice flows out into a relatively deep sea to form an ice shelf which is rapidly broken up by tides and waves (Figure 3.10). This was not the case in the Northern Hemisphere and is the reason that northern ice sheets exert by far the largest control on global sea-level changes.

In North America the Laurentide ice sheet expanded into latitudes (down to 36°N) that were relatively warm and supported forests. It did so because accumulation rates to the north were high enough to balance or exceed ablation rates on its southern margin. This resulted in the ice sheet growing to a thickness of up to 3.5 km. As its surface grew to higher altitudes, the amount of precipitation it received near its summit area was reduced because winds would have already shed the bulk of their moisture at lower levels. This would eventually limit the southward expansion of the ice sheet: a good example of negative feedback in a system.

The isostatic response of the Earth's crust to the accumulation of a thick ice sheet (Box 1.3) may also restrict ice advance. This is because the rates of ice accumulation and isostatic crustal movements are different. Thick ice may build up in a few thousand years, but the continental crust will take several tens of thousands of years to sink into the mantle to compensate for the ice loading. As it does so, it will

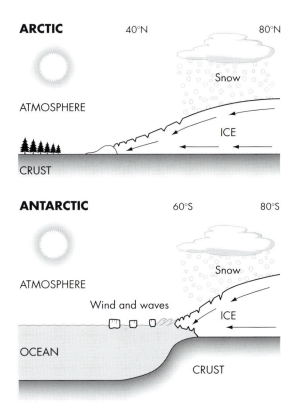

ARCTIC   40°N   80°N

ATMOSPHERE

Snow

ICE

CRUST

ANTARCTIC   60°S   80°S

ATMOSPHERE

Wind and waves

Snow

ICE

OCEAN

CRUST

Figure 3.10 Ice sheet advance in the Northern and Southern Hemispheres. In the north (a), ice sheets advanced south of 40°N into relatively warm areas. A comparable advance was not possible in the south (b), because the Antarctic ice sheet is fringed by ice shelves which are rapidly broken up (i.e. ablated) by waves and tides (cf. Figure 3.3(c)).

reduce the amount of ice available to flow into the zone of ablation, and so shift the equilibrium line away from the ice margin.

The crustal bulge surrounding the ice sheet (Figure 1.10(b)) may prevent meltwaters from the ice sheet draining away freely, and so large lakes may form in front of the ice margin, causing an increase in the ablation rate. Crustal subsidence of an ice sheet bordering the sea may result in the conversion of the edge of a land-based ice sheet to a floating ice sheet, which will also increase the rate of ablation, as wave and tidal action breaks up the ice. Once the ice sheet

discharges into the oceans, a significant cooling of surface waters may occur.

□ Could this have any effect on precipitation rates on surrounding land areas?
■ Yes it could, as evaporation rates would be lowered, and so less moisture would be transported landward, reducing ice sheet accumulation rates.

Such a reduction in precipitation alters the accumulation/ablation balance of an ice sheet, and is yet another potential cause of glacial retreat.

HOW DO ICE SHEETS GET STARTED?

There are no historical records of the initiation and growth of ice sheets. We can only speculate, therefore, about the mechanisms responsible for their inception by proposing theoretical models, three of which are described below (Figure 3.11).

As we saw earlier in this chapter, high latitude mountainous areas are susceptible to glaciation. Areas of Scandinavia, Scotland and north-eastern Canada are on the threshold of being covered by large ice sheets; in fact only Scotland lacks small glaciers today. Slight cooling in these areas would lower the altitude of the snowline/glaciation limit so that small ice caps would develop over plateau areas because summer warmth does not melt all the snow that falls during the previous winter. The albedo feedback mechanism resulting from the growth of these ice caps would lead to further growth, and eventually large ice sheets would develop. This model is known as the snow blitz or instantaneous glacierisation model (Figure 3.11(a)).

In the highland origin, windward growth model, once initiated in the highest areas due to regional or global cooling, the glaciers grow fastest on the windward slopes of the highland areas.

□ Why should ice sheet growth be fastest on windward slopes?
■ Because the highest precipitation rates occur where moisture laden winds rise over the highlands, cool and release snow.

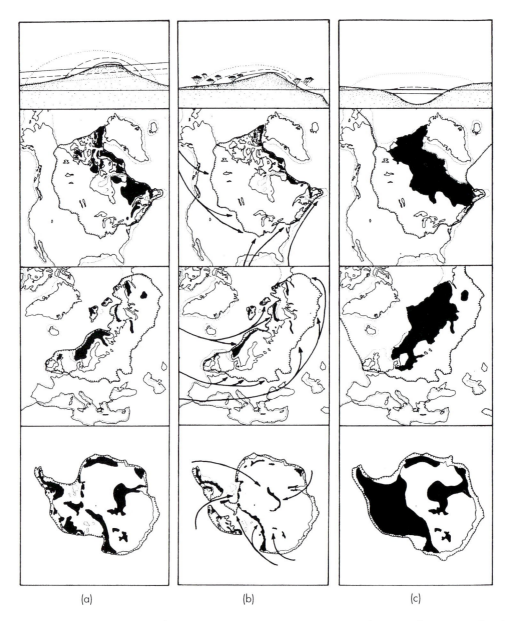

Figure 3.11 Models for the initiation of ice sheets: (a) instantaneous glacierisation; (b) highland origin, windward growth; (c) marine ice transgression. For explanation, see text.

In North America, this resulted in ice sheet growth occurring largely in a south westerly direction from the postulated sites of initiation in Labrador and Baffin Island. Windward advance of newly formed ice

sheets in Scandinavia was restricted by marine ablation, so growth occurred south-eastwards towards the Baltic Sea.

The marine transgression model suggests that the

negative feedback caused by the formation of permanent sea ice results in regional cooling which in turn leads to the formation of thicker sea ice, the grounded parts of which expand into local ice domes. Cooling would also lead to instantaneous glacierisation over surrounding highlands.

The theoretical models outlined above do not explain the cause of cooling that triggered each glacierisation mechanism. In fact, we need to explain the cause of global cooling that led to the first formation of glaciers in Antarctica nearly 50 million years ago, and the growth of large ice sheets in the Northern Hemisphere 2.6 million years ago, and why there have been repeated episodes of glaciation separated by warmer interludes. Possible mechanisms for global cooling since the Cretaceous are discussed in Chapter 7. But what triggered the repeated expansion and retreat of the cryosphere during glacial periods? It is virtually certain that the pacemaker of ice ages is not to be found within the Earth's climate system, but is the result of astronomical effects that cause variations in the amount of solar energy reaching the Northern Hemisphere of the Earth.

THE MILANKOVICH THEORY OF ICE AGES

In the 1920s and 1930s, the modern astronomical theory for the origin of glacial periods was elaborated in detail by the Yugoslav astronomer Milutin Milankovich (Box 3.4). The essence of his theory is that changes in the intensity of the seasons in the Northern Hemisphere controlled the waxing and waning of northern high latitude ice sheets. In particular, he believed that Northern Hemisphere high latitude summer temperatures hold the key to the onset of glaciations. If they were cold enough, winter snows would not completely melt, and so permanent snow fields would grow into glaciers. High latitudes between about 60° and 80° north are now regarded as being 'Milankovich sensitive'. This is because there is a solar insolation minimum in this zone (Figure 3.12) due to the combination of the poleward diminution of insolation received per unit area, and the poleward increase in day length (up to 24 hours) in this direction during the summer months. Changes in the tilt of the Earth's

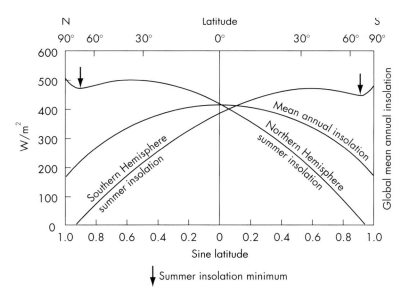

Figure 3.12 Summer insolation today. The graph compares mean annual insolation at different latitudes with that for Northern Hemisphere and Southern Hemisphere summer periods. Note the solar insolation minimum at high latitudes in both hemispheres.

rotation axis (obliquity) change the amplitude of this minimum.

Milankovich's work may be considered in two parts. First, he calculated the periodicities of the three orbital parameters (eccentricity, obliquity and precession, see Box 3.4) that are affected by the gravitational attraction of the other planets. This was essentially a mathematical exercise, which has been checked and refined more recently by computer modelling; changes in eccentricity, obliquity and precession can be calculated for the future as well as the past. Second, Milankovich proceeded to complete the much more difficult calculations of

quantifying the changes of solar radiation received at different latitudes over the last 600 thousand years. He suggested that times of summer insolation minima in northern latitudes coincided with the timings of glacial advances as then known. At the time Milankovich did his work, only four glacial periods were known, and the age of these was not well constrained, but as we shall see in Chapter 5 there is now convincing evidence to indicate that the timing of some fifty episodes of growth and decay of the cryosphere is controlled by northern high latitude seasonal insolation changes as suggested by Milankovich.

Box 3.4 Seasonality, Orbital Parameters and the Milankovich Pacemaker

The seasons

The Earth's hemisphere that is tilted towards the Sun is warmer in summer than in winter because the Sun is visible for more than 12 hours, and also because it rises higher in the sky so that the amount of solar radiation reaching a given area of the surface (insolation) is higher than that received in winter. Midsummer day (i.e. the day with the longest period of daylight) for the Northern Hemisphere occurs on 21 June each year, and is known as the summer solstice. This is midwinter for the Southern Hemisphere. Six months later the situation is reversed, giving the Southern Hemisphere summer and the Northern Hemisphere winter. The winter solstice for the Northern Hemisphere is on December 21. Spring and autumn correspond to intermediate positions of the Earth around the Sun, marked by the equinoxes, when every part of the Earth has 12 hours of sunlight a day.

Eccentricity (Figure 3.13)

The shape of the Earth's orbit changes from near circular to an ellipse over a period of about 95 thousand years with a longer cycle of change of

about 400 thousand years. Put another way, the long axis of the ellipse varies in length over time. Today, the Earth is at its closest (146 million km) to the Sun on January 3rd: this position is known as perihelion; on July 4th it is most distant from the Sun (156 million km) at aphelion (Figure 3.13(a)). Changes in eccentricity cause only a 0.03 per cent variation in total *annual* insolation, but they have significant *seasonal* effects. If the orbit of the Earth were perfectly circular, there would be no seasonal variation in solar insolation. Today, the average amount of radiation received by the Earth at perihelion is $\sim$351 Wm^2 reducing to 329 Wm^2 at aphelion, a difference of more than 6 per cent. At times of maximum eccentricity since 5 Ma the difference may have been 30 per cent. Milankovich suggested that the northern ice caps are more likely to form when the Sun is more distant in summer so that not all the previous winter's snow would melt resulting in the build up of glaciers over many years. As the intensity of solar radiation reaching the Earth diminishes as the square of the planet's distance from the Sun, global insolation falls at the present time by nearly 7 per cent between January and July – a situation that is favourable to snow surviving better in northern

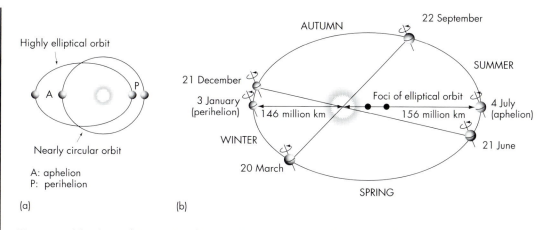

Figure 3.13 The shape of the Earth's orbit around the Sun. (a) The shape of the orbit changes from near-circular to elliptical. The position along the orbit when the Earth is closest to the Sun is termed perihelion and the position when it is furthest from the Sun aphelion. (b) The present-day orbit and its relationship to the seasons, solstices and equinoxes.

rather than southern latitudes through the summer. The more elliptical the shape of the orbit becomes, the more the seasons are exaggerated in one hemisphere and moderated in the other.

Obliquity

The tilt of the Earth's axis of rotation with respect to the plane of its orbit (the plane of the ecliptic) varies between 21.8° and 24.4° over a period of 41 thousand years (Figure 3.14(a)). The greater the angle of tilt, the greater the difference between summer and winter.

Precession (Figure 3.14)

There are two components of precession: that relating to the elliptical orbit of the Earth, and that related to its axis of rotation. The Earth's rotation axis moves around a full circle, or precesses, every 27 000 years (Figure 3.14(a)). This is similar to the gyrations of the rotation axis of a spinning top. Precession causes the dates of the equinoxes to travel around the Earth's orbit, resulting in the Earth–Sun distance, say, during the Northern Hemishere summer to vary (Figure

3.14(c)). The precession of the Earth's orbit is shown in Figure 3.14(b). It has a period of 105 000 years and changes the time of year when the Earth is closest to the Sun (perihelion).

Combinations of different orbital parameters result in a number of other periodicities: precession of the axis of rotation (Figure 3.14(a)) plus precessional changes in the orbit (Figure 3.14(b)) produces a period of 23 thousand years. Combining the cyclical changes in the shape of the orbit (eccentricity, Figure 3.13(a)) and precession of the axis of rotation results in a period of 19 thousand years.

These two periodicities (23 thousand years and 19 thousand years) combine so that perihelion coincides with the summer season in each hemisphere about every 21.7 ka, resulting in the precession of the equinoxes, as shown in Figure 3.14(c).

Combining eccentricity, obliquity and precession

Combining the effects of the eccentricity, obliquity and precession cycles enabled Milankovich to calculate the changes in solar radiation

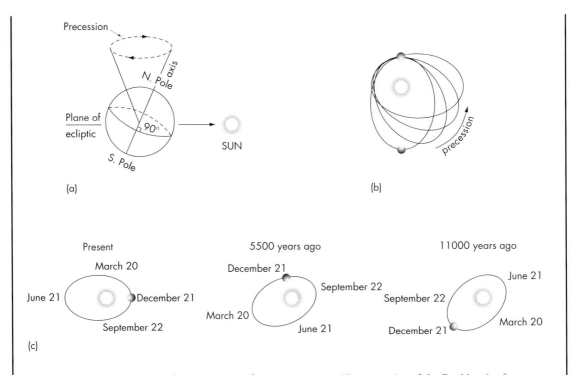

Figure 3.14 The components of the precession of the equinoxes. (a) The precession of the Earth's axis of rotation. (b) The precession of the Earth's orbit. (c) The precession of the equinoxes. For explanation, see text.

received at a given latitude through time back to 600 ka. He found that the maximum change was equivalent to reducing the amount of summer radiation received today at 65° to that received now over 550 km to the north at 77°N.

☐ What might such change in insolation have done to the altitude of glacial limits around the North Atlantic shown on Figure 3.5?

■ It must have significantly reduced their altitude.

In fact, if there was a simple southward shift of ~600 km, this would bring the altitude of the glacial limits seen in mid-Norway today at 65°N (where glaciers still exist) down to the latitude of Scotland, which, as we saw in Chapter 1, was ice covered during glacial stages. Of course, changing insolation may not have shifted the glacial limits in such a simplistic way, for they are controlled not only by latitude, but by climate, and in particular the direction of winds bringing weather systems that result in snowfall. But this simple 'thought experiment' shows how significant insolation changes caused by astronomical effects might be in changing the climate so that glaciers expanded in the Northern Hemisphere.

Figure 3.15 shows the results of a modern calculation concerning how the Earth's orbital parameters have changes through time, and the resultant changes in solar insolation at Milankovich sensitive northern high latitudes.

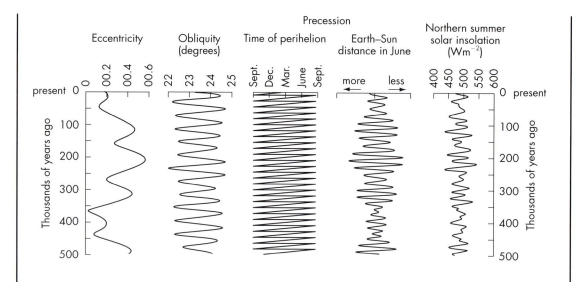

Figure 3.15 Variations in the Earth's orbital parameters: eccentricity, obliquity and precession (precession is shown both as the time of perihelion (see Figure 3.13(b)) and the Earth–Sun distance in June) and the resultant variations in solar energy received during summer at the top of the atmosphere in Milankovich sensitive northern high latitudes.

SUMMARY

1 The Earth's cryosphere consists of land ice (glaciers and permanently frozen ground (permafrost)) and sea ice. Today, about 14 per cent of the Earth's surface consists of land ice, and about 4 per cent sea ice. During glacial maxima, these proportions increased up to 25 per cent and 6 per cent respectively, largely due to growth of the cryosphere in the Northern Hemisphere.

2 Glaciers are masses of moving ice. Radial flow patterns characterise ice domes: ice sheets are continent sized domes, whereas ice caps are much smaller. Small glaciers moving through channels in mountainous areas occur as cirque glaciers occupying amphitheatre-like depressions, and valley glaciers (the former often feed ice into the latter).

3 Glaciers can only be initiated in areas where winter snowfall regularly survives through the following summer so that ice accumulates. The altitude at which glaciers can form – the glaciation limit – is controlled by a combination of factors: latitude, the presence of mountains and the proximity to oceans. The glacier limit rises from sea-level at polar latitudes to ~5000 m at 20° latitude; due to higher precipitation rates, it is depressed slightly at the Equator.

4 Glacier movement is accomplished by internal flow within the ice mass, and by sliding over bed rock. If the underlying material consists of unfrozen unconsolidated sediments, flow within the latter can significantly augment ice movement. Basal flow only occurs beneath warm glaciers, because under cold glaciers, the ice is frozen to the bedrock. The temperature of glacier ice is dependent on ice thickness, surface temperature, and geothermal and basal frictional heating. Changes in the temperature profile of glaciers may lead to changes in flow rates, and even their collapse.

5 Ice sheet expansion may be limited by:
   - availability of moisture
   - snowfall decreasing as the elevation of the surface of the ice sheet increases
   - isostatic subsidence may eventually outpace

the rate at which new ice can be added in summit areas

- the crustal bulge caused by isostatic subsidence may act as a dam to melt waters so that ice front lakes form, causing an increased ablation rate at the ice margin
- marine ablation: ice sheets cannot advance far across continental shelves as they are broken up by wave and tidal action (e.g. the Antarctic ice sheet).

6 According to the Milankovich theory of ice ages, Northern Hemisphere summer temperatures hold the key to the origin of glaciations, because if they are cool enough, winter snows will survive leading to successive annual accumulation of snow and ice. Milankovich calculated changes in solar insolation over Northern Hemisphere high latitudes caused by variations in the eccentricity of the Earth's orbit, the obliquity of its rotational axis, and precessional changes. He suggested that the ages of the four Quaternary glaciations as known when he did his work coincided with periods of lower summer insolation in the Northern Hemisphere.

FURTHER READING

Benn, D.I., and Evans, D.J.A. *Glaciers and Glaciation*. Arnold, London. A very advanced text, best read after Bennett and Glasser (see below).

Bennett, M.R., and Glasser, N.F. 1996. *Glacial Geology: Ice Sheets and Landforms*. John Wiley and Sons, Chichester. A readable follow up to this chapter for readers wanting to know more about glacial processes and associated landforms.

Duff, P.McL.D. (ed.), 1993. *Holmes' Principles of Physical Geology* (4th edition). Chapman and Hall, London. See Chapter 20 on 'Glaciers and Glaciation'.

See also references on figure sources for this chapter (p. 254).

4

THE DEEP SEA RECORD

INTRODUCTION

Our understanding of the Great Ice Age began with descriptions of landforms and deposits found on the continents, and the interpretation of the palaeoclimatic record that they preserve. Large areas of the continental landmasses of the Northern Hemisphere are covered by glacial deposits, but each ice sheet impoverished the record of its predecessors by eroding away earlier sedimentary records. This means that the continental record since 2.6 Ma is fragmentary and discontinuous. The record preserved on the continental shelves is little better, for not only were some of these formerly covered by ice, they were also subjected to successive flooding and emergence as global sea-levels rose and fell with successive contractions and expansions of the cryosphere. The deep oceans, therefore, are almost the only areas where relatively undisturbed and continuous sedimentary records extending back millions of years occur; the only notable exceptions are deposits of dust (loess) in China (Chapter 5) and some sediments deposited in deep lakes. Continuous continental records providing uninterrupted time series extending back to the last and/or penultimate interglacial stages are available from ice sheets and stalactitic material in caves (Chapter 5).

DEEP SEA SEDIMENTS

Deep-sea sediments consist of materials that may be derived originally from erosion of the continents,

from subaerial volcanic eruptions, from the skeletal remains of marine organisms, or through direct chemical precipitation.

The floors of deep oceans are not always tranquil places where a rain of the remains of planktonic organisms settles to the bottom. This rain may be interrupted by density and contour currents, or by debris shed by floating, melting ice. A detailed discussion of deep ocean sediments is beyond the scope of this book. The key depositional and erosional processes that need to be kept in mind when searching for oceanic sites that may yield high resolution and uninterrupted palaeoclimatic records are described in Box 4.1. Figure 4.1 shows the distribution of deep-sea sediments on the floors of the world's oceans. All these deposits may carry palaeoclimatic signals, but the most important are the glaciomarine sediments and the calcareous oozes.

In polar regions glaciomarine sediments are formed. These are characterised by the presence of ice-rafted material of all sizes. The sediments are very poorly sorted sometimes containing a large range of grain sizes – from clay particles to boulders (compare with Figure 1.6(a)). Significant glaciomarine sedimentation around the Antarctic continent, occurred much earlier (almost 50 million years ago) than in the North Atlantic. Deep-sea cores show that from about 2.6 million years ago the extent of glaciomarine sedimentation increased in the North Atlantic and North Pacific. There was an even greater increase in the extent of this type of deposit from about one million years ago, the cause of which will become apparent later in this book.

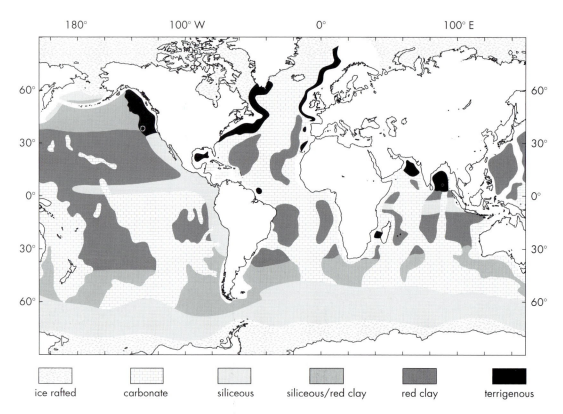

Figure 4.1 Map showing the present day distribution of sediments on the deep ocean floors. Ice rafted: see main text and Box 4.1. Carbonate: these sediments are almost entirely oozes composed mostly of coccoliths, and smaller amounts of foraminifera and other skeletal debris. Siliceous: these sediments are oozes containing the remains of plankton that secrete siliceous skeletons (radiolaria, diatoms). Red clay: these sediments are comprised of fine grained material transported from the continents by wind, or in suspension in seawater. They are deposited in very deep water settings below the carbonate compensation depth (CCD: see main text), so that the rain of the remains of planktonic carbonate material is much reduced or even eliminated. Terrigenous: these deposits are dominated by muddy and sandy sediments transported from the continental shelves by turbidity currents (Box 4.1). Siliceous/red clay sediments show features transitional between siliceous and red clay sediments.

Calcareous oozes are particularly important in the study of climatic changes from the late Cretaceous onwards. This is because the fossil plankton not only enable past sea-surface temperatures to be determined (Box 1.4), but they also contain a record of the oxygen isotopic composition of the oceans which can be used to determine not only past temperature changes, but also past ice volumes. They occupy a large part of the floor of the Atlantic and Indian Oceans and a smaller area of the Pacific. These oozes consist of up to 30 per cent of inorganic clay that also settled out

of suspension. This is the finest grade of sediment derived from continental rocks which remains in suspension long enough to be widely distributed across the oceans. The oozes are principally composed of the skeletal remains of planktonic marine organisms, particularly abundant coccoliths and the tests (skeletons) of foraminifera (Figure 1.12(a)). Coccoliths are the tiny calcareous platelets of a type of unicellular algae, the Coccolithophora, which live in great numbers in surface waters of the oceans (Figure 4.3).

In order to obtain uninterrupted time series data,

sites likely to preserve a continuous record of calcareous oozes must be selected carefully.

☐ What processes could interrupt the deposition of calcareous oozes (see Box 4.1)?

■ The influx of turbidity currents would interrupt

the rain of planktonic debris, and even erode previously deposited oozes. Contour currents may also remove older sediments.

☐ In view of the above answer, where are continuous sequences of calcareous oozes likely to accumulate?

Box 4.1 Some Deep Sea Depositional and Erosional Processes

There are four main processes responsible for the deposition of deep-sea sediments.

1 Settling from suspension of organic and inorganic material (Figure 4.2(a)). The settling out from suspension of the remains of the skeletal parts of calcareous and siliceous planktonic

organisms results in the accumulation of calcareous or siliceous oozes. Inorganic clays may form by the settling out of land derived particles, which may be initially transported by wind, or by surface or deeper ocean currents. Ash and dust from volcanic eruptions also settle out from suspension: they provide useful

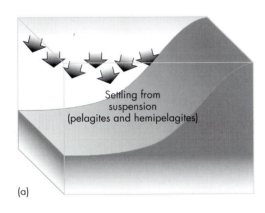

(a)

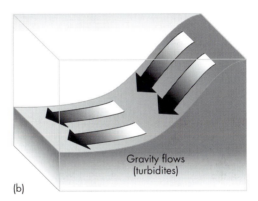

(b)

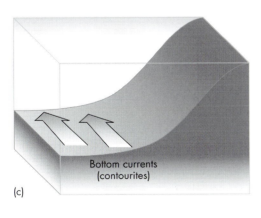

(c)

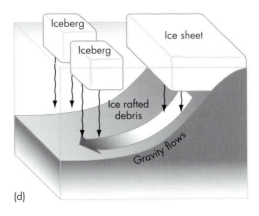

(d)

Figure 4.2 The origins of deep sea sediments.

marker horizons enabling successions in widely spaced areas to be correlated with one another. They may also contain minerals that enable accurate radiometric dates to be determined.

2   Downslope transport of sediments within density currents consisting of a turbulent mixture of sediment and water (Figure 4.2(b)). As the velocity of these flows, known as turbidity currents, diminishes, sediments are deposited to form turbidites. These are characterised by sharp, often erosional, bases overlying older sediments, and an upward diminution in grain size.

3   Erosion, transport, and deposition by deep-water currents that flow parallel to the depth contours along continental slopes (Figure 4.2(c)). The deepest of these relate to global thermohaline circulation (Figure 2.6(a)).

4   Sediment released from shelf ice, and icebergs, or released from land based ice may be transported downslope as density currents (Figure 4.2(d)). Sediment released from melting icebergs consists mostly of silt and clay that settles to the sea floor, but this fine sediment often contains larger particles (up to the size of boulders) that dropped out of the ice. The larger particles are referred to as ice rafted debris.

Erosion associated with processes 2, 3 and 4 results in discontinuous records but even discontinuities may provide evidence for past climatic changes, especially the initiation of thermohaline circulation.

■ Uninterrupted deposition of calcareous oozes occurs in locations where turbidity and contour currents are unlikely to reach. Such areas are most likely to be isolated elevated areas above oceanic abyssal plains, deeper parts of continental margins isolated from the input of land derived sediments (e.g. plateaux and spurs), and along the flanks of ocean ridges.

Over such features the rates of sedimentation vary from several centimetres to less than one centimetre per one thousand years. Provided the nature of the sedimentation has remained the same, cores taken through sedimentary successions in these areas often reveal relatively even rates of sedimentation over thousands of years, spanning both cool and warm episodes of global climate.

Although the remains of calcareous plankton fall to the sea floor throughout the oceans, they are preserved only where the sea floor lies above a certain depth. This is because the bottom waters contain CO_2 in solution which dissolves carbonate skeletal material sinking down from the surface waters. The depth above which carbonate can accumulate is called the carbonate compensation depth (CCD). Below it the rate of solution exceeds the rate of supply of skeletal material. The CCD is shallower in sub-polar regions than in the tropics because the solubility of $CaCO_3$ is higher in cold than it is in warm water.

The mixing-up of sediment by burrowing and feeding organisms is termed 'bioturbation'. It does not occur if the bottom waters lack oxygen (i.e. are anoxic).

☐   How will bioturbation affect the resolution of the climatic record that can be obtained from calcareous oozes?

■   It will tend to mix together sediments deposited at different times, thus reducing the climatic and time resolution of the record.

Bioturbation generally affects the top 4–5 cm of the sediment on the ocean floor, or deeper on occasion, so that with the slower rate of sedimentation the details of the record of events over 5 to 6 thousand years could be blurred and smoothed out. Since climatic changes of importance can take place within 1000 years, even a sedimentation rate of 5 cm per 1000 years will make the determination of the exact time and intensity of such changes very difficult if bioturbation has occurred (Figure 4.4).

☐   What are the advantages of cores with high and low sedimentation rates?

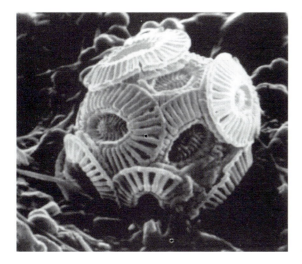

Figure 4.3 Scanning electron microscope photograph of Coccolithophora. Platelets (coccoliths) of calcium carbonate are secreted on the cell walls. Distinctive shapes and patterns exhibited by the platelets are used to identify different species of Coccolithophora, which in turn enables the relative ages of calcareous oozes to be calculated.

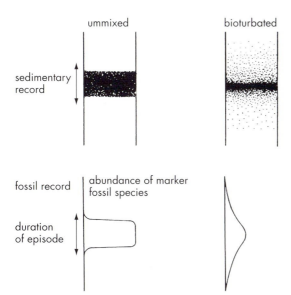

Figure 4.4 Diagram showing the mixing of sediment layers by the churning effect of burrowing organisms (bioturbation).

■ Low sedimentation rates enable a record to be obtained from shorter cores extending much further back in time, showing a more or less continuous geological record covering hundreds of thousands to millions of years. On the other hand, the faster the sedimentation rate the greater the detail that can be obtained from that record, particularly if a long core covers just a few tens of thousands of years.

The last interglacial stage, as recognised in continental sequences, lasted for about 11 thousand years. This would occupy only 11 cm of core at a depositional rate of 1 cm per one thousand years but 55 cm at 5 cm per one thousand years, giving a potentially much fuller and clearer picture of how conditions were changing during that period.

OXYGEN ISOTOPE STUDIES

Isotopes and isotopic ratios are explained in Box 4.2.

In 1946 the American scientist Harold Urey first demonstrated that oxygen isotope ratios of the calcium carbonate laid down in the skeletons of marine organisms are related to the temperature of the seawater in which they lived. This relationship was investigated experimentally by keeping molluscs under different temperature regimes in seawater of known isotopic composition. He showed that the lower the temperature of the water, the greater was the proportion of ^{18}O relative to ^{16}O incorporated in the shell. A quantitative calibration of this effect was obtained which enabled the temperature of growth of samples of planktonic forams to be calculated from the isotopic composition of their tests. This corresponded satisfactorily to the ocean-water temperatures from which they had actually been taken. The results were given as a simple ratio as the parts per thousand (‰) change of one isotope to another, with respect to a standard (Box 4.2), known as the delta (δ) value. With each 1°C fall in temperature, $\delta^{18}O$ was shown to increase by 0.2‰.

The results of oxygen isotope analyses of forams obtained from deep-sea cores are generally presented as oxygen isotope curves, plotted as isotopic changes

Box 4.2 Isotopes

All atoms of the same element have the same number of protons in their nuclei (i.e. they have the same atomic number) but they do not necessarily have the same number of neutrons and so they may have different mass number (mass number of an element = number of protons + number of neutrons). Varieties of the same element with different mass number are known as isotopes.

Isotopes of the same element have identical chemical properties. This is because these properties are determined by the number of electrons around the nucleus, (which is equal to the number of protons in the atom, i.e. the atomic number) and are independent of the number of neutrons in the nucleus.

Hydrogen has two isotopes so that in water 0.02 per cent of the atoms of this element are represented by the heavier isotope deuterium (Figure 4.5).

The third isotope of this element is tritium (having two neutrons in its nucleus). It is unstable, and therefore radioactive, being produced in nuclear reactors or explosions. The other two isotopes (hydrogen and deuterium) are stable isotopes and so do not decay over time. Hydrogen is the only element that has named isotopes.

Isotopes are usually denoted by writing the mass number as a superscript before the chemical symbol of the element (e.g. ^{2}H for deuterium).

Oxygen has three stable isotopes with mass numbers 16, 17 and 18. Over 99 per cent of natural oxygen is made up of ^{16}O, with ^{18}O comprising most of the balance.

The isotopic composition of water melted from ice, and planktonic fossils in sediments, can be measured accurately using mass spectrometers. The proportion of isotopes present is given as a delta (δ) value, which is determined by comparison of the sample with a standard, and results in parts per thousand ‰ or 'per mil'):

$$\delta\,^{18}O = 1000 \times$$
$$\frac{(^{18}O/^{16}O)\ \text{sample} - (^{18}O/^{16}O)\ \text{standard}}{(^{18}O/^{16}O)\ \text{standard}}$$

The standard used to indicate the isotopic composition of water and ice is seawater which has a $\delta^{18}O$ value which is zero and is known as SMOW (Standard **M**arine **O**cean **W**ater). Another standard (PDB: **P**ee **D**ee **B**elemnite) is used to express the isotopic composition of calcite. This is a fossil belemnite (an extinct group of cephalopods related to squids) that Harold Urey used as his standard. It comes from the Pee Dee Formation of Cretaceous age in Carolina, USA.

Variations of $\delta^{18}O$ in deep sea calcareous oozes are caused by changes in the temperature of the waters in which organisms live, and changes in the volume of the cryosphere. The relative contributions of these two factors can be determined, and is explained on p. 71. Changes in salinity also influence the oxygen isotope composition and skeletal remains, but this is beyond the scope of this book.

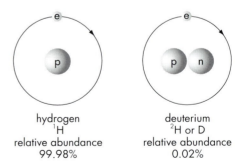

Figure 4.5 Sketch showing the atomic structure of two isotopes of hydrogen: p, n and e represent a proton, neutron and electron, respectively.

($\delta^{18}O$) against core depth for individual species. The curve for a core from the western equatorial Pacific Ocean (Figure 4.6) demonstrates a pattern of marked fluctuations in isotopic composition with time.

☐ From what you have learnt so far about the relationship between $\delta^{18}O$ values and temperature change, what is the range of temperature change suggested by the data in Figure 4.6?

■ As $\delta^{18}O$ varies by 2‰, a 10°C range of temperature is indicated (a 1°C change in temperature results in a 0.2‰ change in $\delta^{18}O$).

This result is surprisingly high - even greater than that thought to have occurred over most land areas.

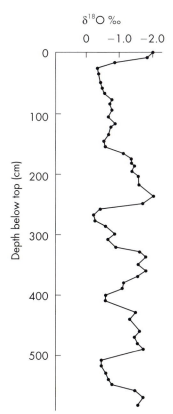

Figure 4.6 Oxygen isotope analysis of shells of a planktonic foraminifera species from a series of samples obtained from a deep sea core taken in the western equatorial Pacific Ocean.

The enormous volume of ocean water acts as a buffer to temperature changes, so that oceanic variations are likely to occur over a much *smaller* range than do those on land which means that temperature cannot be the sole reason for the observed changes in isotopic composition. So why are such large changes in $\delta^{18}O$ observed? The answer is that they are caused by changes in the volume of land ice, as explained below.

The isotopic composition of materials formed chemically or biochemically from seawater depends not only on temperature conditions, but also on the original isotopic composition of seawater. It used to be asserted that the isotopic composition of present-day seawater varies very little from place to place, and that isotopic values have remained essentially uniform for at least 150 million years. It is now known that neither of these statements is true.

☐ Marine organisms, as you have seen, can fractionate oxygen isotopes. What other natural mechanisms might do the same for water derived from the ocean?

■ Evaporation and condensation. This is because the latent heat of 'heavy' water ($H_2^{18}O$) is higher than that of 'light' ($H_2^{16}O$) water, so the latter evaporates more easily. This results in water vapour being isotopically lighter than its parent ocean water. An explanation of the isotopic fractionation involving liquid water, water vapour and ice is given in Box 4.3.

There are two climatic signals combined in oxygen isotope time series obtained from forams extracted from deep ocean cores:

● local ocean surface temperature changes, and
● global changes in the volume of land ice, and therefore sea-level.

Fortunately, the two can be separated. This is because temperature conditions at the bottom of the oceans are very different from those at the surface. Cold, dense, saline water is produced in Antarctic and Arctic regions and sinks to form the major bottom-water currents, flowing towards and even beyond the Equator

Box 4.3 Oxygen Isotopic Fractionation in the Hydrological Cycle

Isotopic fractionation occurs during evaporation and condensation (Figure 4.7). When seawater evaporates from the oceans, water molecules with the lighter isotope ($H_2^{16}O$) evaporate faster, so atmospheric water vapour is relatively enriched in the lighter isotope. When water vapour condenses and is precipitated back into the ocean, the heavier isotope ($H_2^{18}O$) condenses preferentially. Both processes deplete water vapour in the atmosphere in $H_2^{18}O$ relative to $H_2^{16}O$. As water vapour moves towards the poles, it is progressively depleted in ^{18}O, so that the snow that falls there is depleted in ^{18}O relative to the oceans. In polar regions today, snowfall and ice have $\delta^{18}O$ values of $-30\permil$ to $-50\permil$. Therefore, the larger the volume of land ice and ice sheets, the higher will be the relative proportion of ^{18}O in seawater. During glacial stages, when at their maximum extent

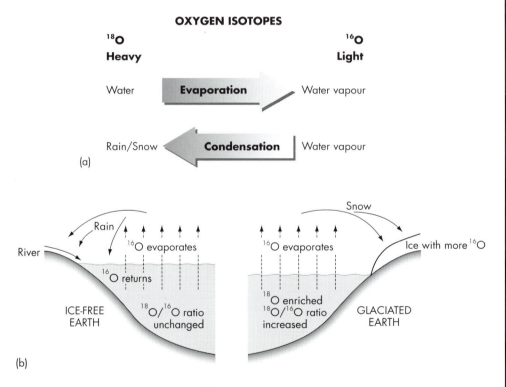

(a)

(b)

Figure 4.7 Oxygen isotope fractionation in the hydrologic cycle during interglacial and glacial periods. (a) Diagram showing how more ^{16}O is incorporated into water vapour when seawater is evaporated: this depletes the ^{16}O content of seawater resulting in it becoming isotopically heavier (i.e. containing more ^{18}O). When water vapour condenses, the remaining vapour becomes even more depleted in ^{18}O. As water vapour moves polewards, it loses progressively more of the heavier isotope. If the polar precipitation results in the formation of ice sheets, these are significantly depleted in ^{18}O. (b) During interglacial periods, most precipitation on land returns to the sea via rivers, so there is only a small amount of additional ^{18}O in sea water due to the Antarctic and Greenland ice-caps. But during glacial periods, ^{18}O enrichment of oceanic water is much greater (i.e. it is isotopically heavier), due to larger amounts of the lighter isotope being locked up in the more voluminous ice sheets.

glaciers and ice sheets covered approximately three times their present area, about 3 per cent of ocean water was abstracted, enriching the ocean waters in ^{18}O. Higher, or less negative, δ^{18}O values in deep sea sediments indicate larger ice caps and lower global sea-level at the time the sediments were deposited.

(Box 2.2). Today, even in the tropics, bottom-water temperatures are only about 1.5°C. They could not have become much colder even during the major glacial stages when the temperatures of ocean surface waters fluctuated by 2–3°C in equatorial regions, but 10–15°C in high latitudes. If both planktonic and benthonic (i.e. bottom dwelling) forams occur together in ocean floor sediments, oxygen isotopic analyses of their tests enable a record of both surface and bottom water temperatures to be obtained.

Figure 4.8 shows the results of oxygen-isotope analyses of forams obtained from planktonic and benthonic forms in the upper part of a core collected in the western equatorial Pacific Ocean. Curve P shows δ^{18}O variation with depth in the planktonic species and curve B the results from benthonic species.

☐ What conclusions can be drawn by comparing plots B and P?

■ The shape of curve B is a very similar form to curve P. So if there was very little or no temperature change in the deep bottom waters in which the benthonic species were living, virtually the whole range of 1.5â isotopic change must be due to changes in ice volume and thus provide a summary record of the growth and decay of the major ice sheets.

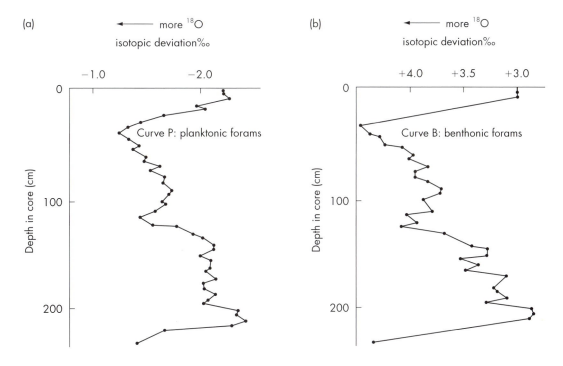

Figure 4.8 Oxygen isotope compositions of (a) planktonic and (b) benthonic forams obtained from a core of ocean floor sediments taken in the western equatorial Pacific Ocean.

The isotopic difference between planktonic and benthonic species in a sample from the top of the core is 5.3‰.

□ What is the reason for the difference of 5.3‰ in $\delta^{18}O$?

■ It is due to the markedly different temperatures at the ocean floor and near its surface. The latter is at 1.5°C, and the surface is 21°C warmer.

If curves P and B are superimposed, using the most recent maxima (i.e. virtually the present day) as the common point (Figure 4.9), the two curves are very similar, taking into account the possibility of sam-

pling error due to reworking of sediment. The two curves track each other quite closely, with the biggest difference between them occurring at the minima at 35 cm depth. This suggests that the range of isotopic fluctuation in curve P as well as curve B resulted entirely from a large change in global ice volume and that there was very little change in the temperature of deep ocean water as the Earth warmed from glacial to interglacial conditions.

□ What is the difference in isotopic composition between the glacial minimum and the interglacial maximum in curve P?

■ It is approximately 1.1‰.

The range of 1.1‰ in $\delta^{18}O$ values shown in Figure 4.8(a) is an index of the change in the total volume of land ice. Dating the sediments from which the record was obtained shows that global sea-levels began to fall at about 120 ka at the end of the last interglacial, reaching their lowest level about 20 ka, after which they rose very rapidly.

As can be seen in Figure 4.10, $\delta^{18}O$ records from the Pacific show that significant ice-volume changes occurred during the past few million years.

□ Ignoring the rapid fluctuations of $\delta^{18}O$ in the benthonic record shown in Figure 4.10 (caused by ice volume changes), what is the overall trend this figure shows over the past four million years?

■ It rises from about 3.3‰ four million years ago, to around 3.5‰ one million years ago.

□ What does this indicate?

■ It could indicate a decrease in the temperature of ocean bottom waters of about 2°C, but it is likely that the trend is due to a significant increase in the volume of the cryosphere, particularly in the Northern Hemisphere.

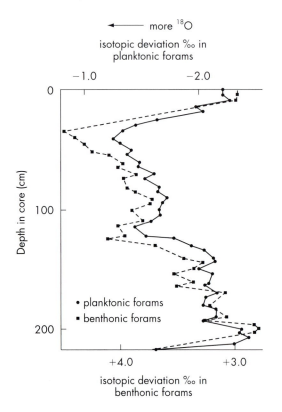

Figure 4.9 The oxygen isotope curves shown in Figure 4.8 plotted on the same scale of isotopic change, but with the zero points differing by 5.3‰ so that the most recent maximum is the common point. This value is the difference between present day planktonic and benthonic values.

The time series shown in Figure 4.10 also contains a long record of global sea-level change and ice volume: but how can we calibrate such curves to show sea-level change?

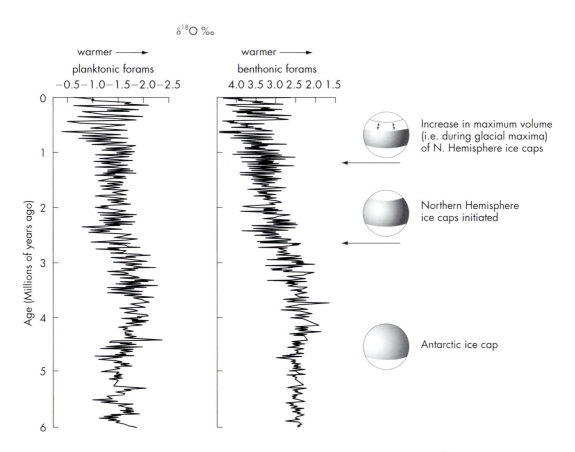

Figure 4.10 A composite plot of records from the western equatorial Pacific Ocean showing $\delta^{18}O$ values obtained from planktonic and benthonic forams. The three key stages in the development of the cryosphere are shown.

GLOBAL CHANGES OF SEA-LEVEL

There is no independent method available to estimate past ice volumes: in places we can map the maximum extent of ice sheets, and make rough estimates of their former thickness based on rates of isostatic uplift, but these data do not permit detailed calculations to be made of variations in ice volume through time. Fortunately there is an independent method of measuring past sea-level changes that enables the isotope proxy data to be calibrated in terms of global sea-level rises and falls. This is based on studies of the elevation of coral reefs, a method first used in Barbados, but subsequently applied in Papua New Guinea where older reefs occur.

Figure 4.11 shows a staircase of terraces on the Huon Peninsular, Papua New Guinea. Each terrace is a fossil coral reef that has been left high and dry by relative changes in sea-level.

☐ Assuming that coral reefs grow at or very close to sea-level, what does the staircase of reefs indicate about changes in relative sea-level in this area?

■ It must have been lowered to leave each reef-terrace high and dry.

The relative drop in sea-level could have been caused by tectonic uplift of the crust, or by global sea-level fall due to land ice formation (i.e. a glacio-eustatic fall), or by a combination of both causes. In fact rapid tectonic uplift occurs along the Huon Peninsula, ranging systematically along the coast from 0.5–3.0 mm per year. It does not seem that in any one place this rate of uplift has varied significantly during the last few hundred thousand years. If global sea-level remained the same, the reefs would not flourish, as despite being ideally situated near the equator, they would be killed off as tectonic uplift raised them above the water level.

☐ What would happen to the fringing reefs if global sea-level rose by a few millimetres per year due to the melting of ice caps?

■ They would be able to survive for longer, as glacio-eustatic sea-level rise kept pace with, or even outpaced, tectonic uplift.

Reefs have no difficulty in keeping up with a relative sea level rise since corals can grow at rates of more than a few millimetres per year. This means that they seldom drown, but they die if exposed to relative sea-level falls. The reef terraces, therefore, mark periods when glacio-eustatic sea-level rise outpaced tectonic uplift. Thus the flights of fossil reefs, which extend up to an elevation of 700 m, record ice-melting episodes. The glacio-eustatic sea-level record can be determined by (1) radiometric dating of the fossil reefs and (2) subtracting the amount of tectonic uplift since their formation from their present day elevation. In this way, the Huon Peninsular reef record enables past sea-levels to be determined, except for the lowest stands, as rapid falling global sea-levels prevent reef growth. The results are shown in Figure 4.12, which also shows how the oxygen isotope record is calibrated in terms of sea-level change. The two plots have remarkably similar shapes. Matching peaks on the two curves enables the $\delta^{18}O$ curve to be calibrated in terms of global sea-level change. This shows that a 1‰ change in $\delta^{18}O$ is equiv-

Figure 4.11 A staircase of coral reef terraces exposed on the Huon Peninsular, Papua New Guinea.

alent to a 10 m change in global sea-level, although later work indicates it may be as much as 11 m.

OXYGEN ISOTOPE STAGES

The analysis of isotope curves we have just undertaken should have convinced you that they can be used as a proxy index of global sea-level and volumes of land ice. The correlation of patterns of isotopic change through time seen in cores taken in many parts of the world is so good that a series of global oxygen isotope stages has been established. The scheme of oxygen isotope stages can be applied world-wide for stratigraphic subdivision of ocean sed-

iments. In fact, it is so well-defined that it is even possible to detect time gaps due to erosion or non-deposition where discontinuities in the isotope curves occur. The stages are based on maxima and minima shown by isotopic records. Maxima are the 'peaks' seen on records such as that on Figure 4.13 at 8 and 120 ka.

☐ What do such maxima and minima represent?
■ The maxima represent interglacials, and the minima successive cooling episodes as land ice sheets grew and drew down global sea-level.

Figure 4.13 shows the isotopic stages interpreted from the δ^{18}O data for the last 150 thousand years. Stage 1 (at the top) represents the Holocene, the present warm interglacial stage. Stage 5e was warmer than today, with even less ice cover; it represents the last interglacial stage. It was followed by a period of growth of land ice with some interruptions. The most positive δ^{18}O values of Stage 2 reflect the most extensive glacial advances of the last glacial stage, and the very steep drop in δ^{18}O between Stages 2 and 1 is an example of a termination (Termination I) marking a period of very rapid deglaciation. A similar feature, Termination II, preceded the last interglacial stage.

It is clear from Figure 4.13 that the slow growth of land ice, indicating global cooling, was interrupted several times by a return to warmer conditions that partially melted ice sheets, as well as colder periods that caused their growth. These intervals of warming and cooling are known respectively as interstadials and stadials.

☐ Which isotope stages and sub-stages are equivalent to interstadials and stadials.
■ Stadials: 2, 4, 5b, 5d. Interstadials: 3 (early part), 5a, 5c.

The melting of land ice at the end of the last glacial stage was clearly very rapid, occurring over a few thousand years in contrast to the slow step-wise cooling that took place over some 90 thousand years. This asymmetry in the cycle of climatic change is one important feature that models of the climate system must explain.

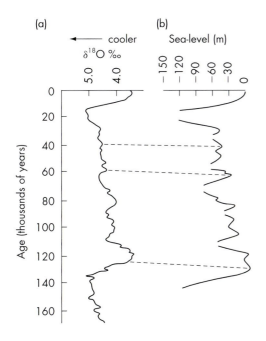

Figure 4.12 The eustatic sea-level record obtained from the coral terraces of the Huon Peninsular (b) compared to the oceanic oxygen isotope record (a). Note that the Huon record does not contain a record of the lowest stands of sea-level. The similarity between the two curves is remarkable. The slight difference in ages of the rises, peaks and falls is probably due to errors in the age dating of the two records. The δ^{18}O record has been smoothed, and displayed in time rather than depth in sedimentary cores.

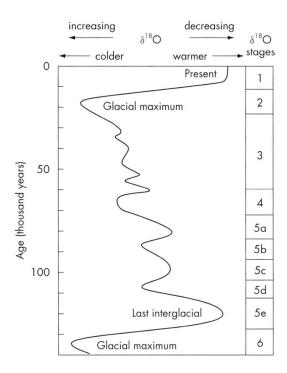

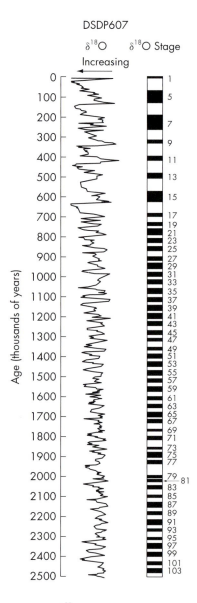

Figure 4.13 Oxygen Isotope Stages for the past 150 thousand years. Odd numbered stages indicate warm periods, and even numbers indicate cold periods.

Deeper cores provide a continuous record right through the Quaternary into the Tertiary. Just over 100 isotope stages are now recognised in the Quaternary (Figure 4.14), and the scheme has been extended back into the Miocene, and has the potential to be continued even further back in time. In the next chapter, we discuss how this long time series provides important clues concerning the forcing factors that may trigger glaciation and deglaciation, but first we need to examine how oxygen isotope studies have shown how the Earth has cooled since the Cretaceous.

POST-CRETACEOUS COOLING

Figure 4.15 shows the progressive decrease in the marine $\delta^{18}O$ record through Cainozoic time. This is generally accepted to indicate global cooling.

Figure 4.14 The $\delta^{18}O$ record from cores taken at Deep Sea Drilling Program Site 607 in the Pacific, showing isotope stages recognised since 2.5 Ma. The black bars indicate interglacials.

☐ Can you identify time intervals over which significant increases in $\delta^{18}O$ values occur? Ignore the relatively short term fluctuations that last 2–3 million years or less.

- Increases occur during the Middle to Late Eocene, the Middle Miocene, and from the Pliocene onwards.

- □ What is the most likely explanation for these isotopic trends?
- Significant increases in the volume of the cryosphere.

Identifying where on the Earth such increases in the volume of the cryosphere occurred requires direct sedimentary evidence of the former presence of ice sheets on land and/or extending over continental shelves. Deep sea drilling has recovered glacial sediments off Eastern Antarctica indicating the formation of an ice sheet in this region during the Middle and Late Eocene. Western Antarctica was glaciated during the Oligocene, but it is probable that an extensive ice sheet only developed during the Middle Miocene, at which time the Antarctic ice sheet grew almost to its present size. There is no sedimentary evidence for extensive Arctic ice sheets at this time, although it is likely that alpine glaciers appeared in northern high latitudes before this time, and that an ice sheet existed in Southern Greenland since about 5 Ma, from the beginning of the Pliocene. Intensification of Northern Hemisphere glaciation in Eurasia and North America occurred between 2.7 and 2.5 Ma.

Having reviewed the timing of significant growth episodes of the cryosphere, we now need to examine in more detail the nature of the changes in ice volume that occurred once the Earth had entered an icehouse state.

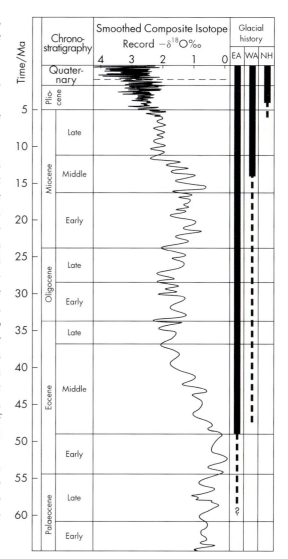

Figure 4.15 Composite marine $\delta^{18}O$ record for the Cainozoic. The columns on the right side of the diagram indicate the glacial history of the East Antarctica Ice Sheet (EA), the Western Antarctica Ice Sheet (WA) and ice sheets in the Northern Hemisphere (NH) based on the occurrence of ice rafted debris (Figure 3.1) and the isotope record. Solid lines indicate strong evidence for the presence of ice sheets in each area, and dashed lines indicate probable early phases of their growth.

SUMMARY

1 Unlike the fragmentary continental sedimentary record of past glacials and interglacials, some successions of deep sea sediments contain an uninterrupted proxy climate record extending back to the onset of glaciation in the Northern Hemisphere at about 2.6 Ma, and even earlier.

2 Ice rafted debris and calcareous oozes carry the most important palaeoclimatic signals.

3 Deep sea sediments formed by a rain of debris formed by the remains of planktonic organisms, and undisturbed by bioturbation, density (turbidity currents) and contour currents, have yielded long proxy climate records. These reveal that there were many more periods of global cooling and warming than suggested by the land based record.

4 Studies of the oxygen isotope composition of planktonic and benthonic forms contained in deep sea calcareous oozes reveal past changes in global sea-level caused by the formation and melting of land ice. This is because water containing the lighter isotope ^{16}O evaporates more rapidly than that containing the heavier ^{18}O. Condensation of water vapour favours the heavier isotope. As water vapour evaporated from sea surfaces in equatorial regions moves polewards, it is progressively enriched in ^{16}O, resulting in high latitude ice sheets being enriched in this lighter isotope. As the ice sheets increase in volume, the world's ocean waters become enriched in ^{18}O. Therefore the amount of enrichment in the heavier isotope is controlled by global land ice volumes.

5 The oxygen isotopic composition of the calcareous shells of organisms is controlled by the temperature of the surrounding water, and its isotopic composition. It is possible to separate the ice volume and temperature signals contained within oxygen isotope records. This is because it is probable that the temperature of deep ocean waters has not varied significantly during the last 2–3 million years. This means that isotopic variations obtained from bottom dwelling (benthonic) forams must be caused by changes in land ice volumes. Removing the land ice signal from isotopic data obtained from planktonic forams enables the palaeotemperature of surface water to be determined. Changes in the isotopic composition of fossil calcium carbonate material is expressed as variations in the value of $\delta^{18}O$ in comparison to a standard.

6 Rapid negative $\delta^{18}O$ excursions characterise glacial terminations, between which lower amplitude changes indicate shorter periods of cooling and warming known respectively as stadials and interstadials.

7 Variations in the value of $\delta^{18}O$ have been calibrated to show changes in global sea-levels. This was achieved by determining past global sea-levels indicated by uplifted coral-reef terraces in several parts of the world, particularly on the Huon Peninsula in Indonesia. The reefs, which only flourish at, or just below sea-level, have been uplifted by Earth movements at a constant rate. As they can be dated accurately by radiogenic methods, subtracting the amount of tectonic uplift from their present day elevation yields the past sea-level (which may be above or below the present day level). The resultant calibration of the $\delta^{18}O$ variations indicates that a 1‰ change is equivalent to a 11 m change in sea-level, and shows that during the last glacial maximum at 18 ka, global sea level was 120 m lower than it is today (allowing for isostatic effects).

8 Global synchroneity in past $\delta^{18}O$ fluctuations makes them a reliable tool for correlating deep-sea sediment successions around the world. Over 100 oxygen isotope stages have now been recognised, extending back to 2.6 Ma.

9 Oxygen isotope studies have documented the post-Cretaceous cooling of the Earth, revealing episodes of more rapid ice sheet growth in Antarctica during the Middle to Late Eocene and during the Middle Miocene. Intensification of Northern Hemisphere glaciation occurred at 2.6 Ma.

FURTHER READING

Open University Course Team, 1989. *Ocean Chemistry and Deep-Sea Sediments*. Pergamon Press, Oxford in association with The Open University. This book provides more detailed information about deep-sea sediments.

See also 'Further reading' section for Chapter 1 (pp. 28–9) and references to figure sources for this chapter (p. 255).

5

REVEALING THE MILANKOVICH PACEMAKER

INTRODUCTION

As explained in Chapter 3 (Box 3.4), early in the twentieth century, a Yugoslav astronomer, Milutin Milankovich, proposed that changes in the amount of solar radiation received at high latitudes in the Northern Hemisphere were the cause of ice ages. His ideas found little favour in Europe, and virtually none in America. Only when, from the 1960s onwards, the oceanic record revealed that there were many more than the four glacial periods recognised from the terrestrial record in Europe, did palaeoclimatologists realise that Milankovich was right after all. One of the pioneers of the oxygen isotope studies described in the previous chapter, Cesare Emiliani, summed up the irony of the revolution in opinion as follows:

> the revolution in our views of the ice ages is a true revolution because it destroyed a canon that had been cast in granite half a century earlier and that had generated a seemingly unassailable mountain of field evidence. To add insult to injury, that revolution was not engineered from a study of glaciers or at least glacial remains like moraines, drift sheets, or loess deposits.[*] It came instead from the deep-sea, which obviously has nothing to do with glaciers, and from mass spectrometry, which obviously has nothing to do with the trusted geological hammer.
>
> C. Emiliani. 1995. Two revolutions in the Earth Sciences. *Terra Nova*, 7: 595–6.

[*] Loess are dust deposits deposited during glacial stages; they are described on page 93 of this Chapter.

This chapter examines four proxy-climate records that show clear evidence for climate change driven by the Milankovich 'pacemaker'. We start by examining in more detail the oceanic oxygen isotope record of global sea-level/ice volume change, and then consider three terrestrial records: cave deposits, loess and ice cores. They all show that changes in the oceans, atmosphere and cryosphere were somehow driven by the Milankovich pacemaker.

THE DEEP-SEA OXYGEN ISOTOPE RECORD

Revealing the Pacemaker

Figure 5.1 shows a $\delta^{18}O$ record for the last two million years.

☐ Can you observe in Figure 5.1 a regular periodicity for the occurrence of interglacial periods during the past million years or so?

■ Interglacials occur at a periodicity of about 100 thousand years.

☐ Does the same frequency occur before about one million years ago? If not, can you determine whether there is still a regular frequency present in the $\delta^{18}O$ plot?

■ Before about one million years ago, the shape and width of the 'teeth' in the sawtooth pattern depicted in Figure 5.1 change. The teeth are narrower, and their asymmetry less evident. The warm peaks and cold valleys on the graph repeat at a frequency of about 40–50 thousand years and are roughly the same size, showing that cold and warm periods were about the same length. This

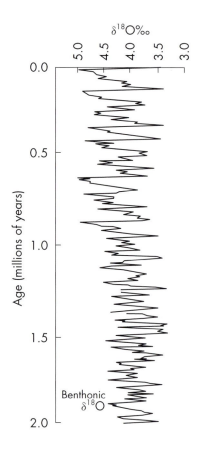

Figure 5.1 A composite deep-sea benthonic δ¹⁸O record for the past two million years. This was constructed by combining records from several cores (or parts of cores) containing high resolution records largely unaffected by bioturbation, and containing no erosional breaks.

contrasts with the ~100 thousand years frequency from about one million years to the present, when interglacial conditions prevailed for only about 10 per cent of the time.

☐ What astronomical phenomena do periodicities of 100 thousand years and 41 thousand years relate to?
■ Changes in the eccentricity of the Earth's orbit around the Sun occur at a periodicity of 95 thousand years, and the variations in the tilt of the Earth's rotational axis show a ~41 thousand years periodicity (Box 3.4).

A more objective analysis of time series, such as variations in oceanic δ¹⁸O or changes in solar insolation through time can be made by determining their frequency spectra. The principle underlying this technique is explained in Box 5.1.

The spectral analyses (see Box 5.1) shown in Figure 5.4 show beyond reasonable doubt that Milankovich frequencies are present in the oceanic δ¹⁸O signal for the past two million years.

☐ How does the amplitude of the different frequencies change from 2 Ma to the present day?
■ The 41 thousand years obliquity signal stays the same, but the precessional (19 thousand years and 23 thousand years) and especially eccentricity (100 thousand years) signals within the isotopic climate record increase.

It must be stressed that the changes in the relative strength of the different orbital signals is caused by feedback mechanisms *within* the Earth's climate system.

Since the 1970s, when the 100 thousand years signal was first recognised in oceanic δ¹⁸O records, palaeoclimatologists have assumed that it was the result of the 105 thousand years and 95 thousand year periodicities (of, respectively, the precession of the eccentricity of the Earth's orbit and change in the eccentric shape of the orbit (Box 3.4)) being blurred into one.

The strengthening of the 100 thousand years signal is accompanied by larger positive excursions of δ¹⁸O that indicate the formation of significantly larger volumes of land ice in the past one million years or so.

So, there is a clear link between past episodes of global cooling and warming, as indicated by the isotopic record of land ice volumes, and the Milankovich pacemaker. Is the signal a Northern Hemisphere one, as predicted (or strictly, retrodicted) by Milankovich? Figure 5.5 shows insolation curves for 60°N and S. The two curves have very similar shapes, but are out of phase: a warming trend in one hemisphere is matched by cooling in the other. The reason that the shape of one curve is not an exact mirror image of the

Box 5.1 Time Series and Frequency Spectra

The record in Figure 5.1 shows variations in temperature with time, and are called (in mathematical terminology) time series. In the case of the time series in Figure 5.1 we interpreted how often the pattern repeats – the period or frequency of the curve. For a simple wave, period is easy to interpret. For example, in Figure 5.2 the period is 20 thousand years (the time between the peaks, or troughs of the wave). Trying to interpret by eye more complicated curves that combine to give the climatic record of Figure 5.1 is even more difficult: a peak about every 100 thousand years back to one million years ago? How about the rest of the pattern?

Complex curves can be made by mixing several simple curves with different wavelengths/frequencies. The mathematical term for such mixing is synthesis: this is why an electronic keyboard is called a synthesiser. The instrument produces a wide range of sound by mixing signals with dif-ferent frequencies. The combination of frequencies that builds a complex natural curve such as that shown in Figure 5.1 can be analysed by the reverse of this process. It is a complex operation, but one which can be done routinely by a computer (an explanation of the details of how it is done are not necessary here).

The periodicity of a wave with a single period, such as in Figure 5.2(a) would be represented by Figure 5.2(b). The single line indicates that there is only one period present in the wave.

A graph such as Figure 5.2(b) is called a frequency spectrum and the mathematical process of producing frequency spectra from time series is called spectral analysis. Frequency is the reciprocal of period: frequency $= \dfrac{1}{\text{period}}$.

Figure 5.3 shows the frequency spectra for the Milankovich orbital changes described in Box 3.4. Unlike Figure 5.2, none of the spectra show a

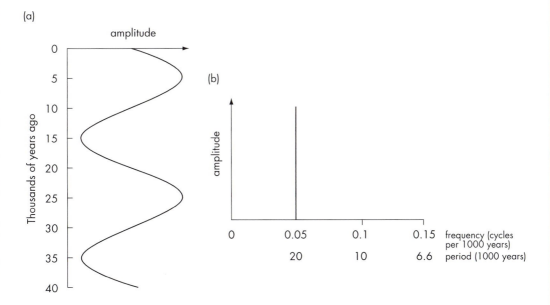

Figure 5.2 (a) A periodic time series. (b) A frequency spectrum of the time series shown in (a). The horizontal scale is linear for frequency (b) but not period (a), and longer periods are towards the left.

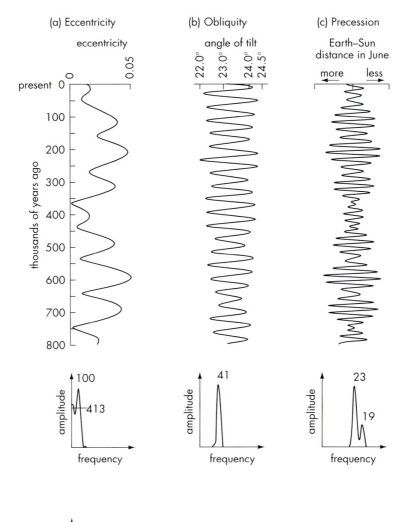

Figure 5.3 Frequency spectra of the Milankovich orbital changes. (a) Variations in orbital eccentricity (the higher the value, the more elliptical the orbit: an eccentricity of zero indicates a circular orbit). Two versions of frequency spectra are shown: one indicates the presence of a strong 100 thousand years cycle and a weaker 413 thousand years one and in the other periodicities are at 400 thousand years (the strongest), 125 thousand years and 95 thousand years. (b) Variations in the angle of tilt of the Earth's rotational axis show a strong 41 thousand years cyclicity. (c) Precession expressed as the Earth–Sun distance in June. This shows a strong 23 thousand years cycle and a weaker 19 thousand years one.

single line, indicating just one period, but instead the line has broadened into one or more peaks, showing that although there are one or more main periods there are also periods of less amplitude with values just above and below the main periods. This is the reason that even the plot for obliquity shows a peak rather than a single line; although 41 thousand years is the dominant tilt period, the time series is modified by other periods close to 41 thousand years. The precession also has two peaks, the main one at 23

thousand years and a lesser one at 19 thousand years.

Two spectral analyses of eccentricity variations are shown in Figure 5.3(a). One is the version commonly shown in text books: it has two peaks at 413 thousand years and 100 thousand years, with the latter being the strongest. But the calculations of Milankovich and later workers showed that eccentricity varies with three different periodicities: 400 thousand years (the strongest), 125 thousand years and 95 thousand years.

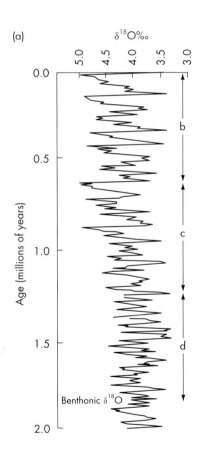

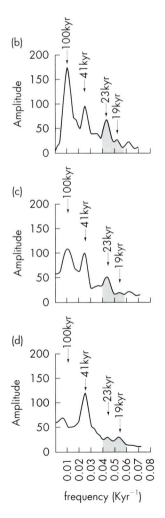

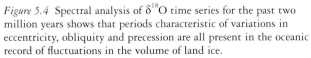

Figure 5.4 Spectral analysis of δ^{18}O time series for the past two million years shows that periods characteristic of variations in eccentricity, obliquity and precession are all present in the oceanic record of fluctuations in the volume of land ice.

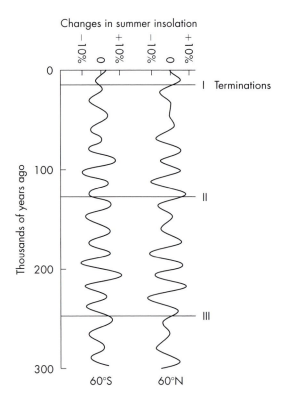

Changes in summer insolation

60°S   60°N

Figure 5.5 Variations in solar insolation in June and December at 60°N and 60°S. The timings of glacial Terminations I, II and III (as determined from oceanic δ^{18}O variations) are shown.

other is due to the eccentricity of the Earth's orbit – today the planet is nearest the Sun during December.

- ☐ How did solar insolation change in both hemispheres at the time of Terminations I and II?
- ■ Warming in the Northern Hemisphere, and cooling in the Southern Hemisphere, with a greater magnitude of change in the north.

So why should global temperature changes during the last million years track summer insolation variations reaching Northern Hemisphere high latitudes? Somehow, the Earth's climate system amplified the Northern Hemisphere eccentricity pacemaker (and, in particular, the near-100 thousand years component of it) to give an asymmetric global signal of ice-

volume changes indicated by the oxygen isotope signal obtained from ocean floor sediments. Figure 5.6 illustrates this 100 thousand year cycle problem. It shows that whereas there is a strong linear relationship between variations in summer radiation received at high northern latitudes related to obliquity and precessional variations, the amplitude of the 100 thousand years variation is only 2 W m^{-2} which is an order of magnitude smaller. Yet the response of the Earth's climate system, as recorded in changes in the volume of land ice through time, is almost twice as large as the linear response to obliquity and precessional changes. This suggests that the 100 thousand years cyclicity is the result of the amplification of orbital variations by mechanisms *within* the Earth's climate system.

Before one million years ago, the repeated cooling and warming of the Earth over periods of about 41 thousand years was symmetric following changes in Northern Hemisphere high latitude insolation caused by variations in obliquity. When the frequency of climatic change switched to 100 thousand years, the pattern became asymmetric, with slow, usually step-like cooling followed by rapid warming. The relationship, therefore, between climatic shifts and orbital forcing changed from being linear (related to obliquity) to *non-linear* (apparently related to eccentricity) around one million years ago. This change is sometimes referred to as the Mid-Pleistocene Revolution: satisfactorily explaining this revolution is a major problem that must be overcome by palaeoclimatologists. Possible solutions are discussed in Chapter 7.

Although the majority of palaeoclimatologists agree that changes in the shape of the Earth's orbit (eccentricity) were the pacemaker that controlled glaciations for the past million years or so, there is another astronomical explanation of the 100 thousand years cycle. This postulates that changes in the inclination of the plane of the Earth's orbit relative to the plane of the solar system drive the climatic shifts. This, it is suggested, dips the Earth in and out of a band of cosmic dust concentrated along the plane of the solar system, somehow triggering glacial episodes due to an as yet unknown mechanism. The

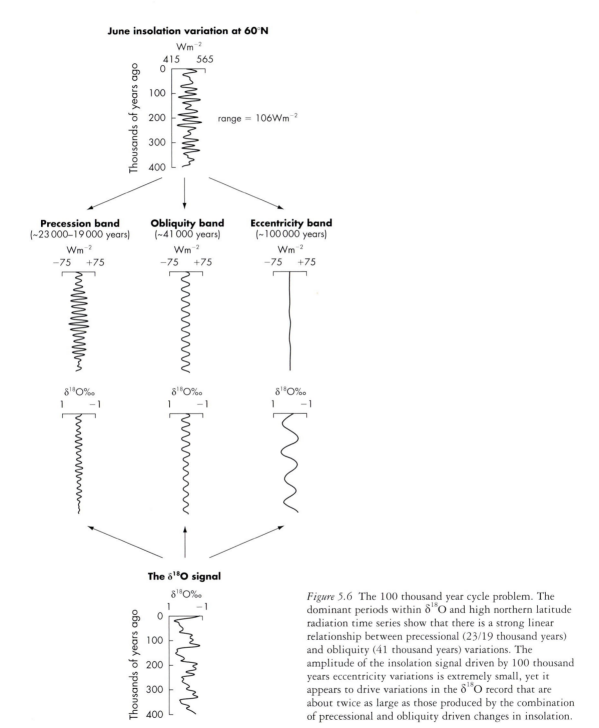

Figure 5.6 The 100 thousand year cycle problem. The dominant periods within $\delta^{18}O$ and high northern latitude radiation time series show that there is a strong linear relationship between precessional (23/19 thousand years) and obliquity (41 thousand years) variations. The amplitude of the insolation signal driven by 100 thousand years eccentricity variations is extremely small, yet it appears to drive variations in the $\delta^{18}O$ record that are about twice as large as those produced by the combination of precessional and obliquity driven changes in insolation.

proponents of this theory base their argument on spectral analyses of the $\delta^{18}O$ time series that show for the past million years only a 100 thousand years cycle is present, and not the three eccentricity periodicities (400 thousand years, 125 thousand years and 95 thousand years) shown in Figure 5.3(a). At this point, readers may be thinking: 'so what – this just adds to the evidence that the 100 thousand years eccentricity cycle is the pacemaker'. Remember, however, that the previous discussion in this book is based on the conclusion reached by many workers who assumed that the 100 thousand year cycle is a combination of the 125 thousand years and 95 thousand years eccentricity components. In fact the 400 thousand years component is the strongest of all, yet does not show up in the $\delta^{18}O$ time series although it is evident in records of plankton productivity in the tropics. Spectral analysis of variations in the orbital *inclination* show just one peak – a 100 thousand years cycle, which matches better with the $\delta^{18}O$ data. Despite this match, the inclination theory has found little favour because it seems unlikely that variations in the amount of cosmic dust would significantly change the amount of solar energy reaching the Earth.

Oxygen Isotope Time Stratigraphy

The global synchroneity of the changes in the isotopic composition (especially the peaks and troughs) of ocean water and calcareous organisms living in it provides a means of global correlation, using the oxygen isotope stages described in the previous chapter.

☐ How rapidly will ice volume changes lead to changes in the deep water $\delta^{18}O$ composition of ocean water throughout the world?

■ The change in composition will, in geological terms, be virtually instantaneous. This is because oceanic circulation takes place far more rapidly than the growth of ice sheets (tens to hundreds of years versus thousands of years; see Table 2.1).

This means that the peaks and troughs in benthonic $\delta^{18}O$ curves obtained from cores of ocean floor sediments recovered from different locations around the

world can be assumed to be of the same age in geological terms. Such correlation does not require that the ages of the peaks and troughs be determined precisely everywhere.

Accurate dates are needed in order to be able to determine the time lags between the response of the climate system to changes in insolation. The problem is that it is difficult to obtain accurate dates of samples of deep-sea sediments older than 40–50 thousand years. This problem has been overcome by 'orbitally tuning' oceanic $\delta^{18}O$ time series. Great care was taken to use data from successions of cored sediment that showed no signs of erosion or bioturbation, or intervals of unusually slow or rapid deposition. This meant that a constant sedimentation rate could be assumed between the sample points for which precise dates were available. For isotope time series such as that shown in Figures 5.1 and 5.4 six dated horizons were known at the time that orbital tuning was undertaken:

● the top of the series (i.e. the present day);

● four radiocarbon dates between the present and 35 ka;

● the age of the last magnetic reversal (i.e. when the polarity of the Earth's magnetic field flipped) at 775 ka. (See explanation given in Figure 5.10.)

Clearly, there is not an even distribution of dates – there is a long (740 thousand years) gap across which ages had to be determined by assuming a constant rate of deposition. Using this initial age calibration, frequency spectra were generated; they did not match perfectly with those shown by the Milankovich pacemaker (cf. Figures 5.5, 5.6). The age calibration was then adjusted many times to obtain a better match. In fact, 120 iterations were needed to obtain the best 'fit' between the frequency spectra of the $\delta^{18}O$ time series and the astronomical spectra. This is analogous to tuning a piano by adjusting the tension of its strings so that the sound they emit matches that emitted by tuning forks: the $\delta^{18}O$ time series equates to the piano strings, and the orbitally induced variations in solar energy to the tuning forks. The process also involved choosing a time lag between calculated insolation changes and the growth and demise of ice sheets.

☐ Why should there be a time lag between the formation and melting of land ice, and insolation changes?
■ Ice sheets take time to form, and so do not respond immediately to any forcing factors, such as insolation changes. This is in marked contrast to the almost instantaneous response of the oceanic $\delta^{18}O$ signal to ice formation or melting.

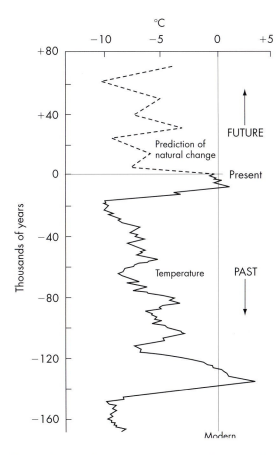

Figure 5.7 A prediction of future climate change based on observations of the past that show the link between astronomically induced variations in Northern Hemisphere high latitude summer insolation and global sea-levels/ice volumes, and predictions of future orbital changes.

You should now be able to appreciate why some caution needs to be exercised when using the ages of the isotope stages. Whereas there can be little doubt that the overall pattern of the standard curve does reflect a truly global signal of ice volume changes through time, only the ages of the most recent isotope stage boundaries have been verified by direct dating of closely spaced samples of deep-sea sediments.

The Future?

Just as past changes of eccentricity, obliquity and precession can be calculated, so can those that will occur in the future. Our knowledge of the relationship between the Milankovich pacemaker and past global sea-level/ice volume changes enables predictions about the future to be made: an example is shown in Figure 5.7. It shows that we are very near the end (in geological terms) of the present interglacial. But will the warming effects of adding CO_2 to the atmosphere by burning fossil fuels, etc. postpone the predicted cooling? As we shall see in Chapter 7, our understanding of the Earth's climate system does not yet enable us to fully explain past changes of climate, let alone predict what the consequences will be of our global experiment in changing the greenhouse gas content of the atmosphere.

CAVE DEPOSITS

Introduction

It may seem surprising that a mere 36 cm thickness of calcite precipitated in a cave in Nevada has, at the time of writing, yielded the best dated record of climate change since 500 ka. This is because small amounts of radiogenic isotopes present permit, in some cases, accurate dates to be obtained from tiny samples collected from individual growth layers in calcitic material precipitated in caves. These isotopes are not usually present in sufficient amounts in deep-sea sediments.

Oxygen isotope analyses of the cave deposits reveal changing $\delta^{18}O$ values that reflect isotopic variations in atmospheric precipitation falling in areas from which the groundwaters were derived. It may take water falling on the ground surface thousands of years to reach sites where cave calcite deposits may be sampled, and so the ages of fluctuations in $\delta^{18}O$ values will be *younger* than the changes in isotopic composition of the parent atmospheric water vapour.

The Devil's Hole Record

Devil's Hole is an open fissure up to 2 m wide developed along a fault, and lined with calcite precipitated by groundwaters. It is situated near the California–Nevada border about 115 km west northwest of Las Vegas on the edge of a large groundwater basin that discharges water along a line of springs on its southwestern side.

A 36 cm long core, obtained by divers working 30 m below the water table, showed that continuous calcite deposition occurred between 50 ka and 600 ka. 221 samples were dated radiometrically. No other record over this period of time has been calibrated by so many radiometric dates. A sampling interval of 1.26 mm along the core represents an average time interval of 1800 years. The plot of the $\delta^{18}O$ values obtained from the samples reveals the familiar saw tooth pattern of the marine record (Figure 5.8).

The importance of the Devil's Hole record is that it is well dated by direct methods, rather than ages being obtained by tuning to the Milankovich pacemaker as described for oceanic sediments. This led authors of the study to question the Milankovich hypothesis. They compared their results to the marine $\delta^{18}O$ record. There is a strong similarity between these records, but when they are examined more closely, there are some significant differences.

☐ Are the ages of glacial terminations the same in the Devil's Hole and marine $\delta^{18}O$ records?

■ Apart from termination IV, the ages of the other terminations are not the same on the two records.

The differences in ages, in thousands of years, of Terminations II and III on the two records shown on Figure 5.8 are as follows:

	Devil's Hole	Marine record
Termination II	~140	~130
Termination III	~253	~244

The larger uncertainty in the dates for Terminations IV and V in the Devil's Hole data means that

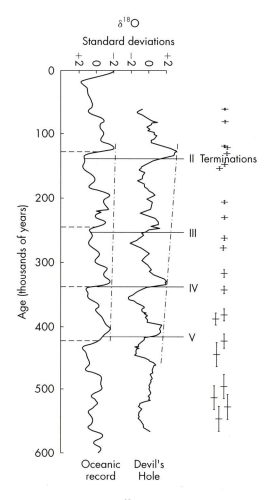

Figure 5.8 Variations in $\delta^{18}O$ values along the 36 cm core obtained from Devil's Hole, Nevada. Horizontal lines to the right of the graphs show where radiometric ages were determined; the vertical lines are error bars (two standard deviations). The $\delta^{18}O$ oceanic record is shown for comparison. The ages of the Terminations shown are determined from the Devil's Hole record; the dashed lines on the left show ages of Terminations interpreted from the oceanic record.

they are not significantly different from those in the marine record.

☐ If an allowance were made for the time it takes water to migrate across the groundwater basin, would this improve the match between the ages of Terminations indicated by the Devil's Hole and $\delta^{18}O$ curves?

■ No, it would make matters worse, as it would increase the ages of the Terminations.

A World Without Milankovich?

Commenting in 1994 on the Devil's Hole results, a leading researcher on Quaternary climatic changes, Wallace Broecker, commented:

> I remain confused. The geochemist in me says the Devil's Hole chronology is the best we have and the palaeoclimatologist in me says that correlation between the accepted marine chronology and Milankovich cycles is just too convincing to be put aside. One side will have to give, and maybe – just to be safe – climate modellers should prepare themselves for a world without Milankovich.
> W.S. Broecker. 1992. Upset for Milankovich theory. *Nature*, 359: 779–80.

The discrepancy between the oceanic and cave records was subsequently explained in two ways. It could be due to changes in the accumulation rate of the cave deposit which would affect the deposition rate of the radiogenic isotopes used for radiogenic dating. Another possible explanation is that the chronology based on radiogenic dating is correct, but that the $\delta^{18}O$ variations reflect regional changes in climate, rather than global changes in oceanic/atmospheric $\delta^{18}O$ content.

Marine records in both the Pacific and Atlantic Oceans indicate that at 150 ka, at the penultimate glacial maximum, oceanic polar fronts were about 5° further south than they were at the last glacial maximum. They also show that significant warming of ocean surface waters to interglacial levels had occurred when the ice sheets began to retreat 145 thousand years ago. The oceanic $\delta^{18}O$ record indicates that there were still extensive ice sheets at this time.

The continued presence of the North American ice sheet would have diverted Pacific low pressure systems across the United States, bringing precipitation derived by evaporation of the warmer ocean to the west. Thus the early warming indicated by the Devil's Hole $\delta^{18}O$ data may record changes in the position of the Pacific oceanic polar front and the extent of the North American ice sheet, rather than pointing to a 'world without Milankovich'.

DUST, MONSOONS AND MILANKOVICH

Introduction

Dusty conditions stem from dryness, windiness, and a lack of vegetation that binds soil. Today, such conditions occur most widely in deserts. As described in Chapter 1, during glacial stages regions equatorward of huge ice sheets were very dusty places for two reasons. First, wind speeds were much higher. This was partly because the temperature gradients between the equator and the poles were higher then than they are today. In addition, very large high pressure areas developed over the extended cold areas at high latitudes, so that strong winds blew from them into warmer lower latitudes. Second, erosion caused by ice (armoured, like sand-paper, by rock fragments) produced huge amounts of rock flour. This was released continuously at melting ice fronts, and the strong winds blew this fine dust over huge areas to form thick loess deposits which today cover about 10 per cent of the Earth's total land area (Figure 5.9). Loess is composed of fine grained sedimentary particles, over 80 per cent of which are usually between 0.005 (5 μm) and 0.5 (500 μm) mm across, with the bulk being at the finer end of this size spectrum.

Chinese Loess Deposits

In north-central China, loess deposits form an extensive (440 000 km³) plateau with an elevation just under 1000 m (Figure 5.9(b)). The deposits are up to 180 m thick, and provide a virtually continuous

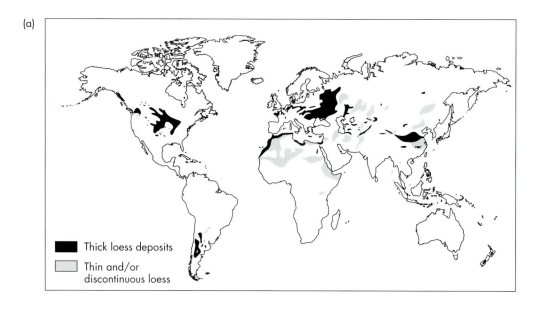

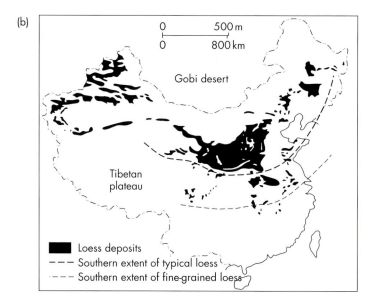

Figure 5.9 Distribution of loess deposits: (a) global occurrences, and (b) occurrences in China

sedimentary record spanning the last 3.5 million years. The loess was transported from deserts in northwestern China during glacial stages. The present day winter wind direction (Figure 1.19) is probably similar to that which prevailed at such times. Today, strong winter cooling results in the development of high atmospheric pressure over Siberia and Mongolia. This high pressure, together with a large low pressure cell over the NW Pacific ocean, causes strong, cold, dry winds to flow from land to sea, forming northwesterly winds over the loess plateau. Evidence that such a wind direction prevailed during the deposition of the loess is provided by the fact that the grain size and thickness of the deposits diminish towards the southeast (Figure 5.9(b)). In contrast to the dry cold conditions of winter, the summer monsoon winds, which blow from the southeast, bring warm and wet conditions to the Loess Plateau between May and September. Most of the region's annual rainfall of about 700 mm falls at this time. This results in a present day climate that is relatively warm (mean annual temperature of around 13°C) and wet.

The loess succession in the southern part of the plateau ranges in thickness from 130 to 180 m, and contains over 30 loess–soil alternations. The alternations are a record of regional climatic fluctuations during the past 2.5 million years. The loess units have a yellowish colour, and a high carbonate content. They are interpreted as having been deposited in glacial periods when the climate was arid and there was little vegetation cover. The soils have brownish or reddish colours, and low carbonate contents, and are finer grained than the loess deposits. They developed during humid periods (similar to the present day conditions in the area) when the influx of dust was reduced and vegetation was more prolific.

The Loess Succession at Baoji

The stratigraphy at one well studied locality, Baoji, is shown in Figure 5.10. The age of the succession is known from palaeomagnetic dating, with the boundary between the Matuyama and Gauss magnetic epochs (2.43 Ma) occurring within the lowermost loess unit.

Thirty-two major soil units have been identified in the field. Each soil unit shows soil features (not discussed here) that are comparable to or better developed than the topmost soil (S0 on Figure 5.10). There are two lines of evidence indicating that the succession at Baoji contains a relatively continuous record of loess deposition, and hence a continuous record of climate change. First, the loess intervals have undergone some alteration by soil forming processes, which suggests the existence of vegetation cover during glacial periods which prevented wind erosion after deposition. The basal parts of the soil horizons show grain size distributions similar to those of the underlying loess intervals, which suggests that only a small part of the loess deposited during a glacial period was subsequently altered by soil processes.

Samples were taken every 10 cm through the succession at Baoji, and grain size analyses conducted on each one. The resultant plot of grain size ratio through the succession is shown on the right side of Figure 5.10.

☐ Do you notice any marked change down the succession in the spacing of the 'spikes' in the grain size ratio plot in Figure 5.10?

■ There is a change in spacing from S6 downwards. Above S6, there is a broad spacing between the spikes, and a narrower one below.

☐ How many major spikes formed by deviations to the right above the base of the Brunhes magnetic epoch are there?

■ Seven, if the subsidiary spike at ∼330 ka is ignored.

☐ What is the average interval in time of the occurrence of these seven spikes above the base of the Brunhes epoch?

■ About 100 thousand years.

☐ What does this time interval remind you of?

■ The 100 thousand years spacing between interglacials seen in oceanic and ice core records, and the periodicity of variations in the eccentricity of the Earth's orbit.

There is a strong resemblance between the grain size variations at Baoji and the patterns of change shown by $\delta^{18}O$ values obtained from ocean sediments (compare Figures 5.10(b) and 5.1).

Spectral analysis of the Baoji grain size variations reveals a major change in the dominant periodicities about 800 thousand years ago. The 100 thousand years eccentricity periodicity dominates back to that time, before which the 41 thousand years obliquity signal predominates. Prior to 1.6 Ma, spectral analysis shows significant peaks at 400 thousand years (eccentricity), 41 thousand years and 55 thousand years (which cannot be matched to any Milankovich cycles). As yet, this earlier change has not been observed in the oceanic record, and so may be a local phenomenon. The succession of loess soils and soil deposits at Baoji (and others in the Loess Plateau area) has the potential to yield a high resolution climate record completely independent from those based on isotope studies of oceanic sediments. This will require high resolution dating which still remains to be done (locating the magnetic reversals within the succession gives only a small number of irregularly spaced dates).

Figure 5.10 The Baoji loess–soil succession: (a) the stratigraphy (black/S: soils) and (b) grain size variations. GSR: grain size ratio. The significance of the grain size changes is discussed in the text. The age of the sediments was determined from the magnetic properties of the sediments. When sediments settle out from suspension, any grains of magnetic minerals within them will become oriented close to the configuration of the Earth's magnetic field at the time of deposition. The Earth's magnetic field is known to have flipped many times in the past – in other words there were periods when a compass needle would have pointed southwards instead of northwards. A timetable of these magnetic polarity changes is now well known, so that epochs of normal (compass needle pointing north: shown black) and reversed (compass needle pointing south: shown white) have been defined (shown in the middle column). The timing of changes from reversed to normal polarity and vice versa has been used to date the Baojii sediments.

Monsoons and Milankovich

The grain size changes at Baoji are a proxy record of variations in the intensity and/or location of high atmospheric pressure over Siberia and the associated monsoonal winds and associated seasonal precipitation, and so provide another insight into the link between Milankovich cycles and components of the Earth's climate system.

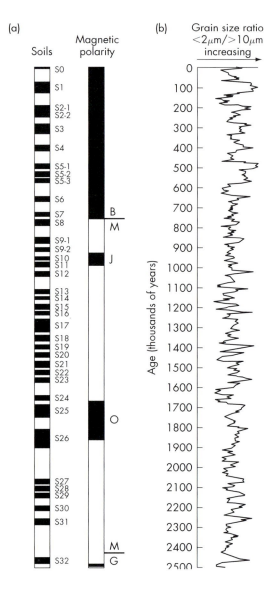

Wet and dry periods in low latitude regions can be discerned in terrestrial and oceanic records, but continuous records spanning the last 150 thousand years only occur in the oceans. The terrestrial record is seldom continuous. This is because lake sediments are mostly deposited during wet periods; they may be eroded by high winds during dry intervals. Deposits of wind blown sediments are only preserved when water tables rise. This is because wet sediments cannot be eroded by wind action.

Lakes only form when the level of ground water (i.e. the water table) rises above the surface of topographic depressions. Such rises reflect increased rainfall and reduced evaporation. The records shown in Figure 5.11(a) are compilations of the ages of lake sediments indicating permanent (i.e. lakes that do not dry up during the summer) freshwater or saline lakes.

As already stated, wind cannot pick up sedimentary grains from wet surfaces. Therefore the amount of dust in ocean sediments is an indication of both elevated wind speeds and aridity in the adjacent land area (Figure 5.11(b)). The presence of freshwater diatoms in oceanic sediments (Figure 5.11(c)) provides another aridity indicator. The diatoms can only be blown out of lake sediments after they have dried up, or when exposed diatomaceous sediments are dry.

Layers rich in organic carbon occur in cores of sediment recovered from the floor of the Mediterranean Sea north of the Nile Delta mark periods when large inputs of freshwater inhibited the mixing of surface and deep waters. Such mixing carries oxygen to the sea bottom, resulting in the oxidation of organic matter derived from dead plankton and other sources. When the River Nile discharge was very high, a layer of freshwater developed at the surface of the eastern Mediterranean, above denser more saline water. This shut down convective circulation, and so oxygen consumed by the oxidation of organic matter on the sea bottom was no longer replaced. This led to the accumulation of organic rich sediments known as sapropels. Four sapropel horizons are known off the Nile Delta (Figure 5.11(d)), and are thought to mark periods of high discharge from the Nile, which were caused by increased monsoonal rainfall in its catchment area far to the south.

Figure 5.11(e) shows variations in the average Northern Hemisphere summer insolation since 150 ka. There is a good match between the major peaks on this curve (times when the summer thermal contrast between continent and ocean would be highest, triggering the strongest summer monsoons) and the proxy climate data shown in the rest of Figure 5.11. The match is particularly good for the organic carbon present in sediments of the Nile Delta, and there is a reasonable match with high lake levels. Indicators of high aridity and wind velocities (dust levels and the abundance of freshwater diatoms in ocean sediments) are mutually exclusive of the wet monsoonal periods.

The proxy climate data for the last 150 thousand years shown in Figure 5.11 demonstrate that four wet monsoonal periods can be detected that coincide with peaks in solar insolation related to precession. These occurred during the present and the last interglacial, and the other two during the early cooling that led to the last glacial.

As shown in Figure 5.11(b) records of past dust fluxes can be obtained from oceanic sediments; they are also present in cores taken through ice sheets in Greenland and Antarctica (Figure 5.16). Everywhere the pattern is broadly the same: dust fluxes increase during cooler stadial phases (i.e. isotope stages 4 and 2). So were the increases in dust flux merely a consequence of global aridity and higher wind speeds, or could dust in the atmosphere contribute to feedback processes that would have contributed to global cooling?

Windblown dust contributes to the aerosol content of the atmosphere. Aerosols are small particles (0.01–10 μm in diameter) that remain suspended in the atmosphere for periods ranging from days to months or a few years. Their residence time in the atmosphere depends on their size, atmospheric circulatory patterns, and the rate at which they are washed out by precipitation. Unlike sulphate aerosols derived from volcanic eruptions and the burning of fossil fuels (Box 7.1), dust aerosols reflect and absorb both incoming (visible) and outgoing (infrared) radiation:

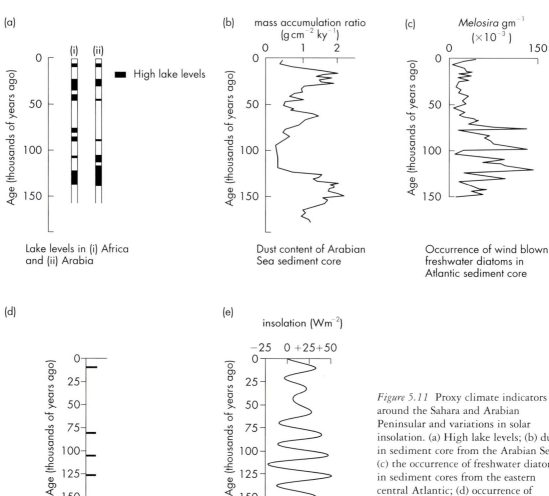

Figure 5.11 Proxy climate indicators around the Sahara and Arabian Peninsular and variations in solar insolation. (a) High lake levels; (b) dust in sediment core from the Arabian Sea; (c) the occurrence of freshwater diatom in sediment cores from the eastern central Atlantic; (d) occurrence of sapropels in sea-floor sediment from the Mediterranean Sea north of the Nile Delta; (e) Variations in summer solar insolation in the Northern Hemisphere.

most sulphate aerosols only reflect radiation. It is probable that the reflection of incoming solar energy is almost cancelled by the absorption of outgoing infrared radiation. This does not mean, however, that dust aerosols cannot affect climate. This is because dust may *reduce* the amount of energy reaching the ground, but *increase* the amount retained in the atmosphere. This may change atmospheric stability in ways not yet understood. Today, about half the atmospheric dust originates from human disturbance of soils in arid regions, and results in strong cooling over the western North Pacific, the North Atlantic and Arabian Sea. This is evidence of regional, rather than global, forcing of climate change due to atmospheric dust loading. Such regional forcing would most likely have been much stronger on a more arid and windy planet. As we shall see later, regional changes in the climate – particularly in the North Atlantic – may trigger global cooling and warming. Atmospheric dust also supplies nutrients to the oceans, and so may affect the rate at which the 'biological pump' exchanges CO_2 between the oceans and atmosphere (Box 5.2).

THE RECORD IN ICE SHEETS

Introduction

Ice is, in essence, frozen atmosphere, containing a record that may, in Antarctica, go back in time half a million years. Ice sheets, therefore, have the potential to yield very high resolution records (at least in their upper parts) of past changes in the atmosphere to a decadal or even annual level of resolution.

As snow is buried it compacts to produce small interlocking crystals of ice that are visible in thin slices using a polarising microscope. As the ice is buried to deeper levels, the crystal size increases, and bubbles of trapped air eventually become diffused into the ice crystals to that they are no longer visible. Analysis of the ice, and its dust and gas content, yields information concerning variations in:

- surface temperature at the drilling site as snow fell;
- the storminess of the atmosphere as indicated by the amount of dust and sea salt preserved in the ice, and the 'dryness' of continental areas at lower latitudes from which the dust was blown;
- the content of greenhouse gases in the atmosphere.

The proxy climate data obtained from ice cores ranges in scale from local (surface temperature), through hemispheric (storminess of the atmosphere, aridity of nearby continents), to global (atmospheric composition). Ice sheets have the potential to yield time series that may indicate whether teleconnections occurred between the Northern and Southern Hemispheres. Teleconnections are linkages between widely separated climatic phenomena. Of particular interest is whether the Antarctic ice sheet contains evidence of the Milankovich pacemaker, and whether, if present, this signal is linked to changes in high latitude seasonal insolation in the Northern Hemisphere, as is

the case with the oceanic isotopic record of global ice volume changes, or with a Southern Hemisphere insolation signal.

Comparing results from ice cores obtained in Antarctica and Greenland is not as straightforward as it might seem. This is not only because ice accumulation rates were markedly different in the two areas, which affects the time resolution that can be obtained, but also because of problems in calibrating core depths in terms of their age.

Figure 5.12(a) shows the location of the ice cores that are discussed in this and the next chapter. As can be seen by examining the age/depth relationships shown in Figure 5.12(b) for two cores drilled in Antarctica and Greenland, ice accumulation rates were much higher in Greenland.

☐ Based on your reading of Chapter 3, suggest why ice accumulation rates were so much higher in Greenland?

■ Annual snowfall today is much higher over Greenland than it is over Antarctica. This is because moist air from the relatively warm North Atlantic frequently blows over Greenland, whereas Antarctica is isolated from the southern oceans by the cold circum-Antarctic ocean current resulting in less moisture being transported from the Southern Ocean to the ice cap.

The different ice accumulation rates result in ice records that have contrasting resolutions. Drilling in Antarctica has yielded a record back to about 400 thousand years ago. The ice record in Greenland yields much higher resolution data (that is discussed in the next chapter) back to about 110 thousand years ago, but the effects of ice-deformation near to bedrock mean that it is difficult to obtain reliable information from ice accumulated during the last interglacial stage, let alone for earlier glacial and interglacial stages.

In Greenland, it is possible to count annual layers of ice, much like counting tree rings, back to 8600 years ago. This has been done at the Dye 3 site, where accumulation rates were even higher than they were at the Summit site. The depth at which 8600 year old ice occurs in the Summit GRIP cores was determined

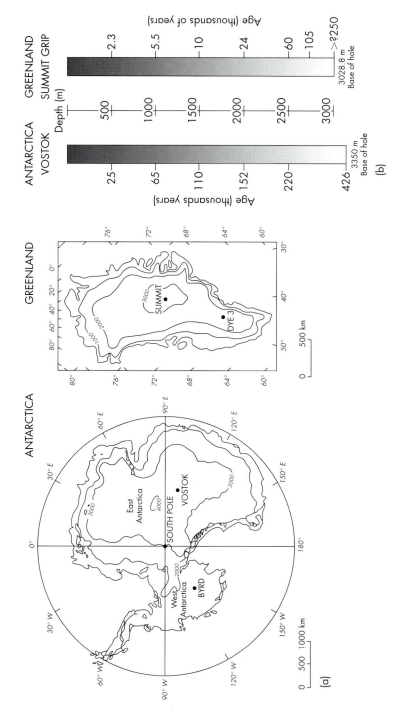

Figure 5.12 Ice cores in Antarctica and Greenland. (a) Map showing the drilling location of ice cores discussed in Chapters 4 and 5. Two holes were drilled near the summit of the Greenland ice cap: one by a European team (the Greenland Ice-core Project: GRIP) and another by an American group (Greenland ice-sheet Project: GISP). (b) Diagram showing the age/depth relationships and location of the ice/bedrock contact for the Vostok and GRIP boreholes.

by correlation with Dye 3 using thin volcanic fall-out layers identifiable in both cores. Seasonal layering, as determined from analyses of dust content, enabled layer counting to be extended back to just over half way down the core to 14 500 years ago (in the GISP cores, similar layer counting was extended to 17 400 years ago). Older, deeper, ice was then dated using ice accumulation and flow models based on the age/depth relationship determined for the younger ice. This had to take into account variations in snowfall during interglacial and glacial stages: it may have been up to 50 per cent higher during interglacials. Radiometric dating of ice is also possible using the isotopic ratio of beryllium 9 and 10 but this assumes that cosmogenic ^{10}Be is added to the atmosphere at a constant rate, which like the addition of ^{14}C (Box 7.2), is not strictly true. Annual layer counting was not possible in the Antarctic Vostok core, so age dating relied exclusively on modelling. The details of the modelling need not concern us, but it is important to remember that nearly all the dates quoted below were not obtained by direct methods, and that their precision diminishes with increasing depth. In the case of the Vostok core, the precision of the model-based dating at ~2000 m depth is probably plus or minus 10–15 thousand years – similar to the duration of interglacials, and longer than stadials and interstadials.

The high resolution data from Greenland ice cores is discussed in the next chapter. Here, the results from cores obtained from boreholes drilled at the East Antarctic station of the former Soviet Union are discussed. This station is at an altitude of 3488 m, with a mean annual temperature of −55.5°C. By January 1996, drilling had reached a depth of 3350m, covering the last four glacial cycles back to about 426 thousand years ago.

Isotopes and Ice

Isotope analyses of ice are used to determine the temperature of the site at the time the original snow fell.

☐ Will snow be isotopically heavy or light compared to the ocean water from which it was originally evaporated?

■ It will be isotopically light, for the reasons explained in Box 4.3.

A temperature change of 1°C at the site of condensation of water vapour results in a δ^{18}O deviation of 0.6–0.8‰ (Figure 5.13). Thus δ^{18}O values obtained from ice cores can be used to determine temperature changes relative to those measured today at the site. However, the δ^{18}O values may be affected by a number of factors for which corrections must be made:

● the isotopic composition of ocean water changes as ice sheets grow, because progressively more lighter isotopes are stored in the latter;
● as the height of an ice sheet increases, temperatures above it will decrease, so that the ice core record indicates cooling through time;
● δ^{18}O values are also influenced by the temperature of the original moisture source. This means that δ^{18}O values in ice could be affected by changes in the temperature of surface ocean waters, or by winds bringing moisture to the ice cap from different parts of oceans characterised by different temperatures.

It is important to remember that increased volumes of land ice result in *higher* δ^{18}O values in deep-sea sediments, and lower local temperatures produce *lower* values in ice cores. Ice sheets are isotopically light, showing δ^{18}O values down to almost −40‰, in contrast to ocean water, which is isotopically heavy (~δ^{18}O 5‰).

The heavier isotope of hydrogen, deuterium (Box 4.2), is also used to determine past temperatures from ice cores in a manner similar to that described for oxygen isotopes. Water vapour evaporated from the oceans is depleted in deuterium, and as it moves inland, it preferentially loses the heavy isotope due to condensation and precipitation. Values for δD are calculated in the same way as for δ^{18}O, giving concentration differences with respect to modern ocean water. They show a good relationship to mean air temperatures at the site of snowfall (Figure 5.13).

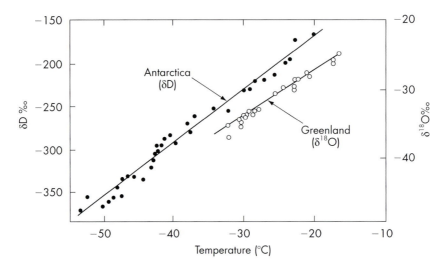

Figure 5.13 The relationship between the isotopic composition of snow and annual average surface temperatures in Antarctica and Greenland.

The Vostok Isotope Record

In the 1980s, drilling had penetrated the Antarctic ice sheet below the Russian Vostok research station to a depth of just over 2.5 km. This enabled an ice core record to be recovered that extended back to the last interglacial period. By January 1996, a fifth hole had been drilled to a depth of 3350m, extending the core record across four glacial–interglacial cycles. As can be seen from Figure 5.14, the isotopic record obtained from the Vostok ice cores shows a sawtooth pattern of slow step-wise cooling containing relatively minor warming events followed by rapid warming that is very similar to the oceanic isotope record of land-ice volume. Having reviewed some of the problems of assigning ages to ice core samples, we now need to extend the discussion to consider the difficulties involved in time correlating between ice and oceanic records. To do this we will examine the results from the Vostok record that spans the period between the last and present interglacial.

Figure 5.15(a) shows the δD record from the Vostok ice core displayed in time (using the chronology based on ice modelling as discussed above), together with eight climatic states (A–H) identified by the Franco-Russian team that analysed the cores.

☐ How much did the temperature rise in the Vostok area at the end of the last glacial, and at the end of the penultimate glacial?

■ A rise of 8°C occurred at the end of the last glacial, and 10°C at the end of the penultimate glacial.

The Vostok temperature data indicates that the early part of the last interglacial was probably some 2°C warmer than the Holocene, as suggested by the oceanic isotopic record.

☐ Does the Vostok δD record track a Northern or a Southern Hemisphere Milankovich Signal?

■ It tracks the Northern Hemisphere signal back to 125 ka.

☐ How well can the Antarctic climatic stages A–H (Figure 5.15(a)) be matched with the isotopic stages 1–6 identified in the oceanic record (Figure 5.15(c)).

■ There is a good match between A and 1 (the present interglacial), B and 2 (the last glacial maximum, although the start of the two stages is adrift by 10 thousand years), C and 3 (but the δD plot shows a pair of double peaks compared to a pair of single peaks in the oceanic record), and D and 4.

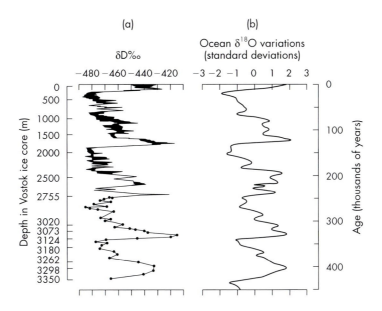

Figure 5.14 Isotopic records of the last four glacial–interglacial cycles. The δD record from the Vostok ice cores (a) compared to the marine δ¹⁸0 record (b) on which the numbers of warm isotope stages (interglacials) are shown. Note that the sample spacing was every metre above 2755 m, but was more discontinuous below this depth.

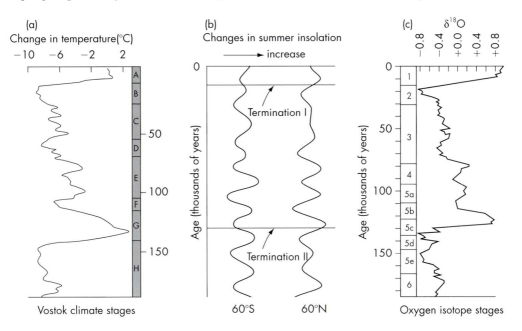

Figure 5.15 A comparison of isotopic records from the Vostok ice core and from the oceans made in 1987. (a) The Vostok δD record (obtained by analysing samples taken every 1.0 metre) calibrated to show variations in temperature from the present day annual mean of −55.5°C. (b) Changes in insolation at high northern and southern latitudes. (c) The composite δ¹⁸O oceanic record, showing isotope stages.

There is a good correspondence between the peaks in E and those defining marine isotope stages 5(a) and 5(c) and the troughs in E and 5(b) and 5(d), but the earlier records do not match well. There is a large difference in the timing of the termination of the penultimate glaciation: the marine isotope record lags the Vostok data by some 15 thousand years. Moreover, the duration of the last interglacial indicated by the two curves is different: at Vostok it is about 22 thousand years (Stage G), much more than the duration of Stage 5(e) in the $\delta^{18}O$ record (15 thousand years, 115–130 ka).

The remarkable correlation between the Vostok and marine records back to 125 ka ago supports the ice-core chronology based on glaciological modelling. But beyond this time there is a major mismatch.

□ How could this mismatch be explained?
■ It could be due to incorrect estimates of the rate of ice accumulation incorporated into the model which was used to produce the Vostok chronology, or be due simply to the lack of precision for dates in the lower part of the core (remember, a precision of only 10–15 thousand years is possible at the base). Another unlikely possibility is that the dating of the oceanic record is wrong. Yet another alternative is that the demise and growth of the Antarctic ice sheet was out of step with ice sheets in the Northern Hemisphere (as we shall see later, it is now accepted that this is the case).

These uncertainties illustrate the problem of calibrating proxy climate records in time. Most palaeoclimatologists would argue that the original dating of the oldest part of the Vostok core is incorrect, and would not subscribe to the view that the dating of the oceanic record over the last interglacial is wrong. By dating the Vostok record by matching it to the Milankovich tuned oceanic record we lose the possibility of having an independent check on whether climatic conditions over Antarctica track Southern or Northern Hemisphere high latitude summer insolation changes. Not surprisingly, the French and Russian authors of the Vostok isotope studies originally cited the close match between their temperature data and changes in annual insolation at high southern latitudes to support their proposed ages for ice older than 125 ka. Although later re-calibration of the age of the Vostok core showed that the original dates were in error, comparison of later high resolution ice core records from the Antarctic and Greenland does indicate that warming out of the last glacial in Antarctica was ahead of that in Greenland (page 126).

Dust: the Key to Calibrating the Age of the Vostok Ice Core

Comparing the dust contents of the Vostok ice cores, and that present in cores from deep sea sediments in the Southern Hemisphere has enabled the Vostok timescale to be recalibrated. Figure 5.16(a) shows the variation in dust content of the Vostok ice core: it is a proxy record of the storminess of the Earth's atmosphere in the region around the Antarctic ice sheet.

□ Do the peaks in the dust content correspond to colder or warmer periods?
■ They correspond to the colder periods (B, D and H), indicating higher wind speeds at these times.

Not surprisingly the dusty intervals 1–4 correspond to cold periods (B, D and H) as indicated by the δD-based temperature measurements. This is because latitudinal temperature gradients between the poles and the equator are higher during cold periods (Chapter 1). Surprisingly, cold period F does not show up as a dust peak. Perhaps wind speeds were not high enough to transport dust onto the Antarctic ice sheet from other continental areas in the Southern Hemisphere, or perhaps the climate was still too humid for significant dust generation to occur. Another plausible explanation for the lack of dust during period F is the fact that global sea-levels may not have dropped significantly at this time, and so continental shelf areas (including those around Antarctica) were not exposed, and so were not yet sources of dust.

Comparing dust records from the Indian Ocean (Figure 5.16(c)) and a new Vostok ice dust record (Fig-

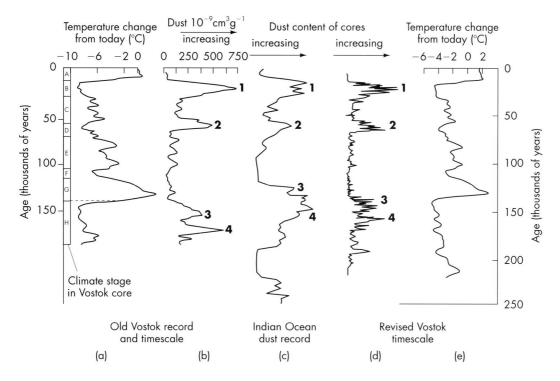

Figure 5.16 Re-calibrating the age of the Vostok ice core record. A–H are the climatic stages originally recognised by the Franco-Russian team who studied the first cores. (a) Temperature changes determined from δD record; (b) variations in dust content of the ice core; (c) A dust record from a sediment core taken in the southern Indian Ocean. Comparison with the Vostok ice core and oceanic dust records suggests that the estimated ages of ice older than 125 ka are 10–20 thousand years too old; (d) is a newer dust record from Vostok, displayed using a timescale which results in the dust peaks matching the age of those from the Indian ocean record in (c). The resultant revised temperature–time record for Vostok (e) shows that the temperature maximum during the last interglacial is >10 thousand years younger than on the old timescale based on ice-flow modelling. Note that in this later plot of temperature changes, values have been reduced compared to (a).

ure 5.16(d)) suggests that the ages of Vostok ice older than 125 ka were over-estimated by 10 to 20 thousand years. Remember that the ages were based on ice flow modelling studies, and not on direct dating methods. Most researchers now agree that the recalibrated Vostok temperature record is more likely to be correct.

There is a better match between the recalibrated Vostok temperature record and the sea-surface temperature record for the Southern Ocean than there is between it and the δ^{18}O record of global ice volume (Figure 5.16). The following differences are evident:

- at the end of the penultimate glacial (Stage 6) warming over the Antarctic ice sheet and of Southern Ocean surface waters leads the decrease in global ice volume by 5–10 thousand years;
- the cooling of the southern records between 110–120 ka leads the increase in global ice volume;
- the southern records show extreme cooling at the end of the last interglacial (~110–120 ka).

This lack of synchroneity between southern high latitude temperature changes and the record of global ice volumes (which largely reflects the volume of Northern Hemisphere ice sheets) is hardly surprising, given that when solar insolation is rising in one hemisphere it is falling in the other (Figure 5.15(b)).

Greenhouse Gases in Vostok Ice Cores

Changes in atmospheric CO_2 have only been monitored systematically and reliably since 1958, although the first measurements were made in the second half of the nineteenth Century. In 1958, the atmosphere contained 315 ppm by volume of CO_2, and in the 1990s the amount had increased to 355 ppm. The atmospheric concentration of another important greenhouse gas methane (CH_4) has also increased dramatically, growing by 1–1.15 per cent per year over the last 25 years from 1500 ppb (parts per billion) to 1700 ppb. How can we judge whether the amounts and rates of these changes are exceptional in terms of geological time scales? Fortunately, continuous sampling of the atmosphere at the surface of ice sheets has occurred naturally for several hundred thousand years in Antarctica. Analyses of samples of the Vostok ice core have revealed how atmospheric CO_2 and CH_4 have changed since the end of the penultimate glaciation. As we shall see, even during the last interglacial, when average global temperatures were 1–2°C higher than today, concentrations of these greenhouse gases were significantly lower in the bubbles trapped in the ice in comparison to their proportion in today's atmosphere.

We have already seen that, unless seasonal layers can be counted, dating the ice record is not straightforward; it is impossible to assign ages with the same precision as historical dates. And dating the air found in ice bubbles is not simply a matter of assuming that the air is the same age as the surrounding ice. This is because air circulates through the ice long after snowfall accumulates at the surface. This circulation ceases once the pore spaces are no longer connected with each other. The age difference between enclosed air and surrounding ice is dependent on the ice accumulation rate, which in turn is controlled by temperature (more snow falls during warmer periods). The age difference is thought to vary between 2.5 thou-

sand years for warm periods (i.e. interglacials) to around 4.3 thousand years for the coldest intervals (i.e. glacial maxima). The data shown in Figure 5.15 has been corrected in this way, using the temperature record derived from the δD analyses of the ice. The sampling interval shown in this figure is approximately 25 m below 850 m. At this depth in the core, this distance equates to a time interval of about 2 thousand years, rising to around 4.5 thousand years at the bottom of the hole. This may seem a rather coarse sampling interval, but when viewed in the light of the fact that each CO_2 and CH_4 measurement represents an average value over several hundred years, it is realistic. The reason that the analyses do not give an annual value, but an average for a several hundred year period, is that all the pores in the ice do not become isolated from one another at the same time. They become pinched off from the atmosphere over a period of time lying between 300 and 750 years, depending on the rate of accumulation. Thus one analysis spans a much longer time period than the direct measurements of atmospheric CO_2 made since the last century.

□ Which of the records shown in Figure 5.17 are likely to better reflect global changes rather than more localised changes over Antarctica?

■ The records of CO_2 and CH_4 reflect global changes in atmospheric composition, as the mixing time of these gases in the atmosphere is very rapid (Table 2.1). The temperature record, however, may carry a more local signal reflecting climate change in the Antarctic. The overall pattern of temperature change from glacial to interglacial is similar to proxy records of temperature obtained from the oceans.

□ How does the CO_2 and CH_4 content of the atmosphere change between the last two glacial maxima, and the last and present interglacials?

■ During glacial maxima, CO_2 levels were around 200 ppm, rising to nearly 300 ppm at the start of the last interglacial, and almost to 280 ppm in the Holocene. CH_4 levels were around 350 ppb (parts per billion by volume, or 0.35 ppm) during

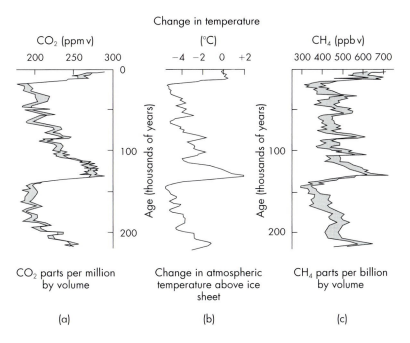

Figure 5.17 Records of CO_2, temperature change and CH_4 from the Vostok ice core plotted using the revised timescale. The shaded bands on the CO_2 and CH_4 plots indicate the degree of uncertainty in the analyses.

glacial maxima, and rose to nearly 700 ppb during interglacials.

It is clear that the anthropogenically induced influxes of CO_2 and CH_4 into the modern atmosphere have significantly changed its composition: methane levels are already more than double their normal interglacial maxima, and CO_2 levels, even using conservative estimates, are set to be nearly double the interglacial levels by the end of the next century.

A superficial inspection of the three plots shown in Figure 5.17 suggests that they are very similar. The terminations at the end of the two glacials are very abrupt and coincide, and there is a similarity in the saw tooth patterns spanning the last glacial. Spectral analysis of Vostok CH_4 and CO_2 records shows a strong link with the 40 thousand year obliquity and 23/19 thousand year precessional cycles. But there are some significant differences between the three plots:

- the CH_4 curve has four peaks rising above a roughly continuous base

- the CO_2 curve shows a gradual decrease after 115 ka, with a marked drop commencing at about 70 ka.

These differences may be explained by the fact CH_4 sources are predominantly land based (particularly from wetlands). This is in contrast to variations in atmospheric CO_2 which are influenced mainly by changes in the strength of the ocean pumps that transfer CO_2 between the atmosphere and the oceans (Box 5.2).

Rises in the CO_2 curve at the end of the penultimate and last glacial track the surface temperature rise at Vostok. However, the major temperature drops at ~125 thousand and ~75 ka are not accompanied by CO_2 decreases: the latter lag the former by about 10 thousand years, and the temperature drop at ~75 ka is followed by a short-lived CO_2 rise. So it seems that whereas a CO_2 greenhouse effect could have amplified the Milankovich signal during rapid warming into interglacials (positive feedback), CO_2-forced cooling seems unlikely.

Box 5.2 The Carbon Cycle and Atmospheric CO$_2$

This box provides a very simplified summary of the way carbon is exchanged between different parts of the Earth's system. It focuses on processes that affect the atmospheric content of CO$_2$, variations of which contributed to climate forcing into and out of glacial periods.

A summary of the global carbon cycle is shown in Figure 5.18. It shows the sizes of each of the key carbon reservoirs, and how carbon is exchanged between them.

As all gases are more soluble in cold than in warmer water, more CO$_2$ can be dissolved in seawater when it is cold. Cooling into glacial periods, therefore, resulted in positive feedback as more CO$_2$

was dissolved; warming into interglacials would also result in positive feedback as the gas became less soluble in warmer oceans. In addition to changing solubility, there are three pumping mechanisms that transfer carbon between different reservoirs. Figure 5.19 illustrates the nature of these processes using very simplified chemical equations.

The oceans contain by far the largest reservoir of carbon that can be transferred as CO$_2$ to and from the atmosphere over timescales appropriate to help explain Milankovich-scale climatic changes. As CO$_2$ diffuses into the oceans from the atmosphere it reacts with water to form three ionic 'species' (bicarbonate, HCO$_3^-$: ~85%,

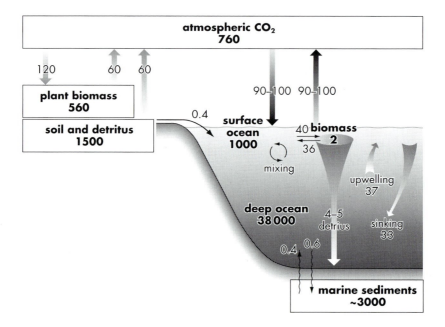

Figure 5.18 Summary diagram of the components of the global carbon cycle and how carbon is exchanged between them. The approximate mass of carbon in each reservoir (shown in **bold**) is given in units of 10^{12} kg of carbon, and the fluxes between different points of the Earth System in 10^{12} kg of carbon per year. The lithospheric reservoir of carbonate rocks includes the skeletal remains of shallow water organisms that lived on the continental shelves, including fossil reefs. Note that the effects of fossil fuel burning, deforestation and agriculture are not included.

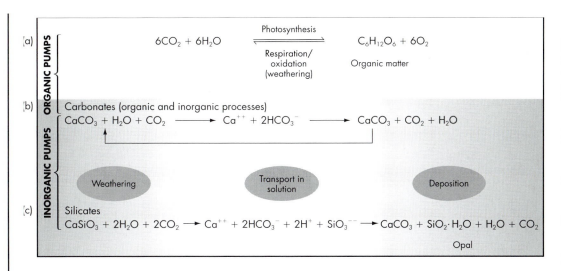

Figure 5.19 Highly simplified chemical equations illustrating the mechanisms involved in biological and inorganic pumps that remove and/or release atmospheric CO_2. Note that only one simple exemplar composition of a silicate mineral is shown in (c); in reality the composition of most silicate minerals include aluminium and several other metals. The mineral opal on the depositional side of (c) is usually precipitated as the hard parts of plankton such as diatoms and radiolarians.

carbonate CO_3^{2-}: 10–15%; carbonic acid, H_2CO_3: < 1%). The formation of these is the reason that CO_2 is the most soluble of all gases in the atmosphere. Biological pumping involves the fixing of carbon into organic matter from carbon dioxide in the atmosphere, or dissolved in the oceans. If the organic matter then accumulates (as plant biomass or in soils on land, or in marine sediments: see Figure 5.18) carbon is removed from the ocean–atmosphere system. Increases in plant biomass during interglacial periods (mainly in tropical rainforest areas) may have reduced atmospheric CO_2 by 80 ppm, resulting in a negative feedback that limited greenhouse warming.

Carbon may also be removed from the ocean-atmosphere system if the skeletons of organisms composed of calcium carbonate are buried permanently. Similarly, any carbonate sediment precipitated chemically and subsequently buried will remove carbon from the ocean and atmosphere, but this process is of relatively minor importance.

The rate at which the biological pump removes carbon from the oceans is related to the productivity of planktonic organisms, which in turn is limited to the availability of dissolved nutrients (especially carbon, nitrogen, phosphorus and iron). Sustained productivity requires the replenishment of nutrients in shallow waters. This occurs largely via ocean upwelling systems delivering nutrients from deeper water, but also through rivers transferring land derived nutrients to the oceans. Marine productivity during glacial periods was probably enhanced because nutrient supply was increased due to vigorous upwelling systems driven by stronger winds. The winds would also have carried to the oceans more dust containing iron.

Figure 5.19 shows that weathering carbonates on land followed by marine precipitation (usually as the skeletal remains of plankton) does not alter the CO_2 content of the atmosphere–ocean system. This is because CO_2 is removed during weathering, but released during deposition. During glacial and interglacial periods, however, carbonate weathering and deposition were not in balance. Eustatic falls accompanying global cooling resulted in the exposure and weathering of carbonate rocks on continental shelves, and the removal of atmospheric CO_2 and its transport in solution to the oceans. This would result in a positive feedback, reinforcing the cooling trend. Global warming and sea-level rise would result in carbonate precipitation over flooded continental shelves, and the release of CO_2 from the oceans into the atmosphere – another positive feedback, but this time reinforcing a warming trend. As carbonate, rocks form about 80 per cent of the reservoir of global carbon (Figure 5.18).

Over longer periods of time, such as the post Cretaceous cooling of the Earth, the weathering of silicate minerals (Figure 5.19(c)) resulted in the removal of atmospheric CO_2. In this case, for every two molecules of CO_2 removed during weathering, only one is returned to the ocean by the associated depositional process in the oceans.

The major natural sources of atmospheric methane today are high latitude and tropical wetlands. Outside interglacial periods, only the tropical wetlands would have generated significant methane, because the former would have experienced low temperatures, and so many would have been permanently frozen. Therefore, it is likely that the methane signal reflects stronger monsoonal phases at low latitudes. Such a link is indicated by the correspondence between high atmospheric methane levels and high lake levels in the Sahara (Figure 5.20).

Changes in atmospheric methane levels have only a very small forcing effect on global temperatures. The changes observed in the Vostok core would result in changes of around $0.1°C$. But the rises in CO_2 of 100 ppm at the end of the penultimate and last glaciations would have contributed significantly to global warming – perhaps by as much as $1.5°C$. But the CO_2 and temperature records at Vostok do not track one another during cooling episodes: falls in CO_2 lag behind temperature decreases.

Palaeoclimatologists agree that past changes in the proportions of atmospheric CO_2 (apart from those caused in the last two centuries by the burning of fossil fuels) resulted primarily from variations in the rate of exchange of the gas between the atmosphere and oceans. This is because the oceans contain the largest global reservoir of carbon that can be rapidly (on a millennial scale or less) exchanged with the atmosphere.

Carbon contained in CO_2 is moved between the oceans and atmosphere, and within the oceans, by three mechanisms: diffusion, the organic carbon pump and the precipitation of $CaCO_3$ by marine organisms (Box 5.2). The effectiveness of the biological pump is limited by the availability of nutrients, without which planktonic organisms cannot thrive. These may be supplied by upwelling of deeper waters along continental margins. As upwelling is linked to surface ocean currents driven by winds, it would have been more prevalent in the oceans of a glacial world, causing a reduction in atmospheric CO_2 levels. This negative feedback may have been enhanced by wind-

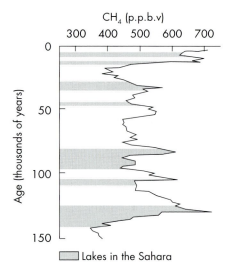

Figure 5.20 Atmospheric methane levels measured in the Vostok ice core and periods of high lake levels in the Sahara. The latter are proxy indicators of periods of enhanced monsoonal rainfall which would result in the expansion of tropical/subtropical wetlands, causing an increase in the generation of methane.

blown dust. Experiments in fertilising equatorial Pacific waters with soluble iron compounds have achieved a fourfold increase in planktonic productivity for short periods of time (hours to days) , and consequent CO_2 drawdown from the atmosphere. The results of these experiments suggest that the increased availability of iron carried in windblown dust during glacial periods may have been partially responsible for low atmospheric CO_2 levels. A more significant negative feedback was probably caused by rougher colder glacial oceans being able to dissolve more CO_2.

SUMMARY

1 Spectral analysis of $\delta^{18}O$ variations shows that they contain signals of insolation variations at high northern latitudes relating to all three orbital parameters: eccentricity (100 thousand years), obliquity (41 thousand years) and precession (23 thousand years and 19 thousand years).

2 The timing of glacial terminations coincides with rapid increases in the calculated solar insolation values at high *northern* latitudes, confirming Milankovich's theory of ice ages. This states that changes in Northern Hemisphere high latitude insolation are critical in determining whether snow survives through successive summers to permit ice-accumulation.

3 Eccentricity driven changes in high latitude insolation are only ~ 2 W m^{-2}. This is an order of magnitude smaller than those caused by obliquity and precessional changes. Yet the 100 thousand years signal within the $\delta^{18}O$ record is twice that produced by the combination of obliquity and precessional changes. This is the '100 thousand years problem'. It is unlikely that the 100 thousand years cyclicity is directly related to astronomical forcing: the Earth's climate system must amplify it in some way yet to be determined.

4 Direct dating of deep-sea sediments older than 40 ka is difficult. Therefore time calibration of oxygen isotope stages is achieved by 'orbitally tuning' the $\delta^{18}O$ variations to fit the astronomically calculated Milankovich pacemaker. As this involves assuming a given time lag between insolation changes and volume changes in ice sheets, it is difficult to independently assess the duration of such time lags.

5 Calcitic cave deposits at Devil's Hole, Nevada, have yielded the best radiometrically dated record of climate change spanning the last 500 thousand years. This record shows that warming into Terminations II and III occurred ~ 10 thousand years earlier than shown by the oceanic $\delta^{18}O$ record. This difference may not call into question the link between climate change and high northern latitude insolation forcing. It may reflect local climate changes such as the position of the Pacific oceanic polar front, and the North American ice sheet. The latter would have diverted southwards low pressure systems that brought water vapour from the Pacific.

6 Loess-soil couplets in the Chinese Loess Plateau deposits show grain size variations caused by changes in the intensity of winter monsoonal winds. The deposits enable a proxy climate record to be obtained extending back to 3.5 Ma, which shows that winter monsoonal winds intensified during glacial periods.

7 The periodicity of the loess climate record changes from 100 thousand years to 41 thousand years at around 800 ka, paralleling the oceanic record. The shift in periodicity at 1.6 Ma from 41 thousand years to intervals of 400 thousand years, 55 thousand years and 41 thousand years is not seen in the oceanic record, and therefore may be a local phenomenon.

8 Ice sheets are, in essence, frozen atmosphere. Analyses of samples taken from boreholes drilled through them yield information concerning the surface temperature at the drilling site when snow accumulated, the storminess of the atmosphere indicated by the dust content of the ice, and the content of atmospheric greenhouse gases at the time of ice accumulation.

9 In Greenland, counting annual layers of ice has been extended up to 17.4 ka. The age of older,

deeper ice is estimated based on the age/depth relationships determined for the younger ice. This is not possible in Antarctica, where age estimates in the Vostok core relied entirely on ice flow modelling studies. This led to the age of ice older than 125 thousand years being overestimated by 10–20 thousand years. The overestimates were determined by correlating the peaks in the dust content of the Vostok ice cores with oceanic dust records.

10   The surface temperature record at Vostok shows a similar pattern of change to the oceanic $\delta^{18}O$ record, with an 8°C rise at the end of the last glaciation, and 10°C at the end of the penultimate glaciation. The last interglacial was about 2°C warmer than the present one. However, at high latitudes in the Southern Hemisphere warming into and cooling out of the penultimate interglacial leads the $\delta^{18}O$ signal of global ice volume.

11   Measurements of CO_2 and CH_4 trapped in air bubbles in the ice enable the scale of recent anthropogenically driven changes in the proportion of these greenhouse gases to be compared to past naturally driven changes. Even when, during the last interglacial, global mean temperatures were up to 2°C higher than they are today, CO_2 levels were 280 ppm compared to today's value of ~360 ppm, and CH_4 levels were ~700 ppb, compared to 1700 ppb today. Peaks in CH_4 that occur during the last glacial were probably related to increased monsoonal rainfall in tropical latitudes, which increased the amount of vegetation decaying in wetlands. Variations in CO_2 levels were probably caused by changes in the rate of exchange of the gas between the atmosphere and the oceans.

FURTHER READING

See Lowe and Walker (1997) and Williams *et al.* (1993) in reading list for Chapter 1 (pp. 28–9), and in references to figure sources for this chapter (pp. 255–6).

6

EVIDENCE FOR RAPID CLIMATE CHANGE

INTRODUCTION

So far we have dissected the anatomy of climate change (Figure 6.1) on a timescale of $10^3–10^6$ years. Long records of proxy climate data from the oceans and continents show not only that there were many more than the four ice ages originally postulated in the nineteenth century, but that the pattern of climate change is not a simple one of alternating cold and warm episodes. During the last million years, global climate was for most of the time somewhere in-between the extreme cold of glacial maxima and interglacial warmth (Figure 6.1(e)), and present day interglacial conditions only existed for about ten per cent of this time.

It is likely that the pattern of change characteristic of the last few hundred thousand years will be repeated in the future as it is driven by the Milankovich pacemaker (Figure 5.7). This thought is a disturbing one, but it is not an immediate threat to us, or even to the next few generations. Our immediate concern is the possibility that greenhouse warming, due to human activities over the last two centuries, will cause climatic and sea-level changes in the world that our grandchildren will inherit.

What if the geological record revealed that rapid climate change – at the decadal to millennial scale – was not an exceptional but a routine feature of the behaviour of the global climate system? If it did, we would be very concerned about how such changes would affect us if they were to begin now. We also need to know the extent of natural climate variability on shorter timescales so that we can assess whether the current warming trend is linked to anthropogenically driven rises in atmospheric CO_2 content, or is a natural characteristic of the global climate system.

In the last two decades of the twentieth century, the ability to obtain high resolution palaeoclimate records from ice sheets and oceanic and lacustrine sediments has revealed a record showing that rapid changes were common. There is some consolation that these records show that the last 10 thousand years – our interglacial – has been relatively stable in contrast to the previous glacial period. These records do show, however, that some rapid changes did occur during the last interglacial, and that climatic variability, albeit of a lesser magnitude, characterises the present interglacial.

This chapter reviews some of the key findings of high resolution palaeoclimate research. This is a rapidly developing and very topical field in which there is sometimes the temptation to jump to conclusions that link past changes to speculation about our future.

NORTH ATLANTIC ICEBERG ARMADAS

In 1988, a German Oceanographer, Hartmut Heinrich, published the results of a study that showed that six times during the last glacial stage, huge armadas of icebergs were launched from Canada into the North Atlantic. As they melted, they released rock debris that was dropped into the fine grained sediments on the ocean floor (see Figure 4.1(d)). Much of this ice rafted debris consists of limestones very

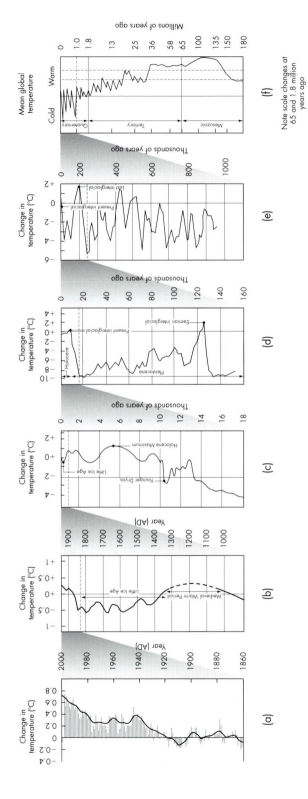

Figure 6.1 The anatomy of past climatic changes: from tens to millions of years.

similar to those exposed over large areas of eastern Canada today. The Heinrich layers, as they have become known, extend 3000 km across the North Atlantic, almost reaching Ireland.

What could have caused the Heinrich events? Clearly they resulted from periodic surges in the flow of the ice sheet covering eastern Canada, but were these surges triggered by climatic changes, or by the dynamics of the ice sheet which resulted in surging unconnected to links between ice, the atmosphere and/or the oceans. The nature of and differences between these two models, known respectively as the MacAyeal (or 'binge–purge' model: Box 6.1) and Denton models, are shown in Figure 6.2 and Table 6.1.

For a while, the MacAyeal 'binge–purge' hypothesis was favoured, until it was discovered that some of the Heinrich layers contain material that could only have come from separate ice sheets covering not only Canada, but Greenland, Iceland, the British Isles and

Table 6.1 A comparison of the MacAyeal and Denton models for the origin of Heinrich events

	MacAyeal 'binge–purge' model	Denton model
Cause	Internal: ice sheet failure	External: ice sheet and/or ice shelf expansion
Forcing	*Geothermal and frictional heat build-up:* enough is trapped periodically beneath the ice sheet so that the frozen base melts causing the ice sheet to fail catastrophically.	*Global cooling:* possibly due to harmonics of the orbital parameters causing ice sheet and/or ice shelf expansion which releases more icebergs.
Points in favour	1) Explains rapid initiation and termination of Heinrich events. 2) Explains the varying time between the Heinrich events (7–13 thousand years) as this will be dependent on the size of the ice sheet. 3) Explains the large amount of debris found in each Heinrich layer in the North Atlantic.	1) Evidence for cooling in the North Atlantic prior to the Heinrich events. 2) Explains the coeval surging of South American glaciers and other global responses without requiring complicated teleconnections. 3) Climate cycles of ~1.5 thousand years also occurred during the past 10 thousand years (see p. 124) although they are an order of magnitude smaller than Dansgaard–Oeschger cycles (pp. 121–3), they suggest external forcing.
Problems	1) Requires a mechanism (e.g. deep water circulation) to transfer the North Atlantic signal around the globe. 2) Does not explain cooling prior to the Heinrich events.	1) Ice sheet response is too slow taking between 1–10 thousand years. 2) The Heinrich events do not have a regular cyclicity. Early during the last glacial period they occurred every 13 thousand years but later became more frequent, occurring every 7 thousand years.
Conclusion	The jury is still out!	

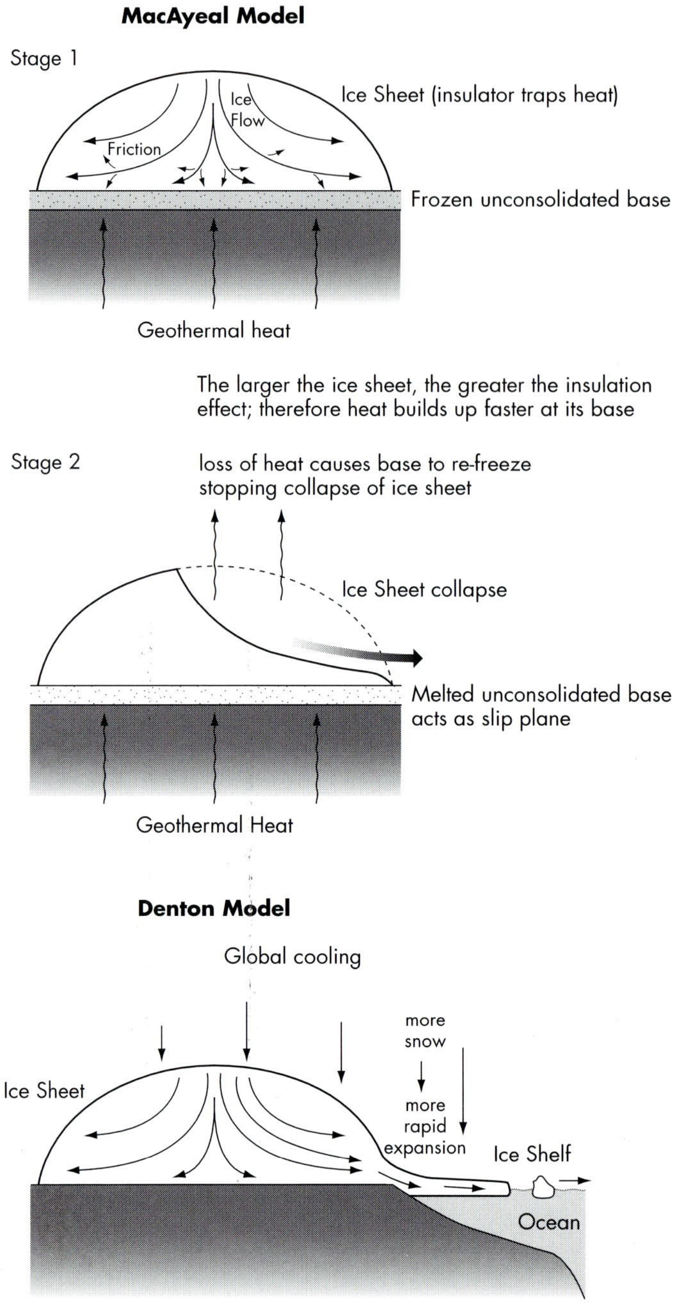

Figure 6.2 Sketches showing the features of the MacAyeal and Denton models for the origin of Heinrich events.

Scandinavia. For example, fragments of basaltic glass could be matched with lavas in Iceland, and in the top two Heinrich layers, abundant coccoliths of Upper Cretaceous age pointed to ice erosion of chalk exposed in the British Isles and beneath the North Sea. It would be most unlikely that separate ice sheets would surge simultaneously *unless* triggered by an external, climatic change, although a rise in sea-level consequent on the melting of Laurentide ice might increase the marine ablation of other ice sheets. None the less, westward thickening of the Heinrich layers across the Atlantic, and the continuation of this trend through the Labrador Sea towards Hudson Bay indicates that much of the floating ice was sourced from the Laurentide ice sheet. So perhaps its break-up triggered a response by other ice sheets.

Studies of landforms and sediments in the Lake Superior area have shown that the Laurentide ice sheet gradually advanced and then rapidly retreated three times towards the end of the last glacial (Figure 6.3). Carbon dating of these events indicates that glacial advances culminated immediately before

Heinrich events 1 and 3. The collapse of the ice sheet caused by the rapid discharge of ice through the Hudson Strait into the Atlantic would have reduced the southward flow of ice into the Lake Superior area, resulting in the rapid retreat shown on Figure 6.3(a).

Research in the Andes and New Zealand indicates that small ice sheets in these areas grew, and then collapsed synchronously with the ice-rafting pulses recorded in the North Atlantic (Figure 6.3(b)). This evidence from strong inter-hemispheric coupling of changes in temperature indicate a global rather than regional forcing of climate change.

High resolution studies of ocean cores have now revealed that the iceberg calving events occurred even more frequently at intervals of 2 to 3 thousand years (Figure 6.4) between 10 thousand years and 38 thousand years. Some of these events were shown to be related to changes in surface water temperatures as determined by the proportion of the foraminifera *Neogloboquadrina pachyderma* showing left handed coiling (sinistral) in the total population of fossil foraminifera (see Box 1.4). Many of the peaks in the

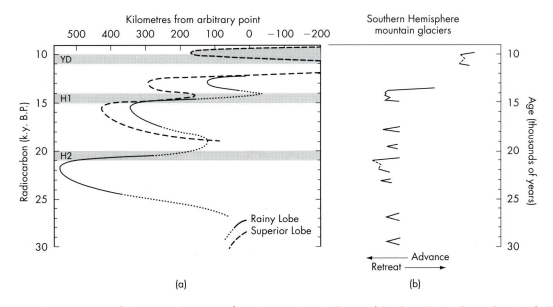

Figure 6.3 The timing of advances and retreats of ice sheets in the Northern and Southern Hemispheres, showing their relationship to Heinrich events. (a) Two ice lobes on the southern margin of the Laurentide ice sheet in the present day Lake Superior area. (b) Maximum advances of mountain glaciers in the southern Andes and New Zealand. YD: Younger Dryas : a ~1000 year long cold period that interrupted the warming trend into the present interglacial.

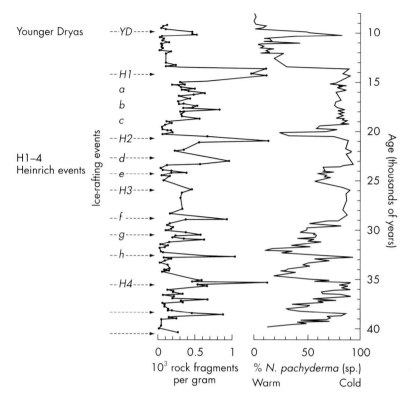

Figure 6.4 Evidence for North Atlantic iceberg armadas. (a) A plot of the amount of rock fragments obtained from samples taken from a core of ocean floor sediments in the North Atlantic shows the youngest four Heinrich events (the oldest (5 and 6) are not shown). The Younger Dryas is also characterised by a peak in the amount of rock fragments, and 10 subsidiary peaks are also recognised. (b) The proportion of the left hand coiled (sinistral) variety of the foraminifera *Neogloboquadrina pachyderma* indicates warm and cold conditions. Cold surface waters are characterised by fossil foraminifera assemblages containing over 90 per cent of sinistral *N. pachyderma* (see Box 1.4).

Box 6.1 A Binge–Purge Model for Iceberg Armadas

A detailed discussion of the conditions necessary for glaciers and ice sheets to grow and retreat is given in Chapter 3.

Movement of ice occurs in two ways: by deformation of the ice mass, and by movement across rock or sediments beneath (Box 3.3, Figure 3.7). If the base of the ice is below the freezing point of water it is effectively glued to the underlying material, and so the ice mass will move only slowly via internal deformation. However, if free water occurs at the base of the ice, its lubricating effect permits much more rapid movement by

basal sliding. The nature of the substrate also has an effect, for the hardness of the underlying material causes friction which retards sliding. Ice flow over crystalline igneous and metamorphic rocks is slower than movement over soft water impregnated sediments.

The author of the binge–purge model, D.R. MacAyeal, suggested that the Laurentide ice sheet (LIS) – the great ice mass that once covered much of eastern Canada – grew whilst its base was frozen solid to both the crystalline rocks, now exposed on land, and softer sediments that exist

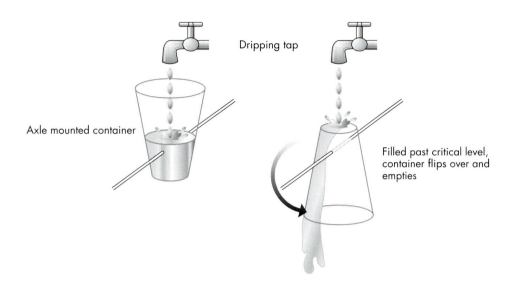

Dripping tap

Axle mounted container

Filled past critical level, container flips over and empties

Figure 6.5 A kitchen-built binge–purge analogue for the cyclic growth and collapse of the Laurentide (or any other) ice sheet.

beneath Hudsons Bay and the area to the south of the ice sheet. When the base of the ice sheet was warmed sufficiently by geothermal heat to permit the sediments to thaw, the rapid ice movement of the purge phase occurred. Frictional heating would have accelerated the rise in temperature, resulting in a positive feedback that accelerated ice movement still further. However, the greater friction at the ice/crystalline rock interface prevented total collapse of the ice sheet.

The author of the binge–purge hypothesis, D.R. MacAyeal, described the model as follows:

> It is instructive to describe a simple kitchen-built experimental device that captures the behaviour qualities needed to explain Heinrich events. Consider the axle-mounted container [sketched in Figure 6.5]. Initially, this container sits upright on the axle, because its center of mass is assumed to lie between the bottom of the container and the axle. As water drips slowly into the container, the center of mass slowly rises to the point where it exceeds the level of the axle. At this point, the container becomes unstable and flips upside down to purge its contents onto the floor of the kitchen. Once the container has emptied, it flips back to the upright position and begins again to fill slowly with water. As long as the slow trickle of

water is maintained, this simple device will continue to cycle between binge (filling) and purge (emptying) behavior.

> The simple oscillatory action of this kitchen-built oscillator captures the essence of the behavior required of the LIS to produce Heinrich events. The points of comparison are listed as follows: (1) The axle-mounted container represents Hudson Bay. (2) The slow filling of the container represents the accumulation of ice over Hudson Bay when the subglacial bed is frozen. (3) The vertical climb of the container's center of mass represents the slow warming of the basal ice temperature in response to the geothermal flux. (4) The flip and purge of the container represents the fast ice stream discharge that ensues once the subglacial bed in Hudson Bay and Hudson Strait has developed a thawed connection to the Labrador Sea. Finally, (5) the recovery of the axle-mounted container to an upright orientation represents the refreezing of the subglacial bed brought on by the increased vertical heat flux associated with ice stream flow.

> The purpose of introducing this simple kitchen-built oscillator is to reinforce the idea that systems can oscillate and display violent changes in state even under steady, time independent forcing. To produce the binge/purge oscillations of the axle-mounted container, all that was necessary was the slow trickle of water.

> D.R MacAyeal. 1993.

Binge/purge oscillations of the Laurentide ice sheet as a cause of the North Atlantic's Heinrich events. *Paleoceanography*. 8: 775–84.

The 'binge–purge' model highlights the fact that natural systems can display rapid changes caused by a forcing factor that does not change with time. In the case of ice sheets, the model explains how rapid ice surges can occur independently of the external forcing (including the Milankovich pacemaker). This is extremely important, as it illustrates the difficulty of making predictions about the possible future effects of climatic forcing factors.

occurrence of rock fragments shown in Figure 6.4 coincide with >90 per cent proportions of the cold water foraminifera, e.g. unnumbered events at before 35 thousand years ago (H4, h, H3, d, H2, H1, YD).

☐ What happens to sea surface temperatures immediately after many of the lithic peaks?
■ There was a rapid rise in temperature, as indicated by significant rises in the proportion of left-hand coiled *N. pachyderma*.

This shows that the release of iceberg armadas coincided with low North Atlantic sea surface temperatures, indicative of stadial periods, which were followed by rapid warming into interstadials. As we will see in the next section of this chapter, these temperature changes are paralleled by temperature changes detected in cores taken through the Greenland ice sheet.

RAPID EUSTATIC SEA-LEVEL CHANGES

In Chapter 4 we saw how oceanic $\delta^{18}O$ variations provide a proxy record of the volumes of land ice and the accompanying changes in eustatic sea-level. This record shows that for about the last million years, ice volumes and sea-levels change over two time scales: ~100 thousand years (glacial–interglacial) and <~20 thousand years (stadial–interstadial). For the last glacial–interglacial cycle, observations of the elevation of fossil coral reefs in Papua New Guinea enabled the amplitude of past sea-level changes to be determined, thus providing an independent calibration of the marine isotopic record (Figure 4.12). Since

this work was undertaken in the late 1980s, a high resolution record of sea-level change has been obtained from fossil coral reefs in Barbados (Figure 6.6(a)).

The curve shown in Figure 6.6(a) is not a simple sinusoidal one. There are changes in the slope of the sea-level rise recorded between 10 and 15 thousand years ago.

☐ How can this change of gradient be explained?
■ The rate of discharge of meltwaters into the world's oceans must have changed. The two steepest parts of the curve (at ~11 ka and ~14 ka) must have been times when the discharge was significantly higher: they are separated by an interval over which a lower discharge rate occurred.

The different slopes of the cumulative sea-level curve on Figure 6.6(a) can be used to calculate how discharge rates varied over time. The results show that there were two major pulses of meltwater, the first peaking at ~14.5 ka and the second at ~11 ka (Figure 6.6(b)), 25 and 17 times that of the present day Mississippi River. In Greenland, these melting events followed rapid rises in temperature (see Figure 6.9). As these temperature rises have not been found in Antarctic ice core records, it is probable that the pulses in meltwater discharge originated from Northern Hemisphere ice sheets, a conclusion corroborated by changes in ocean salinities only found in the North Atlantic.

Almost all the rise in sea-level shown on Figure 6.6(a) took place during the rise in Northern Hemisphere summer insolation driven largely by precession. The lowest sea-level, marking the last glacial maximum, is not well constrained by the coral reef data, but certainly cannot be later than 22 ka, and must have occurred between 22 ka and 25 ka. This time interval is coincident with the solar insolation minimum when the Earth was furthest from the sun in the Northern Hemisphere summer.

GREENLAND ICE CORES

Introduction

Ice coring began in Greenland in the late 1970s, but only in the early 1990s were two holes drilled to bedrock near the summit of the ice dome (Figure 5.12(a)). The first borehole to reach bedrock was that undertaken by the European funded Greenland Ice-core Project (GRIP). It did so in the summer of 1992, and publication of the first results a year later prompted much excitement, because two significant discoveries concerning rapid climate change were made: one occurring during the last glacial, and the other – which turned out to be premature – during the last interglacial.

As discussed in the previous chapter, ice accumulation rates are, and were, much higher in Greenland than in Antarctica. This means that much higher resolutions are possible in Greenland – in fact seasonal layer counting by visual means has been extended back to around 50 ka, and to 110 ka using seasonal proxies (changes in the chemical composition of ice due to its dust content and the influx of sea salt).

Numerous Interstadials

In Chapter 4, it was shown how the $\delta^{18}O$ record from oceanic sediments enabled three warmer interstadials (isotope stages/substages 3, 5a, 5c) to be identified within the last glacial stage. The plot of $\delta^{18}O$ from the GRIP core reveals 24 interstadials, 21 of which are shown in Figure 6.7(a).

The repeated episodes of rapid warming/cooling are known as Dansgaard–Oeschger events after their discoverers. They start with rapid increases in temperatures over the Greenland ice sheet that occurred

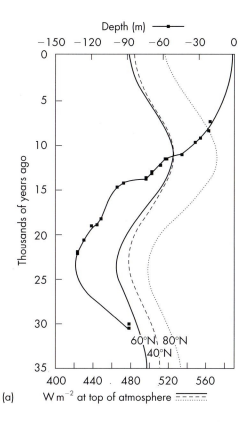

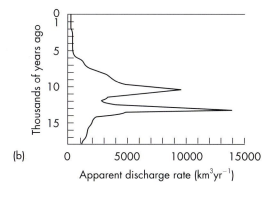

Figure 6.6 Changes in the eustatic sea-level determined from fossil coral reefs in Barbados. (a) Sea-level changes determined from the elevation of fossil coral reefs in Barbados; summer solar insolation curves for latitudes in the Northern Hemisphere are shown. (b) Rates of meltwater discharge calculated from the changes in slope of the sea-level curve shown in (a).

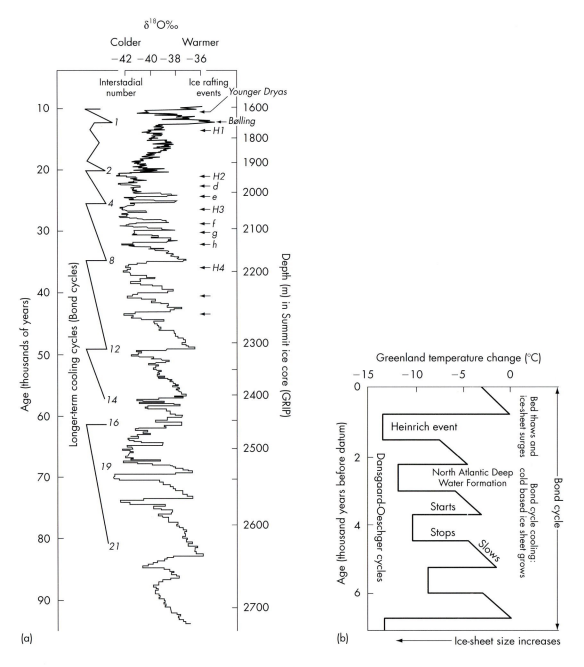

Figure 6.7 Dansgaard–Oeschger and Bond cycles in the GRIP ice core. Figure 5.12(a) shows the location of drilling site. (a) δ¹⁸O record showing numbered interstadials and the ages of the Heinrich events/lithic peaks (H1, H2 etc.; a–h) depicted on Figure 6.4. (b) Diagrammatic representation of the relationship between Dansgaard–Oeschger and Bond cycles and their possible link with changes in the formation of North Atlantic Deep Water is described in Box 6.2.

over a hundred years or less. Relatively slower cooling then follows until the next warming event.

☐ How do the Dansgaard–Oeschger events relate to iceberg armadas indicated by peaks in the occurrence of ice-rafted debris occurring in North Atlantic ocean sediment cores (their ages are shown on Figure 6.4(b), including the larger Heinrich event peaks)?

■ With the exception of H1 they all occur at the terminal coldest part of each event.

Considering the evidence presented in Figures 6.4 and 6.7(a), it is clear that the following rapid changes occurred during the last glacial in North Atlantic area:

● slow cooling of the atmosphere immediately above the Greenland ice sheet and North Atlantic surface waters. A pattern of moderate cooling followed by rapid cooling is evident in events 21, 20, 19, 12 and 8.

● iceberg armadas coincide with maximum cooling and are immediately followed by rapid warming.

These shorter events are bundled together into longer cooling cycles characterised by a steady drop between successive peaks in the $\delta^{18}O$ values. These bundles are known as Bond cycles after their discoverer, and are shown on Figure 6.7(a). This pattern of slow cooling followed by rapid warming is similar to the 100 thousand years glacial/interglacial pattern described in Chapter 4. The large ice rafted debris peaks at the end of each Bond cooling cycle are Heinrich events. The cooling through each Bond cycle may have been caused by the downwind effect from the growth of the Laurentide ice sheet. The Heinrich events at the end of each cycle could have resulted from geothermal heat trapped by the increasing thickness of ice leading to basal melting and consequent surging of the ice sheet.

The collapse of the Laurentide ice sheet cannot be the only cause of the cycles shown in Figure 6.7, for, as we have seen, there is evidence that this ice sheet was not the only source of the iceberg armadas although it was probably the dominant one. Some

external factor must have triggered the almost synchronous collapse of all the northern ice sheets. Changes in North Atlantic oceanic circulation could have been the cause, as explained in Box 6.2.

Spectral analysis of variations in the chemical composition of Greenland ice cores have not only revealed obliquity and precessional cycles but also periodicities of 11 100, 6 100 and 1 450 years the shorter of which could pace the changes summarised in Figure 6.7(b). All of the sub-Milankovich cycles could be overtones generated within the climate system from the longer Milankovich cycles, much as a string on a musical instrument produces overtones when plucked.

The 11 100 year cyclicity may be related to precessional driven climate changes in low latitudes. Precession results in hotter summers alternating between the Earth's hemispheres every 23 thousand years. This results in a continent straddling the tropics, such as Africa, experiencing temperature maxima every ∼11 thousand years, which could affect the amount of dust delivered to high latitudes and incorporated into the Greenland ice sheet.

Seven of the twelve peaks in the 6 100 year cyclicity in the Greenland cores coincide with Heinrich events, and two more with cooling during the mid-Holocene and the Little Ice Age. The 6 100 year overtone could be amplified by ice sheets. They would have reacted so slowly that their response time would be as long as or longer than the cyclicity itself, and so they would amplify the 6 100 year signal in a manner similar to a child's swing being pushed at the top of each swing.

The 1 450 year periodicity observed in the Greenland ice cores is too short to influence the sluggish behaviour of ice sheets, and so it is thought to reflect reorganisation of atmospheric circulation. Changes in solar output have been suggested as the cause, because the same cyclicity is observed in tree ring records of carbon-14. This isotope is produced in the upper atmosphere by the bombardment of cosmic rays, the intensity of which vary with changes in solar activity (Box 7.2). There is, however, no consensus at the time of writing concerning the cause of the 1 450 year periodicity.

Variations in the amounts of ice rafted debris have

also been discovered in Holocene ocean floor sediments of the North Atlantic and show a cyclicity of 1450 years. They are about a tenth of the size of the fluctuations in the amounts of rock fragments characteristic of the ice rafted layers formed during the last glacial stage, but none the less suggest that during colder periods more icebergs were able to travel further. This is but one part of a growing body of evidence that suggests a greater degree of short term climatic variability during the Holocene than had previously been detected.

Re-analysis of the Vostok ice core $\delta^{18}O$ record revealed nine interstadials during the last glacial stage. Each one is characterised by slow warming and cooling of about 2°C, so they are not identical to corresponding events over Greenland. They occurred in

---

Box 6.2 Opening and Shutting the North Atlantic Door

The boundary between cold polar and warmer Atlantic water masses in the North Atlantic is termed the polar front. It is a kind of door that opens and shuts between cold (interglacials and interstadials) and cold (glacials and stadials) (Figure 6.8). During the present interglacial (the Holocene) it has been open, allowing warmer subtropical surface waters caused by the North Atlantic Drift (NAD) to flow northwards, warming NW Europe (Figure 6.8(a)). The evaporation and cooling of the NAD forms cooler saline water which sinks and flows southwards at deeper levels; this is known as the North Atlantic Deep Water (NADW: see Figure 2.6(a)). During the last glacial stage (and probably earlier ones as well), the northward drift of warm surface waters was drastically reduced, and so the door was shut during cold stadials. But it was ajar after each ice sheet collapse that ended each Dansgaard–Oeschger and Bond cycles, allowing some warmer water to penetrate further north in the Atlantic (Figure 6.8(b)). It slowly closed again as the ice sheets grew during the slower cooling phase of the cycles. But what forced the door to open and close? As will be discussed in the next chapter, it is likely that NADW is shut down when a freshwater 'lid' is placed over the northern North Atlantic by the melting of icebergs released by Heinrich events. Modelling studies suggest that ice melting events confined to the Nordic Seas between Greenland and Norway, which may be linked to the Dansgaard–Oeschger cycles, weakened NADW formation, but did not completely shut it down.

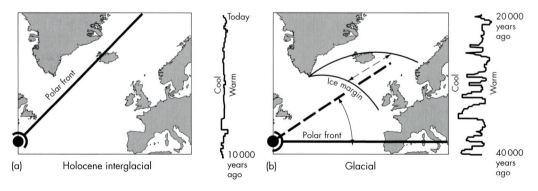

Figure 6.8 The North Atlantic door: the changing position of the polar front during (a) the Holocene and (b) the last glacial.

Antarctica whenever those in Greenland (i.e. the Dansgaard–Oeschger events) lasted longer than two thousand years. The longer interstadial events in Antarctica correlate with minima in the $\delta^{18}O$ record obtained from benthic foraminifera.

Methane

The likely global extent of the numerous interstadials of the last glacial stage was reinforced by the discovery that fluctuations in atmospheric methane between 20 ka and 40 ka are coincident with interstadials revealed by the GRIP $\delta^{18}O$ record (Figure 6.9). It is unlikely that high latitude wetlands would have been a significant source of methane. This is because methane levels reach interglacial values during the Bølling interstadial – a time when continental records indicate that ice sheets were still extensive over northern Canada and Europe, covering areas which are today (in an interglacial) the sites of extensive northern wetlands. It seems likely that the bulk of the methane was generated as a result of increased precipitation stimulating plant growth and decay in tropical regions. Therefore it is likely that interstadial events may have been global climatic phenomena involving rapid coupling between the atmosphere, oceans and cryosphere. Three additional lines of evidence support the northern hemispheric extent of the Dansgaard–Oeschger events: from high resolution loess records in China (p. 131), from ocean sediments deposited in the Santa Barbara Basin off California (p. 132), and from pollen records in lake sediments in Europe (described in Chapter 8).

Figure 6.10(a) shows the methane concentrations in the GRIP and Vostok ice cores for the 7 000 year time interval spanning the end of the last glacial into the present interglacial.

☐ Would you expect the variations in atmospheric methane content to be the same in northern and southern latitudes?

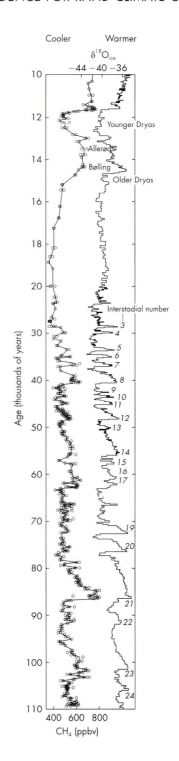

Figure 6.9 The methane and $\delta^{18}O$ record for part of the GRIP ice core. Note change in scale between 10–20 and 20–110 thousand years ago.

■ Yes, they should be synchronous, because gaseous diffusion results in very rapid global dispersal of gases around the world.

□ So why is it that the two records in Figure 6.10(a) are so different?
■ One of the timescales used must be incorrect.

As the GRIP timescale is based on counting annual layers within ice (yielding an accuracy of ± 200 years over the time interval shown in Figure 6.10), variations in methane concentrations in the two cores can be used to revise the Vostok timescale. Figure 6.10(b) shows the revised timescale achieved by matching the Younger Dryas troughs in methane concentration which reveals that temperature changes (as indicated by isotopic analyses) in Greenland and Antarctica were not in phase at the millennial scale, as shown in Figure 6.10(c). As Greenland warmed during the Bølling interstadial, the warming trend in Antarctica was reversed. Even more striking is the fact that during the cold snap of the Younger Dryas in Greenland, Antarctica warmed. It is clear, therefore, that temperature changes in Antarctica lead Greenland by about 1 800 years.

Another method using the GRIP ice core timescale to calibrate the Vostok record is to match variations in the $\delta^{18}O$ values of air trapped in bubbles in the ice at both locations. Note that $\delta^{18}O$ values of *air* ($\delta^{18}O_{air}$) are used, and not those obtained from ice samples ($\delta^{18}O_{ice}$).

□ Why should the isotopic ratio of atmospheric oxygen vary over the last ∼130 000 years?
■ Because the major factor that caused changes in $\delta^{18}O$ values for atmospheric oxygen was fluctuations in the oxygen isotopic content of sea water, which in turn was driven by changes in the volume of land ice (Box 4.3)

Correlation of the GRIP and Vostok records using the $\delta^{18}O_{air}$ method show that warming in the Antarctic commenced ∼3 thousand years before the start of the Bølling interglacial in Geenland (Figure 6.11). Likewise, the proportions of atmospheric CO_2 and

CH_4 also began to rise two to three thousand years before warming in the Northern Hemisphere.

The Present and Penultimate Interglacials Compared

The Vostok ice core record and the oceanic $\delta^{18}O$ records both indicate that during the last interglacial stage (oxygen isotope Stage 5e/the Eemian) was probably a little warmer than the present one (Stage 1/the Holocene). Studying records of the last interglacial has a special fascination, because it may offer clues to climatic changes that might happen as the present one draws to a close.

□ Look at Figure 6.12 and compare the isotopic records in the Vostok and GRIP cores for the last interglacial. How do they differ?
■ The Vostok record shows a rapid rise in $\delta^{18}O_{ice}$ into the last interglacial from 140 thousand to 125 thousand years ago, then there is an initial rapid fall after three to four thousand years, but the values then fall more gradually to a minimum at 110 thousand years ago. The pattern is very different in the GRIP core. There are higher $\delta^{18}O_{ice}$ values between 140 and 115 thousand years ago, but there are rapid fluctuations (7–8‰) quite unlike those at Vostok.

Taken at face value, the changes in $\delta^{18}O_{ice}$ in the GRIP core over the last interglacial indicate temperature changes at the surface of the ice of up to 10–12° in a matter of decades. This is a totally different pattern to the relatively minor temperature fluctuations indicated by the isotope record from the Vostok core during the last interglacial, and is quite unlike anything that occurred during the last 10 thousand years (the present interglacial). When announcing these results, the scientists concerned were not reticent in pointing out the significance of their results for the debate about possible climate changes facing us in the future. They wrote as follows:

> Given the history of the last 150 kyr, the past 8 kyr has been strangely stable; only during the final ∼2 kyr of the warmest stage of the Eemian do our data

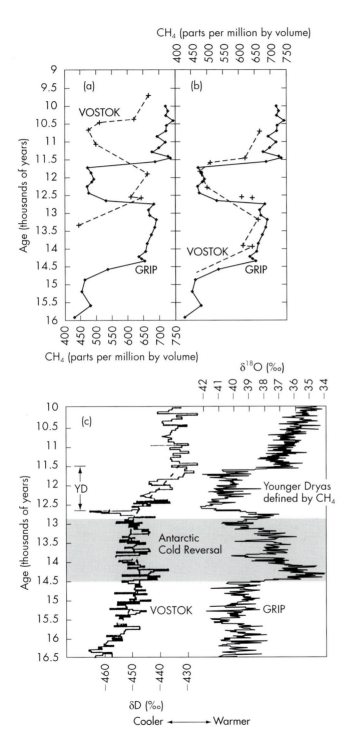

Figure 6.10 Using methane concentrations to revise the Vostok ice core timescale. (a) Variations in methane concentration plotted using original timescales for the Vostok and GRIP cores. (b) Shifting the trough in methane concentration in the Vostok core down to match the Younger Dryas trough in the GRIP core enables the Vostok timescale to be revised. (c) Using the revised Vostok timescale, temperature changes in Greenland and Antarctica, as indicated by $\delta^{18}O$ and δD respectively, do not match. Cooling and warming events in Antarctica lead those in Greenland by about 1 800 years.

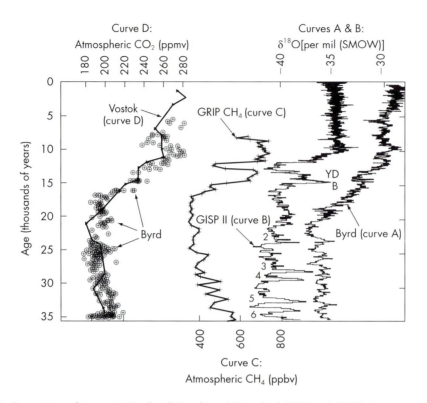

Figure 6.11 Comparison of Antarctic (Byrd and Vostok) and Greenland (GISP and GRIP) ice core records for the last 35 thousand years. Curves A and B show the $\delta^{18}O_{ice}$ curves for the Byrd and GISP cores: these show that (i) Antarctica began to warm ~3 thousand years before Greenland; (ii) the Dansgaard–Oeschger events (cool stadials: 2–6 on curve B) are 'antiphased' in Antarctica by smaller amplitude warming events; (iii) the Bølling interstadial and Younger Dryas stadial (B and YD respectively on curve B) are not present in the Antarctic record. The curves (C and D) and scatter plot showing the changes in atmospheric CO_2 and CH_4 all show rises that precede the rise in temperature in Greenland by 2–3 thousand years.

demonstrate a similar period of stability. The unexpected finding that the remainder of the Eemian period was interrupted by a series of oscillations, apparently reflecting reversals to a 'mid-glacial' climate is extremely difficult to explain. Perhaps the most pressing question is why similar oscillations do not persist today, as the Eemian period is often considered as an analogue for a world slightly warmer than today's . . .

Greenland Ice-core Project (GRIP) Members.1993. Climate instability during the last interglacial period recorded in the GRIP ice core. *Nature*, 364: 203–7.

We find that climate instability was not confined to the last glaciation but appears also to have been marked during the last interglacial . . . and during the previous . . . glacial cycle. This is in contrast to the extreme climate stability of the Holocene, suggesting that recent climate stability may be the exception rather than the rule.

W. Dansgaard, S.J. Johnsen, H.B. Clausen, D. Dahl-Jehnsen, N.S. Gundestrup, C.U. Hammer, C.S. Hvidberg, J.P. Steffensen, A.E. Sveinbjörnsdottir, J.J. Jouzel, and G. Bond. 1993. Evidence for general instability of past climate from a 250-kyr ice-core record. *Nature*, 364: 218–20.

The implication of these results was that the North Atlantic door described in Box 6.2 was partially shut several times during the Eemian, rather than remaining permanently open as it has done during the last interglacial. The first paper concluded: 'Man is already perturbing one of the factors that may be involved, the greenhouse gases, and our first tentative measurements indicate that they may be linked to

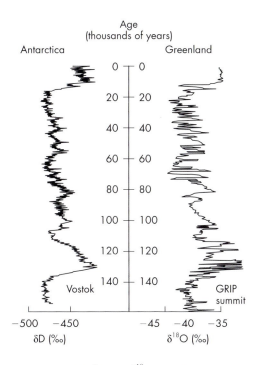

Figure 6.12 The δD and δ^{18}O records for the Vostok and Greenland GRIP ice cores.

climate change within the Eemian.' By the conservative standards of scientific literature this was heady stuff, designed for government funders as much as scientific peers.

Two weeks before the GRIP paper appeared, an American funded Greenland ice sheet Project 2 (GISP 2) completed drilling another hole to bedrock 28 km west of the GRIP site. This location was not quite at the summit, but the ice record here, like that at the GRIP site, was not expected to have been deformed in any major way by flow over bedrock topographic features. By December 1993, early published results from GISP 2 undermined the story that there had been wild temperature fluctuations during the last interglacial, and suggested that the δ^{18}O excursions were not evidence favouring major climatic fluctuations, but an artefact of deformation at the base of the ice sheet, resulting in slices of heavier δ^{18}O ice being emplaced. But these heavier values

could indicate that during the Eemian, the ice cap almost melted away: this is consistent with the coral reef evidence that indicates sea levels at this time were several metres higher than they are today.

Later work indicates that there was some climate instability during the Eemian, but not on the scale incorrectly suggested on the basis of the GRIP ice core results. Oceanic records from the Norwegian Sea, particularly the proportion of the polar left coiling planktonic foram *N. paychyderma*, indicate three cooling episodes during the Eemian (at ~127–126 ka, ~122–121 ka and 117 ka). One of these, (~122–121 ka) has been detected in the Atlantic south of Iceland, off West Africa, and appears to be mirrored by cooling episodes revealed by pollen evidence from France and Germany.

Dust Records at the Annual to Millennial Scale

Electrical conductivity measurements (ECMs) are an indirect method of measuring the amount of wind-blown dust present in the ice. To conduct an electric current, ice must contain acids. These are derived largely from atmospheric SO_2 (from volcanoes) and NO_x (nitric oxides). If the acids are neutralised by dust then the ice will not conduct electricity.

☐ So does a decrease in ice conductivity correspond to an increase or decrease in the amount of dust it contains?

■ A decrease indicates more dust. This is because dust neutralises the acids present, reducing the conductivity of these.

Measurements of electrical conductivity of ice cores can be made at the drilling site. Two separate electrodes in contact with the core are drawn along it, and measurements made continuously. The resolution that can be achieved is between seasons and millennia, and the method is much less time consuming and cheaper than conducting isotope analyses. Remember that low readings indicate dusty intervals.

Figure 6.13 shows δ^{18}O$_{ice}$ and ECM records for the GRIP ice core.

☐ How do the warm intervals (interstadials) of the Dansgaard–Oeschger cycles show up on the ECM record?

■ They show increases in conductivity, as less dust is present.

☐ Why should less dust be blown into the Greenland ice sheet during warmer intervals of time?

■ Because wind speeds are reduced at such times.

Figure 6.14(a) shows the record that was obtained between 10 and 42 ka from the GISP ice core. The time scale shown was obtained by counting annual layers in the ice core back to 17.4 ka, achieving an accuracy estimated at 3 per cent. Further down the core, annual layers were counted over a one metre section at every 10 or 20 m, and the resulting accumulation rates used to calibrate the age/depth relationship.

The proxy dust record shown in Figure 6.14 shows a previously unrecognised signal of rapid climate change on the scale of <5–20 years – this is the spiky pattern shown on Figure 6.14(b–d). The record at the beginning of the present interglacial (Figure 6.14(b)) indicates relatively stable conditions, but the glacial record seems to flicker rapidly between glacial and almost interglacial conditions (Figure 6.14(d)). The record for the warmer interlude that preceded the Younger Dryas (Figure 6.14(c)) appears to be intermediate in character between interglacial (Figure 6.14(b)) and glacial (Figure 6.14(d)). The scientists who studied these patterns referred to them as the 'flickering switch' of climate change. They interpreted this flickering to be caused by changes in atmospheric circulation patterns in the north Atlantic area, rather than changes in the areas of dust sources. They reasoned that the latter, being controlled largely by ice retreat or advance, would be much too slow to cause the rapid change in dust influx into the ice-record. The ECM record shows that the climatic change from the Bølling interstadial to the Younger Dryas stadial occurred in less than a century, and that the change back into the present interglacial may have occurred in just a few decades. Such rapid changes are too fast to have been triggered

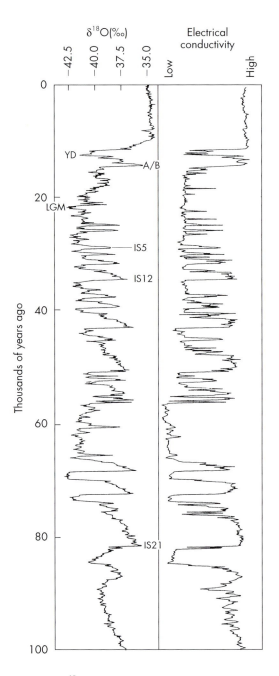

Figure 6.13 $\delta^{18}O_{ice}$ and Electrical Conductivity Records (ECM) from the GRIP ice core. YD: Younger Dryas; A/B: Allerød/Bølling interstadial; LGM: Last Glacial Maximum; IS: interstadial.

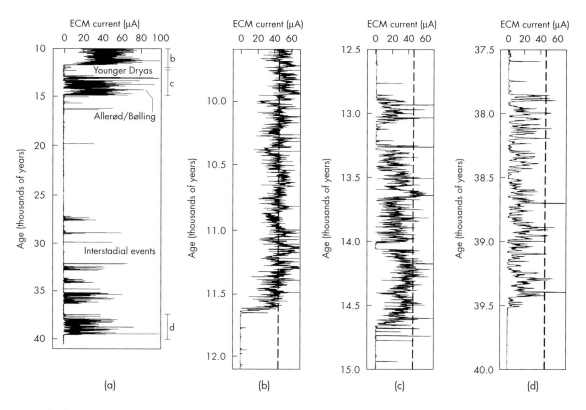

Figure 6.14 The GISP2 ECM (electrical conductivity measurements) record (a) between 10 and 41 ka, and (b–d) detailed records over the intervals shown on (a). Zero readings indicate that high levels of alkaline dust have neutralised the acidity of the ice (caused by the presence of dissolved SO_2 and NO_x). The dashed lines indicate the mean value for the present interglacial.

by ice sheet movement, and are at the limit of the rapidity of possible changes in oceanic deepwater circulation.

MILLENNIAL SCALE EVENTS IN THE LOESS RECORD

As described on pp. 93–6, the alternation of wind-blown dust (loess) and fossil soils (palaesols) provides a proxy record of fluctuations in the Asian monsoon system since 2.5 Ma. Increases in grain size in these sedimentary successions indicate increasing aridity and strengthening of winter monsoonal winds, whereas the fossil soils indicate increased precipitation which favoured soil forming processes. Figure

6.15 shows the grain size variation in a 1–2 m thick succession from the Loess Plateau at Luochan in China. This succession has been dated using cathodoluminesence techniques (not described in this book) and by comparing the fluctuations in grain size to the oceanic $\delta^{18}O$ record. The figure shows that there is a strong correlation between the peaks in the Loess Plateau grain size variations and the timing of North Atlantic Heinrich events. Thus it seems likely that the Chinese loess deposits, which are – and were – downwind of the North Atlantic, record the same high frequency events (Bond cycles) present in ice cores and oceanic sediments far to the west. This relationship is consistent with climate models for glacial periods that indicate more vigorous atmospheric circulation, and the diversion of storm tracks to the south of the Laurentide and Scandinavian ice sheets across southern Europe and central Asia.

MILLENNIAL SCALE EVENTS IN THE NE PACIFIC

Sediments that accumulated during the past 60 thousand years in a small sedimentary basin – the Santa Barbara Basin – off the California coast record changes in oceanic circulation and oxygenation that correlate to 19 out of 20 Dansgaard–Oeschger events. There are two main reasons why such short-lived events can be detected. One is that sedimentation rates are more than ten times higher (>120 cm per thousand years) than those typical of ocean floors. The other reason is that the connection between the basin and the Pacific Ocean is largely via two sills (Figure 6.16(a)), between which oxygen depleted bottom

waters occur today (Figure 6.16(b)). Water below about 300 m depth is derived from intermediate depth Pacific waters which have low dissolved oxygen contents. The anoxic conditions in the Santa Barbara Basin result from the fact that the Pacific water entering it is further depleted in oxygen as planktonic material is oxidised as it falls through the water column, and after it is deposited on the ocean floor. Between the sills, there is no recirculation of bottom water, and so anoxic conditions occur. This means that there are no benthonic organisms present on or within the sediment, and so no bioturbation occurs, resulting in the preservation of finely laminated organic rich sediments.

The sediments deposited in the Santa Barbara Basin were cored in 1993 by the Ocean Drilling Program. Cores of Holocene sediments are largely laminated, but older sediments are alternately laminated and bioturbated.

☐ What is the significance of the bioturbated intervals?

■ They indicate higher ocean floor oxygen levels compared to the oxygen deficient conditions of today.

A semi-quantitative bioturbation index was devised to describe the cores, and is illustrated in Figure 6.16(c), which shows how the indices relate to sea bottom oxygen levels. Figure 6.16(d) shows that the ages of laminated intervals beneath the Holocene sediments closely match the interstadial events identified in Greenland ice cores.

Why did oxygen levels on the floor of the Santa Barbara Basin alternate between high and low levels? The key to answering this question is the age of the bottom water. The age of water masses can be determined by radiocarbon dating of carbon in solution. Surface waters will yield negligible ages, but deeper water masses may be over one thousand years old. The age relates to the time at which the water was at the surface and acquiring planktonic material which was the source of the dissolved carbon. Today, water flowing over the sill into the Santa Barbara Basin is about 1300 years old, which suggests that it originated out-

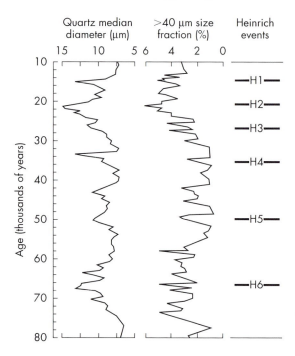

Figure 6.15 Grain size variations in the topmost part of the loess–palaeosol succession at Luochang on the Chinese Loess Plateau compared with the ages of North Atlantic Heinrich events.

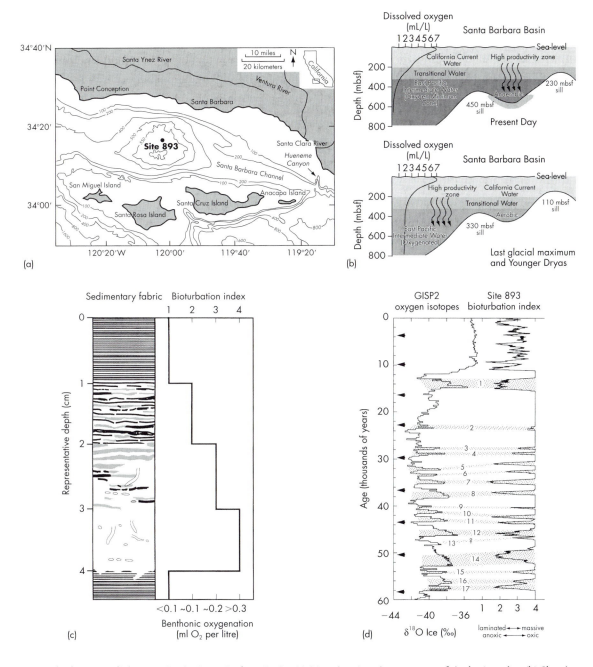

Figure 6.16 Interstadial events in the Santa Barbara Basin. (a) Map showing the geometry of the basin today. (b) Sketch cross sections across the basin showing how the development of anoxic and oxic conditions was related to warmer (Holocene, Bølling interstadial) and cooler intervals (Younger Dryas, last glacial maximum). (c) The relationship between the proportion of laminated and bioturbated sediments (the bioturbation index) and oxygen levels in bottom waters. (d) Comparison of the bioturbation index in the Santa Barbara Basin cores, and the $\delta^{18}O$ record from Greenland ice cores, showing how laminated sediments developed during interstadials.

side the Pacific (circulatory systems within the Pacific recycle deep water on a shorter timescale than this age). The likely source is the 'deep water conveyor' driven by thermohaline circulation (see Figure 2.6) initiated by evaporation of the North Atlantic Drift. By comparing the radiocarbon ages of benthonic and planktonic foraminifera in the Santa Barbara sediments, it is possible to determine whether the bottom waters at the time of deposition were 'young' or 'old'. This is because the radiocarbon ages of the forams are modified by the isotopic composition of the dissolved carbon in the water. If there is no age difference between the benthonic and planktonic forams, the water mass is young, but if there is a difference of a few hundred years or more, the bottom water must have been older than the surface water. This technique has shown that the laminated sediments were deposited beneath relatively old bottom waters, and the bioturbated sediments beneath younger waters. The older waters, which are thought to partly origi-

nate from outside the nearby Pacific basin, contained less oxygen and so could not support bottom dwelling faunas, whereas the younger, oxygen richer waters supported bottom dwelling burrowers. It is impossible to determine whether variations in the oxygen content of the Pacific intermediate depth water supplying the Santa Barbara Basin was controlled by circulatory changes *within* the Pacific Ocean, or by the influx of water from elsewhere. But there can be no denying that there is a temporal link between bottom water oxygen levels in the Santa Barbara Basin and events in the North Atlantic area (Dansgaard–Oeschger cycles and iceberg armadas). As there are no significant time lags between the two records (most of their ages agree to within one per cent) it seems likely that changes in ocean circulation in the two areas were tightly linked.

Foraminifera within the Santa Barbara Basin sediments reveal some remarkable variations in $\delta^{13}C$ (Box 6.3). Benthonic forams in the sediments show large

---

Box 6.3 Carbon Isotopes

Carbon consists of a mixture of two stable isotopes: ^{12}C (carbon-12) and the much rarer ^{13}C (carbon-13). In photosynthesis fixation of the lighter $^{12}CO_2$ is favoured over that of the heavier $^{13}CO_2$, because $^{12}CO_2$ is taken up and diffuses into cells more rapidly and also reacts more readily. As a result of this isotope fractionation, organic matter produced by photosynthesis is enriched in ^{12}C and depleted in ^{13}C relative to the carbon in the inorganic carbon pool in the atmosphere and hydrosphere (mainly present as CO_2 gas, but also as carbonate and bicarbonate ions in solution). Further depletion of ^{13}C occurs when organic matter is decomposed by methanogenic bacteria to produce methane. Enrichment or depletion of ^{13}C is expressed in terms of a $\delta^{13}C$ value which is calculated as follows. The $^{13}C/^{12}C$ ratio for the sample being investigated is given as a ratio to that of a carbonate standard in which the $^{13}C/^{12}C$ ratio is 1/88.99. The number one is subtracted from this value and the whole is multiplied by 1000 to give a $\delta^{13}C$ value in terms of parts per thousand or per mil (â) of ^{13}C relative to the standard. This can be expressed in the following formula:

$$\delta^{13}C = \left[\frac{(^{13}C/^{12}C)\ \text{sample}}{(^{13}C/^{12}C)\ \text{standard}} - 1 \right] \times 1000$$

☐ Using this formula, will the value of $\delta^{13}C$ be negative, positive or zero when the $^{13}C/^{12}C$ ratio is (a) equal for standard and sample, (b) greater in the sample, (c) lower in the sample?

■ Because the number one is subtracted from the ratio of ratios, for (a) the value will be zero, for (b) it will be positive, and for (c) it will be negative.

Present day marine bicarbonate has a $\delta^{13}C$ value close to zero, whereas the value for atmospheric CO_2 is ~8.

Box 6.4 **Methane Clathrate**

Sea-floor sediments often contain organic matter that will naturally decay to form hydrocarbons. Methane is constantly produced on the sea floor by methanogenic bacteria. Unless it is trapped in some way, this methane bubbles up to the surface and escapes into the atmosphere.

Today, the deep-sea bed is at a temperature of around 1–2°C and a pressure of several hundred times atmospheric pressure. Under these conditions, a frozen form of methane in water called gas hydrate can form. Appropriate temperatures and pressures are found on the sea bed where water depths are greater than about 400 m (Figure 6.17(a)), which in turn means that gas hydrates can form over most of the ocean floor.

Gas hydrates form freely in the oceans, but they are not found as a carpet on the ocean floor. This is because gas hydrates are lighter than sea water (they have a density of 850 kg m^{-3}). Normally as soon as they form, they float upwards, turning back into methane and water in the lower pressures and warmer temperatures of the upper layers of the ocean.

However, within the sediments just beneath the ocean floor, crystals of gas hydrates form and are buoyed upwards. Frequently the rising crystals form a 'log-jam' within the pore spaces of the sediment. Once this occurs, more gas hydrate crystals will become trapped in the pore spaces of the sediment underneath. The downward extent of gas hydrates is limited by geothermal heat, and so they accumulate as a distinct layer within ocean floor sediments.

Methane hydrate crystals readily absorb methane into their lattice to form clathrates (a clathrate is a compound in which a gas is enclosed within a crystal structure). Fully saturated methane clathrates can hold up to 200 times their

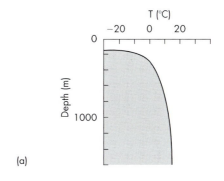

(a)

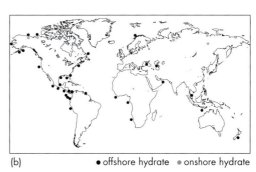

(b)   ● offshore hydrate   ● onshore hydrate

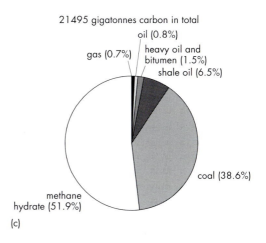

(c)

Figure 6.17 The occurrence of methane clathrate. (a) Shaded area indicates pressure and temperature under which methane clathrates are stable in ocean floor sediments. (b) Known global occurrences of methane clathrate. (c) Percentage of the total carbon in methane clathrate and other fossil fuels. The deep-sea bed is not the only repository of methane clathrates. Permanently frozen ground which gives permafrost conditions can also be suitable for clathrate formation.

own volume of methane within their crystal structure. Eventually, the spaces between grains of sediment become completely filled by gas hydrate crystals saturated in methane. Saturated gas hydrate crystals are denser than seawater, so the zone as a whole is gravitationally stable.

The best conditions for forming gas hydrates are to be found in areas of the ocean where relatively thick carpets of sediment are being deposited and where large quantities of methane are also being produced by the decay of organic matter.

Current estimates suggest that gas hydrates might contain some 10^6 tonnes of carbon (Figure 6.17(c)), or slightly more than the amount of carbon that is locked into all the other fossil fuels added together.

changes in $\delta^{13}C$ values of up to 5‰: low values characterise the forams occurring in laminated sediments (which were deposited in warmer waters) and high values occur in the bioturbated sediments (deposited in cooler conditions). In addition, over several short intervals, large negative $\delta^{13}C$ excursions of up to 3.5‰ occur at or near the beginning of the Dansgaard–Oeschger events. This means that at the start of at least some of the warming events, the whole of the water column in the Santa Barbara Basin experienced a rapid negative carbon isotope excursion. The authors of the isotopic study suggest that the excursion was caused by the rapid release of methane from the seabed. The mechanism whereby this gas can accumulate in continental margin sediments is described in Box 6.5. As pressure and temperature are critical for its accumulation, changes in water temperature and sea level may trigger its release. As methane is a powerful greenhouse gas, it is possible that the release of large amounts of it may have significantly contributed to greenhouse warming at the start of each Dansgaard–Oeschger cycle.

SUMMARY

1 Climate changes at the millennial scale or less occur on timescales that humanity can relate to more easily than those driven by the Milankovich pacemaker. Understanding these possible causes may help us model future climate changes resulting from anthropogenic effects, especially those resulting from the burning of fossil fuels.

2 Layers of ice-rafted debris within ocean floor sediment cores recovered from the North Atlantic show that there were repeated ice sheet surges that released iceberg armadas. Six such Heinrich events were initially discovered, and as the layers of sediment they left behind contain limestone fragments that match those exposed today in eastern Canada, it was suggested that surging of the Laurentide ice sheet (LIS) was responsible. Later research showed that rock debris derived from Iceland and NW Europe occurred in some Heinrich layers, indicating that separate ice sheets surged at the same time. The surges could have been triggered by climatic events (the Denton model) or by ice sheet dynamics (the MacAyeal model).

3 High resolution studies of ocean cores have revealed that these iceberg armadas were launched every 2–3 thousand years into the Atlantic between 10 and 38 ka. Some of the ice rafted debris was shown to originate from ice sheets covering Iceland and NW Europe. As it is unlikely that ice sheet dynamics would cause separate ice masses to surge simultaneously, an external climatic change may have been involved. The abundance of left hand coiled forams in the cores also showed that the iceberg armadas were launched at the end of periods of slow cooling, and were followed by very rapid warming events.

4   Recent evidence suggests that during the Holocene there were regular changes in climate with a periodicity of about 1.5 thousand years. These changes are reflected in peaks in the delivery of ice rafted debris to the North Atlantic, which are about a tenth of the size of the Heinrich layers.

5   Variations of $\delta^{18}O$ in Greenland ice cores reveal 24 warmer interludes (interstadials). The pattern of cooling and warming is asymmetric, with slow cooling followed by rapid warming; it closely matches the North Atlantic ocean floor sediment record. The repeated cooling/warming episodes are known as Dansgaard–Oeschger cycles, and are bundled together into longer Bond cycles.

6   A possible explanation of the virtually identical pattern of temperature changes recorded in ocean floor sediments and the Greenland ice cap involves changes in the position of the oceanic polar front. This front is the boundary between warmer subtropical waters of the North Atlantic, and cold polar waters. The front is a kind of door that opens and closes. When it is open, the warmer waters penetrate further north, and vice versa. It is fully open today, but during stadials, it was closed; the warmer interstadials occurred during short periods of time when it was ajar.

7   Variations in the methane content of gas bubbles trapped in the Greenland ice cores track the temperature changes indicated by $\delta^{18}O$ data. The proportion of methane present increases during the warmer interstadial intervals. As it is unlikely that high latitude wetlands would have been a significant source of gas (because even during interstadials they would have been frozen), the methane peaks suggest higher rainfall in tropical regions, leading to increased plant growth and subsequent decay.

8   The Vostok ice core record and oceanic $\delta^{18}O$ data both indicate that the last interglacial stage was a little warmer than the present one, with global mean temperatures 1–2°C higher than today's value. Initial $\delta^{18}O$ results from Greenland ice cores suggested that surface temperatures over the ice cap during the last interglacial fluctuated by as much as 10–12°C in a matter of decades – quite unlike the stable temperature record of the Holocene. Later results indicated that the $\delta^{18}O$ fluctuations were caused by mixing of different layers of ice by folding near the contact with the underlying rock.

9   The dust content of Greenland ice cores tracks temperature variations recorded by $\delta^{18}O$ fluctuations, increasing during stadials. Electrical conductivity measurements (ECMs) provide very high resolution proxy climate data from ice cores. As the dust content of the ice neutralises its acid content, low conductivity values indicate high dust contents. The ECMs taken along Greenland ice cores show that the interstadial record is characterised by a rapid flickering, at <5–20 year intervals, between glacial and almost interglacial conditions. They also show that the climate change from the Bølling interstadial to the Younger Dryas stadial occurred in less than a century, and that the change back into the present interglacial may have occurred in just a few decades.

10 The stadial and interstadial events of the North Atlantic region are mirrored by changes in the structures displayed by ocean floor sediments in the Santa Barbara Basin off California suggesting that they are at least hemispheric in extent. Sediments deposited during the present interglacial, and during interstadials are laminated, whereas those deposited during stadials are heavily bioturbated. The different features of the sediments were caused by fluctuations in the amount of dissolved oxygen in the bottom waters, with the laminated sediments being deposited in anoxic conditions in which dwelling organisms were unable to live. The change from anoxic to oxic bottom waters is caused by changes in the circulation patterns of deep and intermediate waters in the Pacific. During warmer periods, warmer oxygen deficient intermediate depth waters, ultimately derived from the deep Atlantic, reached the area, whereas during cold periods, oxygenated water supplied the bottom of the Santa Barbara Basin.

11 Negative excursions in the $\delta^{13}O$ values obtained
from planktonic and benthonic forams in the
Santa Barbara Basin sediments occur at or near
the beginning of the rapid warming phase of the
equivalents of the Dansgaard–Oeschger events.
This suggests that the entire water column experienced a rapid negative carbon isotope excursion.
This could have been caused by the release of
methane from clathrates contained in sea floor
sediments.

FURTHER READING

See Lowe and Walker (1997) and Williams *et al.* (1993) in the
reading list for Chapter 1 (pp. 28–9), and references to figure
sources for this chapter (pp. 256–7).
Since this chapter was written, the synchroneity of ice sheet collapse around the North Atlantic has been questioned. See
Dowdeswell, J.A., Elverhøi, A., Andrews, J.T. and Hebbeln, D.
1999. Asynchronous deposition of ice-rafted layers in the
Nordic Seas and North Atlantic Ocean. *Nature*, 400, 348–51.

7

EXPLANATIONS

INTRODUCTION

By now you should have been left in no doubt that the Earth's climate system is extremely complex. Our understanding of it is still not good enough for computer models to be constructed that can simulate changes observed over historical timescales. Yet we are faced with explaining changes over much longer timescales (Figure 6.1). These range from tens of millions of years as the Earth cooled from a greenhouse state during the Cretaceous to the icehouse condition of today, through tens of thousands of years as it oscillated between glacial and interglacial conditions, to changes spanning a few millennia or less such as stadials and interstadials, the Medieval warm period, and the Little Ice Age and subsequent global warming.

So why should we attempt to explain climatic changes that affected the Earth long before humanity began changing many parts of the Earth's system, including, in particular, the greenhouse content of the atmosphere? The answer is simple: understanding the past may help us to comprehend how the present climate system works and possibly predict future climatic changes. We must use historical and geological records of change as the test bed for climate models. If such models can simulate past climatic changes, then we will be more confident in the predictions they make about the future – with or without the continuing release of greenhouse gases caused by human activities.

In this chapter, our search for explanations of past climatic behaviour focuses first on post-Cretaceous cooling, and then on the more rapid climatic fluctuations characteristic of the Great Ice Age. In both cases, neither our knowledge of climatic change through time and in different parts of the world, nor our present understanding of the Earth's climate system enables a single explanatory model to be formulated. It is likely that the cause of post-Cretaceous cooling is to be found *within* the Earth's system, and is not due to astronomical forcing factors as is most probably the case for the origin of climate changes during the Great Ice Age.

POST-CRETACEOUS COOLING

Introduction

There are three broad categories of processes that may have contributed to post-Cretaceous global cooling:

- magmatic processes: volcanism, atmospheric CO_2 and aerosols, ocean crust formation and global sea-level changes;
- continental drift: continental configuration and the opening and closing of oceanic gateways; latitudinal distribution of continental area;
- tectonic uplift of continents: modification of atmospheric circulation; increased rates of weathering lowering atmospheric CO_2; uplift of land areas above regional snow line.

Magmatic Processes

About 120 million years ago, during the Cretaceous, there was a significant increase in the rate of formation of ocean crust along ocean ridges, and in outpourings of lavas that formed huge oceanic plateaux in the Pacific and Indian Oceans. If these plateaux were built above sea-level, large amounts of CO_2 would have been released into the atmosphere, which would account for estimates that atmospheric CO_2 levels may have been between 3 and 15 times the pre-industrial level of 280 ppm. This could have caused greenhouse warming of the planet of between 3 and 8°C. Moreover, the higher atmospheric temperatures would have increased the amount of water vapour in the atmosphere by as much as 150 per cent, so that the resultant positive feedback would have caused further warming because water is the most effective greenhouse gas of all. At the present time, however, the greenhouse effect of water vapour is confined largely to low latitudes because cold air holds very little water vapour (the amount that can be held doubles with every 10°C increase in temperature). This was probably not the case during the Late Cretaceous when polar regions were much warmer, and the atmosphere at high latitudes could have held as much as 1000 per cent more water vapour than it does today, contributing significantly to greenhouse warming in these regions. A decrease during the Cainozoic in the amount of water vapour held in the atmosphere could have played a significant role in global cooling. During this time, the main source areas of atmospheric water vapour – the tropical and sub-tropical oceans – decreased significantly in area, particularly as the Tethys ocean closed (see Figure 7.4 on page p. 144).

Late Cretaceous sea-levels may have been up to 300 m above present levels due to a reduction in the volume of the ocean basins caused by increased rates of sea floor spreading (which results in higher, wider ocean ridges) and the formation of oceanic lava plateaux displaced more water onto the continents. Flooding of land areas and their rearrangement due to continental drift would also have contributed to global warming. Climate modellers have suggested that the net effect of all these changes in the Cretaceous may have resulted in global warming between 8 and 13°C compared to today. This is in good agreement with the 6-16°C of warming inferred from the distribution of shallow marine and plant fossils.

Once magmatic activity diminished, cooling from the greenhouse world of the Cretaceous was inevitable. Not only would a reduction in atmospheric CO_2 levels have resulted in cooling, but lowered sea-levels would have increased the seasonality of climates, particularly in high latitudes, leading eventually to glaciation.

The link between volcanic eruptions and more recent climate changes is controversial. Some researchers postulate that once the Earth had cooled, volcanic eruptions might have triggered the onset of glaciation, but this idea is controversial. Although ashes ejected from volcanoes form the spectacular plumes that spread across large areas of the Earth, the ash particles do not have significant climatic effects. The effects of CO_2 emissions and the formation of sulphate aerosols (Box 7.1) from sulphur dioxide are much more important. It has been suggested that the huge eruption of Tambora, in the northern Indian Ocean, some 74 thousand years ago may have accelerated the cooling from the last interglacial into the last glacial. Measurements following recent major eruptions, such as Pinatubo in 1991, show that global temperatures may decline by up to 0.3°C for a few years after the eruption. High latitudes may suffer greater cooling – as much as several degrees. Cooling is thought to result primarily from the formation of sulphate aerosols in the atmosphere, which, when spread around the globe, increase planetary albedo.

Many workers have discounted the link between long term global cooling and volcanism on the grounds that the sulphate aerosols only remain in the atmosphere for a few years. However, deep-sea drilling results in the northern Pacific suggest that there may be such a link when there is an upsurge of volcanic activity on a regional scale. Figure 7.1 shows that an increase in the ash thickness in the sediments cored coincides with an increase in the amount of ice rafted debris at 2.6 Ma, suggesting that enormous eruptions in the Kamchatka-Aleutian region could have triggered the growth of large polar glaciers so that they reached the sea. It is not suggested that increased volcanism was the sole agent that triggered Northern Hemisphere glaciation, but it could have

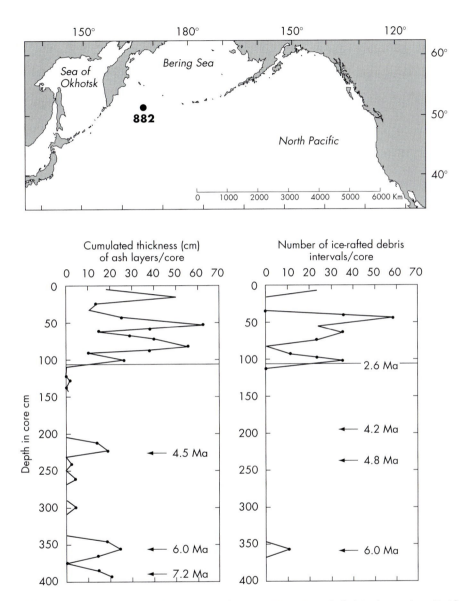

Figure 7.1 Ash layers and ice-rafted intervals (see Figure 4.2(d)) in Hole 882A drilled in the northern Pacific. Each core length is approximately 10 m. Note the coincidence in the increase in ash layer thickness and number of ice-rafted intervals per core at 2.6 Ma, which is the age of onset of Northern Hemisphere glaciation.

been a 'threshold phenomenon' that finally tipped the climate system into a truly global glacial mode. Another interpretation of the coincidence in the increase in the amount of ice rafted debris and volcanic ash thickness is that the onset of glaciation trig-gered increased volcanism due to ice-loading of the crust.

A significant part of NADW consists of deep water spilling over the Greenland–Scotland ridge. This topographic feature separates the colder more saline

Box 7.1 Sulphate Aerosols

Large volcanic eruptions can increase the Earth's albedo, and so cause global cooling. It is not the ash particles (which range in size from 1-2 μm to pebble-sized pieces of pumice) that cause the most significant climatic effects, as most of them fall to the ground in a matter of days. Volcanic gases are far more important as climatic forces.

In addition to the large amounts of carbon dioxide released by volcanoes, large quantities of sulphur dioxide (SO_2) are released. The SO_2 reacts with water vapour in the atmosphere to form tiny airborne droplets (aerosols) of sulphuric acid. As these droplets are so small — (0.1–1.0 μm in diameter) they remain suspended for months or even years. The light scattering by aerosols to produce wonderfully coloured sunsets around the world is a visible manifestation of their effect on the Sun's radiation. More importantly, they scatter a proportion of radiation back into space increasing the Earth's albedo.

water of the Arctic Ocean and Norwegian–Greenland Sea from the warmer water of the Atlantic to the south. Changes in the depth of water above the ridge, caused by glacioeostatic sea-level changes, or tectonic uplift and subsidence, exert an important control over the rate of NADW flow. It is probable that the ridge subsided sufficiently for the northern waters to spill southwards during the early Miocene, initiating Atlantic deep water flow that upwelled around Antarctica. This upwelled water was relatively warm in comparison to 'local' water and so provided a source of moisture that enabled a large ice sheet to grow over Antarctica, as indicated by the increase in $\delta^{18}O$ values of oceanic sediments during the Middle Miocene.

The depths of the oceanic sills along the Greenland–Scotland Ridge were affected by three processes. One of these — glacioeostatic sea-level change — became important once Northern Hemisphere ice sheets were established 2.6 million years ago. The other two are related to processes occurring in the Earth's mantle and crust. The cooling of newly formed oceanic crust as sea-floor spreading moves it further away from the Mid-Atlantic Ridge results in thermal subsidence of the sea bottom. This causes a smooth but decreasing rate of subsidence, superimposed on which are shorter term fluctuations caused by variations in mantle plume activity beneath Iceland. Mantle plumes are huge columns of magma that rise from the core–mantle boundary and impinge on the crust, causing major intrusive and volcanic activity and regional doming. In the case of the Iceland plume, the doming had a radial extent of 500–1 000 km and so affected the elevation of the Greenland–Scotland Ridge, causing it to rise and fall in an irregular fashion. As shown in Figure 7.2 periods of high Atlantic deep water formation correspond to times when the ridge was depressed during periods of low plume activity, and vice versa. This correlation does not, however, establish an unequivocal link between the two, but it does suggest yet another possible link between magmatic and climatic events.

Continental Drift and Oceanic Gateways

Today, the burden of transporting heat from low to high latitudes is shared almost equally by the atmosphere and oceans, but was this necessarily so when the Earth was much warmer? Modelling studies have been conducted to find out the amount of oceanic heat transport necessary to produce global mean surface temperatures determined for past geological periods. As indicated in Figure 7.3 they show that the full range of Cretaceous greenhouse to present day icehouse conditions can be accounted for by changing ocean heat transport by just over a factor of two. Of course this does not prove that changing oceanic circulation was the only cause of post-Cretaceous cooling, but it does suggest that it is likely to have played a significant role.

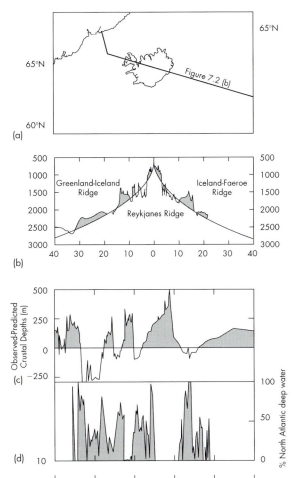

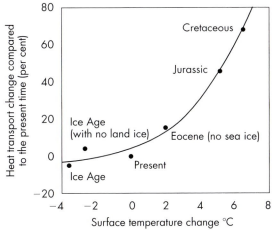

Figure 7.2 The link between the estimated depths of the Greenland–Scotland Ridge and the formation of North Atlantic deep water. (a) Map showing the location of the transect across which observations and predictions shown in (b) and (c) were made. (b) Crustal depth versus age plot across the ridge. The smooth curve shows the depths predicted if the ridge had subsided due only to cooling effects as parts of it moved away from the axis of the Reykjanes Ridge due to sea-floor spreading. The irregular line shows the difference between the actual and predicted depths of the ridge. Note that the actual depths measured are the depth below sea-level of the top of oceanic crust, and not to the sea-floor. (c) The difference between predicted and observed depths of the ridge. (d) Variations in North Atlantic deep water formation determined from the $\delta^{13}C$ content of benthonic forams in the world's oceans (a methodology not described in this book).

Figure 7.3 Model results showing the change in oceanic heat transport from low to high latitudes necessary to produce changes in global mean surface temperature inferred for different time periods.

☐ How could the effectiveness of ocean heat transport have been reduced since the Cretaceous?

■ Changes in the configuration of the continents may have progressively isolated high latitude oceanic areas from warmer equatorial waters.

Figure 7.4 shows the changing configuration of the continents during the Tertiary Period, which resulted in the opening of one key oceanic gateway and the closing of another. During the Palaeocene and most of the Eocene, Antarctica was covered by temperate forests similar to those that occur in Chile today.

Glaciers probably appeared for the first time in the middle Eocene (Figure 4.15). Cooling of Antarctica may have been caused partly by the continued closure of the Tethys Ocean (which began in the late Cretaceous), which reduced the amount of warm water reaching Antarctica from the tropics, but the crucial event was the opening of Drake Passage between Antarctica and South America about 35 million years

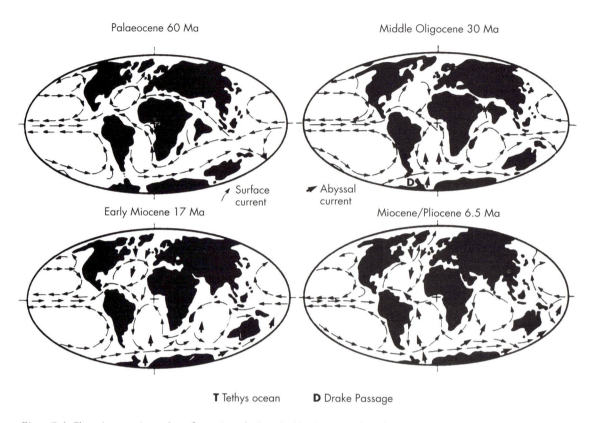

Palaeocene 60 Ma

Middle Oligocene 30 Ma

↗ Surface current

✦ Abyssal current

Early Miocene 17 Ma

Miocene/Pliocene 6.5 Ma

T Tethys ocean   **D** Drake Passage

Figure 7.4 Changing continental configurations during the Tertiary period, and postulated ocean currents (surface: small arrows; deep: large arrows). The maps show two key events: (i) the opening of an oceanic gateway between South America and Antarctica 25 to 30 million years ago permitting a circum Antarctic current to develop, and (ii) the closing of the gateway between Central and Southern America during the Pliocene, isolating the Atlantic and shutting down equatorial circulation between it and the Pacific.

ago. This new gateway resulted in the former continent being ringed by an oceanic current which effectively isolated it from warmer waters to the north and acted like a refrigerator pulling heat away from the polar continent. This resulted in glaciers reaching the sea and depositing significant amounts of ice rafted debris. However, a full Antarctic ice cap did not develop for another 15 million years.

The closure of the gap between North and South America between 4.6 and 2.5 million years ago may have initiated Northern Hemisphere glaciation. This is because closure of this gateway would have increased the salinity of the Caribbean Sea, which in turn would have strengthened the Gulf Stream and

led to more deepwater formation in the North Atlantic. This would have resulted in more moist air reaching colder high latitudes because, as explained in Chapter 3, an adequate moisture supply is crucial for the formation of large ice caps. Strengthening of the Gulf Stream and North Atlantic Drift, could, however, have delayed the onset of glaciation by warming areas around the North Atlantic. Other evidence (not discussed in this book), however, suggests that the formation of deepwater may have decreased between 3.6 and 2 million years ago. So all we can be sure about is that the role that the formation of the Panama Isthmus may have played in the glaciation of the Northern Hemisphere is uncertain.

Tectonic Uplift

Uplift of large areas of continental crust has the potential to modify climate by:

- raising areas above the regional snowline/glaciation limit enabling glaciation to be initiated;
- modifying global atmospheric circulation patterns;
- increasing weathering rates resulting in the removal of CO_2 from the atmosphere.

The centres of major Quaternary ice accumulation in the Northern Hemisphere lie close to areas bordering the Atlantic Ocean and Labrador Sea that were uplifted by as much as two kilometres during the Tertiary (Figure 7.5(a)). When such areas are elevated above the regional snowline (see pages 54–5), or the regional snow line is depressed (due to global cooling), snow accumulating on them will not completely melt during the summer months. Uplifted areas of Labrador, Greenland, Britain and Scandinavia are all within the key latitudinal belt for Milankovich forcing (i.e. at latitude 60–80°N (Figure 3.12). They experienced two episodes of uplift. The first one occurred at about 60 million years ago and was associated with extensive volcanism in NW Scotland and East Greenland, that heralded the opening of the North Atlantic. The second period of uplift occurred towards the end of the Tertiary during the Miocene.

The North Atlantic margin areas have developed a distinctive topography due to the combination of tectonic uplift and glacial erosion. They consist of tilted plateau surfaces, the seaward margins of which have high relief, and which are cut by deep glacial valleys, which today often form lochs or fjords (Figure 7.5(b)). Once global cooling lowered the snow line below the summits of the tilted plateau, their oceanic setting resulted in ideal conditions for ice-sheet growth: a plentiful supply of moisture and summer temperatures low enough to prevent complete melting of the previous winter's snowfall.

Late Tertiary mountain building may have significantly modified atmospheric circulation patterns. The Himalayas and Tibet are particularly important in this regard. Climate models have been used to simulate the effect of elevating the mountains of the

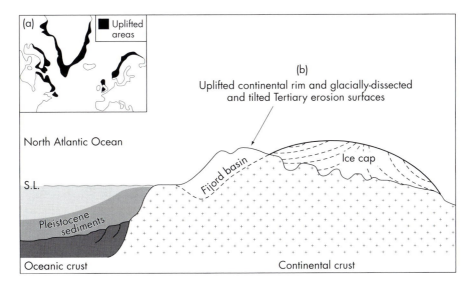

Figure 7.5 Nurseries for northern ice sheet growth: areas of Tertiary uplift around the North Atlantic. (a) Uplifted coastal areas of Labrador, Greenland, Britain and Scandinavia. (b) Idealised cross section across an uplifted area, showing the development of an ice-cap and associated deep glacial valley on the seaward side. Note that the Greenland ice cap is flanked on its eastern and western sides by uplifted areas.

western United States, and the Colorado and Tibetan plateaux (Figure 7.6). Formerly moist areas would become drier, and northern latitudes colder. Uplift of the Tibetan plateau also intensifies the Asian monsoonal winds, driven by increased temperature differences over the plateau in winter and summer. Winter cold is accentuated by the outward flow of cold air from the high plateau. In North America, colder winters resulted from a shift from predominantly westerly winds to northerly ones bringing south cold polar air.

An order of magnitude increase in the dust flux recorded in sediment cores from the North Pacific occured at 3.6 Ma, suggesting an increase in aridity in Asia at this time caused by significant uplift of the Tibetan plateau. The tenfold increase in the dustiness of the atmosphere in northern high latitudes at this time could have caused significant surface cooling, leading to the build-up of northern ice sheets between 3.6 and 2.6 Ma.

Tectonic uplift results in high rates of weathering and erosion. The chemical weathering of rocks into solutions and solid residual material consumes atmospheric CO_2 in large amounts (Box 5.2). The area extent of the Tibetan plateau and its elevation of some five kilometres suggest that its uplift was an exceptional event in Earth history. The wet humid climate experienced on its southern and southwestern flanks makes it an exceptionally large weathering machine which may have significantly reduced atmospheric CO_2 levels, thus contributing in a second way to global cooling. The huge amount of sediments removed from the Himalayas and Tibet were carried to the Indian Ocean by the Ganges and Indus rivers. The resultant rapid sedimentation rates in the Indian Ocean led to the burial of large amounts of organic matter, resulting in the removal of yet more carbon from the atmosphere.

OVER THE THRESHOLD

It seems clear that global cooling since the Cretaceous and the accompanying increase in the thermal contrast between the equator and the poles had two main causes. The first was a reduction in the proportion of greenhouse gases (particularly CO_2 and water vapour) in the atmosphere, and the second resulted from

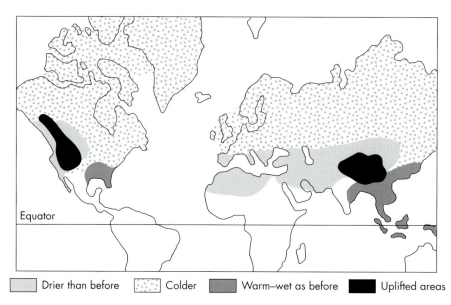

Drier than before   Colder   Warm–wet as before   Uplifted areas

Figure 7.6 Climate changes, predicted by computer models, consequent on the late Tertiary uplift of the Tibetan plateau and western North America.

continental drift causing changes in oceanic and atmospheric circulation patterns. Perhaps we will never know precisely what finally tipped the Earth's climate system into an icehouse state in which fluctuations in Northern Hemisphere insolation were the pacemaker of climate change. It is likely that a combination of events were responsible, rather than single ones. One possible scenario for the final stages of cooling that led to extensive Milankovich-paced Northern Hemisphere glaciation is illustrated in Figure 7.7. Although the details of the timing and magnitude of the events shown in this Figure are open to dispute, the scenario illustrated shows once again the importance of considering climate changes in the context of the many linked components of the Earth's climate system.

The tectonically driven scenario may have been too slow to account for the rapidity of the onset of northern hemisphere glaciation. Some workers argue, therefore, that tectonically driven changes brought the climate to a critical threshold 3 million years ago, after which the relatively rapid Milankovich driven variations in insolation were the final trigger for extensive glaciation. Figure 7.8 (b) shows the calculated variations in insolation during July at 65°N between 3.0 and 2.5 Ma.

☐ Over which of the following periods would the Northern Hemisphere have been most susceptible to glaciation: 2.65–2.75, 2.75–2.85 Ma?

■ 2.65–2.75 Ma, because over this time interval summer insolation reached its lowest levels. This,

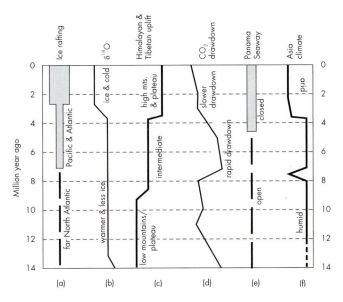

Figure 7.7 Diagram to show possible events leading to extensive glaciation in the Northern Hemisphere. Significant cooling in northern high latitudes which commenced in the Middle Miocene is indicated by seasonal ice-rafting of debris into the North Atlantic (a) and oceanic oxygen isotope records (b). The main phase of uplift of the Himalayas and Tibet may have begun 9 million years ago (c) initiating monsoonal climates and causing a large increase in the rate of chemical weathering which in turn caused an increase in the rate of CO_2 drawdown from the atmosphere (d). The commencement of ice-rafting in the Northern Atlantic and Pacific (a) may be a direct result of cooling caused by the reduction in atmospheric CO_2. The closing of the link between the Pacific and Atlantic as the Panama Isthmus formed in the Early Pliocene (e) diverted equatorial Atlantic surface waters northward, bringing more moisture to the nurseries for northern ice-sheet growth (Figure 7.5). Further Tibetan uplift at this time (c) caused the aridification of central Asia (f) and consequent atmospheric cooling due to a tenfold increase in atmospheric dustiness (the cause of the dust spike at ~7.5 million years ago is unknown). Significant land ice build-up followed the increase in dustiness and is indicated by the ramp in oceanic oxygen isotopes (b) leading to major glaciation in the Northern Hemisphere at 2.6 Ma (a).

according to the Milankovich theory of ice ages, would mean that summers were cool enough to prevent winter snows from melting, thus initiating positive feedback whereby increased albedo caused further cooling. Glaciation, therefore, was unlikely to have been initiated between 2.75 and 2.85 Ma.

As can be seen by examining Figure 7.8(c) there is a pronounced increase between 2.75 and 2.70 Ma in the amount of ice rafted debris present in a deep sea sediment core from the Western Pacific. At this time the amplitude of July solar insolation variations at 65°N increased significantly, so that cooler summers would

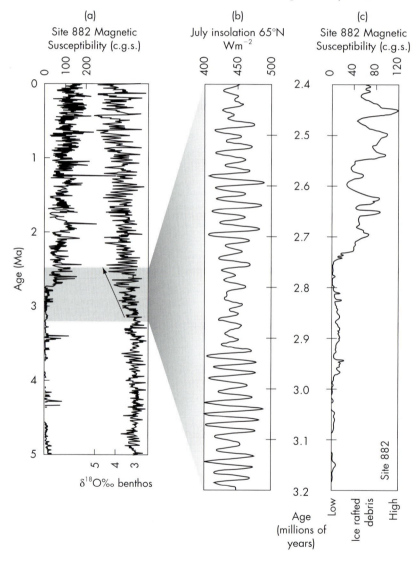

Figure 7.8 Northern Hemisphere summer insolation and the onset of glaciation. (a) Benthonic $\delta^{18}O$ plot and variations in the amount of ice rafted debris in the northern Pacific showing evidence for cooling and the growth of ice sheets between 2.4 and 3.2 Ma. (b) July insolation at 65°N between 2.4 and 3.0 Ma. (c) Variations in the amount of ice rafted debris in a core from the northwest Pacific over the same time interval shown in (b). The variations in ice rafted debris in (a) and (c) were measured by a magnetic susceptibility meter (increased readings indicate more ice rafted debris), the workings of which are not explained in this book.

have occurred every 20 thousand years. This change, therefore, may have been the final trigger that took the climate system over the glacial threshold.

THE MID-PLEISTOCENE REVOLUTION

Introduction

The change from a 41 thousand year rhythm of glacials and interglacials to a 100 thousand year one at about 800 ka (The Mid-Pleistocene Revolution) has perplexed palaeoclimatologists since its discovery some 30 years ago. This is because the changes in solar insolation caused solely by eccentricity related variations in the Earth–Sun distance are considered to be climatically insignificant. Hypotheses attempting to explain this paradox also need to address not only what caused the change from a 41 thousand years to a 100 thousand years cyclicity, but also what drives both of them.

Ice sheet models

Attempts to explain the 100 thousand year cyclicity have been made by developing time-dependent ice sheet models. The main features of one of these models are shown in Figure 7.9. It is designed to simulate changes in ice volumes caused by Milankovich forcing. It is a two dimensional model constructed along a north–south line parallel to the flow of a northern ice sheet. It ignores any east–west flow, and variations in the way the ice may flow along the bedrock due to temperature variations. The northern margin of the

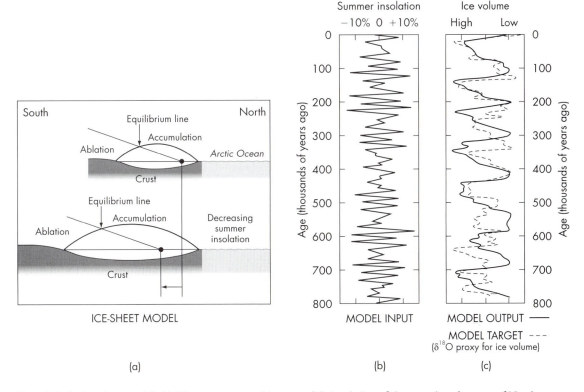

Figure 7.9 An ice sheet model. (a) Diagrams summarising a model simulation of the growth and retreat of Northern Hemisphere ice sheets. The summer insolation variation input at 65°N (b) to the ice sheet model shown in (a), and (c) the resultant ice volume change compared with the $\delta^{18}O$ record. See text for explanation.

ice sheet remains fixed at the edge of the Arctic ocean. The ice sheet model is linked to geophysical models which simulate isostatic adjustment of the crust to ice-loading (Box 1.3). This is important, as accumulation and ablation rates are strongly dependent on altitude. The Milankovich pulse of summer insolation changes (Figure 7.9(b)) is fed into the model by migrating the inclined equilibrium line (see Figure 3.4) to the north (when summer insolation increases) or south (when summer insolation decreases). The success of the model is judged by the extent to which the simulated ice volume variations track the ocean $\delta^{18}O$ curves (which represent the actual ice-sheet fluctuations during glacial cycles).

Early versions of the model produced results that tracked the precession and tilt cycles, but failed to produce a 100 thousand year response. This was because no time lag for isostatic compensation was included in the model. Adding time lags produced better results, but failed to reproduce the glacial terminations. Refining the isostatic response still further produced an output that traced the $\delta^{18}O$ curve fairly well over the past 400 thousand years (Figure 7.9(c)).

The modelling studies indicate that the depression in continental crust caused by isostatic adjustment to ice load is crucial in terminating the ice sheet. This is because the larger increase in solar insolation that coincides with actual terminations causes the model ice sheet to retreat into the isostatic depression resulting from ice loading. This results in a positive feedback, because ablation is dependent on altitude, and the altitude of the surface of the ice sheet is lowered as it retreats into the depression. Increased ablation results in the southern ice slope becoming steeper, which in turn increases southward ice flow. This brings more northern ice into the depression where it rapidly melts. In reality, other positive feedback effects, not simulated by the model, may come into play. These include the role of proglacial lakes in increasing ablation, and the reduction in dust levels (a dusty atmosphere prevents solar heat reaching the Earth's surface) caused by the presence of larger amounts of meltwater in front of the ice edge.

The model results show that the crustal response to ice sheet loading produces a 100 thousand year glacial cycle from higher frequency orbital summer forcing at high latitudes, but is this combination amplifying the 100 thousand year Milankovich signal? Perhaps it is, because after nearly 100 thousand years, the crustal depression results in a runaway positive feedback effect triggered by the much greater change of summer insolation that occurs at the end of this time interval. This positive feedback almost eliminates northern ice sheets. The ice sheets do not immediately return to their previous volumes at the next insolation minimum because the crust has yet to rise to its pre-glacial elevation. Canada, Scandinavia and northern Britain are still rising after the last glaciation (Figure 1.11): as these regions continue to rise, they will become more susceptible to glaciation.

By now, you could be forgiven for thinking that the model results just described go a long way to explaining global cooling and glaciation. They do not, however, explain the change from 41 thousand year to 100 thousand year cyclicity about 800 thousand years ago. Nor do they explain the 'Stage 11 problem' namely why, at about 400 ka, was the rapid glacial termination (Termination V) and subsequent interglacial of marine isotope Stage 11 not associated with elevated levels of solar insolation at northern high latitudes that are typically associated with other terminations (Figures 3.15, 7.11).

THE CHANGE FROM 40 THOUSAND YEAR TO 100 THOUSAND YEAR CYCLICITY

Why did the growth and melting of the cryosphere change from a 41 thousand year (obliquity) cyclicity to a 100 thousand year one about a million years ago, about 1.6 million years after ice sheets were established in the Northern Hemisphere? There must have been a change in one or more of the boundary conditions of the climate system. Oxygen isotope records not only document the change in cyclicity (Figure 5.4) but they also show that the time averaged volume of the cryosphere during glacial periods significantly increased about one million years ago. Yet tills were deposited by the Laurentide ice sheet about two million years ago south of the location of the margin of this sheet as it existed at the last glacial maximum. This suggests that Northern Hemisphere ice sheets were as extensive before one million years ago as they were in later times.

☐ So how can the increase in volume of the cryosphere during glacial maxima since about one million years ago be explained?

■ The ice sheets must have been much thicker.

So why should ice sheets have become thicker about one million years ago? One suggestion is that the bed across which the ice sheets flowed changed from being entirely soft-bedded to a mixture of hard and soft material. This change could have resulted from successive glacial advances eroding away a thick mantle of weathered material that had developed by chemical weathering over millions of years before the onset of glaciation.

☐ Can you recall from Chapter 3 the three mechanisms involved in glacier movement?

■ Ice flow, basal sliding and deformation of soft sediment beneath the ice (Figure 3.7).

Ice flows faster over soft sediments than it does over hard rock, and faster flow rates usually result in thinner ice sheets. This relationship means that once soft weathered material had been eroded away, ice sheets would have become thicker, a scenario that has been modelled for the Laurentide ice sheet as shown in Figure 7.10.

The change in the length of the glacial–interglacial cycles could have been a direct result of the development of thicker ice sheets. These are more likely to survive warming episodes related to the 40 thousand year obliquity and 21 thousand year precessional cycles, but once the crustal depression beneath them deepens, the runaway positive feedback effects triggered by high summer insolation result in deglaciation after ~100 thousand years.

Another explanation for the change in periodicity of the glacial–interglacial cycles suggests that a long term reduction in atmospheric CO_2 levels due to tectonic uplift and associated increased weathering rates (Box 5.2, and page 146) was the cause. This would decrease the greenhouse effect, allowing ice sheets to expand into areas that were previously too warm for ice to survive through successive summers. This permitted 100 thousand years old ice to grow during unusually long periods of low summer insolation that are shown in black on Figure 7.11(a). These long periods are required because it seems unlikely that ice sheets as large as those that formed during the last

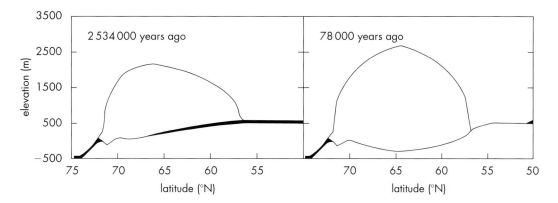

Figure 7.10 Cross sections produced by modelling the behaviour of Laurentide ice as it flowed over different substrates. In (a), the ice sheet is underlain by sediments (shown in black) which result in faster ice flow rates, whereas in (b) when the sediments have been eroded away by earlier glaciations, flow rates are slower, and so the ice sheet is thicker, and the depression in the bedrock is deeper.

million years could not grow during just one cold half (~11 thousand years) of a single precessional cycle. Once such large ice sheets are established, however, they last until the next large increase in summer insolation, when they melt catastrophically. As can be seen from Figure 7.4, the length of glacial–interglacial cycles is either four or five precessional cycles, and that very rapid changes from full glacial to full interglacial conditions only occurs after the accumulation of unusually large amounts of ice. It must be stressed that the long term reduction in CO_2 shown on Figure 7.11 has not been verified in the geological record.

A Multiple-State Climate Model

None of the models described so far is able to replicate the $\delta^{18}O$ record of ice volume changes with reasonable accuracy, especially those around isotope Stage 11. As we saw earlier in the book there is evidence that the Earth's climate system exists in different states characteristic of glacial and interglacial and stadial and interstadial interludes. Similarly ocean circulation, particularly in the North Atlantic, may shift from one state to another. The multi-state model described below is based on the idea that the climate system may function in more than one way.

The model assumes that the climate system operates in three different regimes, depending on northern high latitude insolation and ice volume. The three regimes are interglacial, mild glacial and full glacial – only three transitions between these regimes are permitted by the model and occur across three key thresholds as summarised below:

- interglacial → mild glacial: summer solar radiation at high northern latitudes falls below a critical threshold;
- mild glacial → full glacial: insolation stays below a critical threshold for an extended period;
- full glacial → interglacial: insolation increases above a critical threshold significantly higher than that for the interglacial → glacial transition.

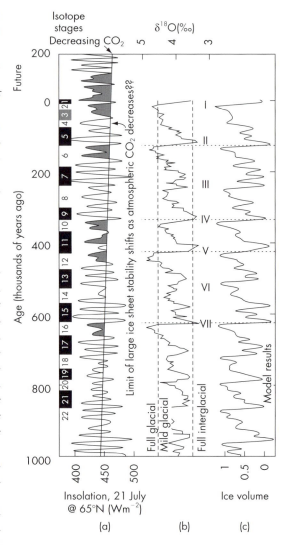

Figure 7.11 The 100 thousand year cyclicity problem. (a) Changes in insolation received on July 21st at 65°N. The inclined line is a postulated level, related to a hypothetical reduction in atmospheric CO_2 levels caused by tectonic uplift and weathering. The black areas indicate times when insolation levels were, and may be in future, low enough for long enough for large 100 thousand years old ice sheets to form. (b) The normalised marine oxygen isotope record for ice volume. The horizontal dashed lines define boundaries between full and mild glacials, and interglacials – three regimes used in a multistate climate model, the output of which (as ice volume) is shown in (c).

As can be seen on Figure 7.11(c), the model produces a good match with the marine $\delta^{18}O$ record, including the problematic isotope Stage 11.

It seems likely that the 100 thousand year cyclicity began when CO_2 concentrations in the atmosphere were lowered past a critical threshold which resulted in northern high latitudes being cold for long enough for very large ice sheets to develop. When such a threshold is added to the model it successfully simulates the change in periodicity from 41 thousand years to 100 thousand years that occurred at about 800 ka. The length of glacial–interglacial cycles that have occurred since that time varies from about 80 thousand to 120 thousand – that is between four and six precessional cycles. So it seems that the periodic build-up of very large ice sheets occurred largely during periods of lowered summer insolation resulting from the interaction of precession with eccentricity, and to a much lesser extent obliquity. In fact, it seems surprising that before the Mid-Pleistocene Revolution, glacial–interglacial cycles were not paced by precession, given how summer insolation at high latitudes changes significantly about every 20 thousand years. So perhaps the '41 thousand year world' that characterised the first two million years or so of Northern Hemisphere glaciation is an even more perplexing mystery than the later '100 thousand year world'.

MILLENNIAL-SCALE CLIMATE CHANGES AND REORGANISATIONS OF OCEAN CIRCULATION

Thermohaline Seesaws

Chapter 6 reviewed the evidence for sub-Milankovich scale changes in climate during the last glacial. With the discovery of proxy records of these changes in Greenland ice cores and in variations in the amount of ice rafted debris in North Atlantic sea-floor sediments, it was thought, initially, that these climatic shifts were confined to areas surrounding this ocean. This North Atlantic focus also pointed to changes in deep thermohaline circulation driven by NADW

formation as being the cause of the changes. Today, we know that these rapid climate changes affected most parts of the world, leading to periods of increased storminess over Asia during intense cold intervals, and increased methane production in low latitude wetlands during milder interludes. Even glaciers in the Southern Hemisphere advanced and retreated synchronously with those in the North. Moreover, ocean circulation was affected beyond the Atlantic as recorded in sediments deposited in the Santa Barbara Basin of California.

For some time, changes in the rate of deep water formation in the North Atlantic were regarded as the likely cause of the globally synchronous millennial scale changes in climate. Two discoveries changed this view and led to the proposition that there may have been major re-organisations of oceanic thermohaline circulation that involved sources around Antarctica as well as in the North Atlantic.

One of these discoveries was that during the transition between the last glacial and the current interglacial, temperature changes in Antarctica and Greenland were anti-phased. In other words, as Greenland cooled, Antarctica warmed (Figure 6.10). Evidence that this antiphasing of temperature changes was caused by re-organisations of thermohaline deep ocean circulation came from varved (i.e. seasonally laminated) sediments in the Cariaco Trench off Venezuela (Figure 7.12). Despite the fact that the Younger Dryas (a significant cooling event that interrupted warming out of the last glacial) lasted for nearly one thousand years (Figure 7.12(a)) changes in the carbon isotope content of organic rich sediments of this age preserved in the Cariaco Basin off Venezuela indicate that deep water circulation only shut down for about the first two hundred years of this period of time. If it had done so for one thousand years there would have been an even larger rise in the proportion of the heavier ^{14}C isotope present in the sediments as newly formed ^{14}C atoms would not have been removed from the atmosphere by deep water circulation (Figure 7.12(b) and Box 7.2). The Cariaco Trench record does show a rapid rise in ^{14}C for two hundred years at the beginning of the Younger Dryas, but this is followed by a gradual return to normal

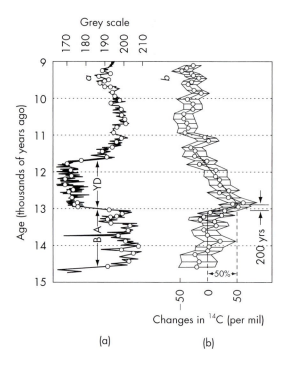

Figure 7.12 The ¹⁴C record in sediments of Cariaco Trench for the period between 15 and 9 thousand years ago. (a) Colour changes exhibited by the sediments have been shown to be related to warm and cool phases, with darker more organic rich sediments being deposited during warmer phases. YD: Younger Dryas cold period (stadial) B/A: Bølling Allerød interstadial. (b) Variations in the ¹⁴C content of the sediments. During the first 200 years of the Younger Dryas cold period, newly produced ¹⁴C atoms accumulated in the atmosphere and upper ocean, indicating a shut-down in deep thermohaline circulation in the oceans (Box 7.2). Reduction of the ¹⁴C content of the sediments during the subsequent 800 years must have occurred as the isotope was removed from the atmosphere-shallow oceanic reservoir by deep water formation. This ocurred in the Southern Ocean rather than in the Atlantic because if it occurred in the latter it would have warmed the climate in the Cariaco Trench region, and there is no record of this in the sediments.

values over the remaining part of the cold event. As the climate around the North Atlantic remained cold during the Younger Dryas the implication is that removal of ¹⁴C from the atmosphere into the deep ocean resumed somewhere else. The antiphasing of temperature changes in Antarctica and Greenland

suggests that deep water formation increased around Antarctica, and so drew warmer low latitude waters southwards, thus warming the continent. This is strong evidence for a seesaw between North Atlantic and circum-Antarctic deep water formation – a phenomenon that has come to be known as the bipolar seesaw of thermohaline circulation. This is not the first time we have encountered a seesaw or oscillating component within the global climate system: ENSO events in the Pacific are another example, though they only last for a few years.

A Role for the Tropical Pacific?

The way in which the relatively energy starved northern high latitudes appear to drive glacial and interglacial cycles rather than the energy rich tropics is a paradox that is as yet unexplained. Climatologists are increasingly turning their attention to the role that events in the tropical Pacific might play in triggering global-scale climate changes. The oscillation between El Niño and La Niña states (Box 2.1) are clear examples of the atmosphere and oceans acting as one system. The effects of ENSO events are felt around the globe. They spread to higher latitudes in the Pacific and they affect other oceans because changes in the pressure field over the Pacific cannot occur independently of other pressure systems. Even the strength of the Asian monsoons is affected by the state of the Southern Oscillation in the Pacific. Parts of North America warms by as much as 6° during El Niño events. Hudson Bay Trading Company records indicate that in El Niño years, the ice retreats early from Hudson Bay. So could it be that when the central Pacific is warmer (as in El Niño years today) the Laurentide ice sheet would have retreated? Oscillations between pressure systems similar to ENSO events are found today in the tropical Indian Ocean and the tropical Atlantic. The pressure difference between the low pressure normally over Iceland and the high pressure normally centred over the Azores in mid-Atlantic oscillates like that between the Indonesian Low and the South Pacific High. It has been found that when the pressure difference is large, westerly winds over the Atlantic are stronger and flow in the

Box 7.2 Radiocarbon and Deep Water Circulation

In addition to the stable isotopes of carbon (^{12}C, ^{13}C), an unstable isotope – Carbon 14 (^{14}C) is constantly being formed in the upper atmosphere as cosmic rays impact on atomic nuclei nitrogen (Figure 7.13). Deep-water circulation removes some of it from the atmosphere – shallow ocean reservoir. If this did not occur, there would be a steady build up of ^{14}C in the atmosphere and surface ocean waters because cosmic rays replenish 1 per cent of the World's total radiocarbon every 82 years. The rate of build up would result in the ratio of ^{14}C to the stable carbon isotopes rising by a third in a thousand years. If deep water circulation had been significantly reduced in the past, it would make radiocarbon dating almost impossible. This dating method is based on two assumptions: that the amount of ^{14}C incorporated into plant material is proportional to the content of the isotope in the atmosphere, and that the rate of ^{14}C formation has remained constant through time.[*] If this is the case, then the smaller the amount of ^{14}C remaining in fossil material, the older it must be. If thermohaline circulation had shut-down or been greatly reduced during cold periods, more ^{14}C would have accumulated in the atmosphere and upper ocean. Plant material growing at such times would have taken up more ^{14}C, and so could have yielded younger radiocarbon dates, thus appearing to be contemporary with organic material that grew during the succeeding warmer

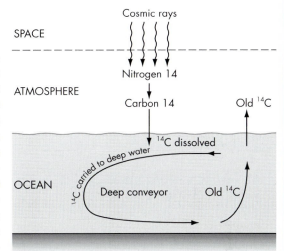

Figure 7.13 Radiocarbon in the atmosphere and oceans.

period! That no such anomalies have been found indicates that no significant reductions in global deep water circulation occurred. Yet there is no doubt that thermohaline circulation in the North Atlantic did periodically shut down during exceptionally cold periods during the last glaciation. The fact that there are no major ^{14}C anomalies suggests that thermohaline circulation increased in other areas (probably the Southern Ocean) when it shut down in the North Atlantic.

[*] In fact, it is known that relatively small variations do occur, due to changes in solar activity affecting the intensity of cosmic ray bombardment.

Gulf Stream (and hence northwards transport of heat in the Atlantic) is greater than usual; when it is low, westward flow of winds and ocean currents are less than usual. Such changes could have contributed to the retreat and growth of northern ice sheets.

Perhaps there were periods of time when climatic oscillations that last for a few years at the present time persisted for much longer intervals. Wallace

Broecker suggested that there might be some evidence for such a possibility:

Immediately after the last Heinrich event about 14 000 years ago, Lake Lahontan in Nevada achieved its greatest size, an order of magnitude larger than today's remnant. Supporting such a large body of water requires immense amounts of precipitation, of the magnitude experienced during the record El Niño winter of

1982–1983. One way of thinking about the impact of these earlier occurrences, then, is as changes in the pattern of ocean circulation that led to El Niño's lasting 1000 years.

W.S. Broecker. 1995.
Chaotic climate. Scientific American, 267: 44–50.

These speculations require much more research to be substantiated, but they indicate that the North Atlantic is now not the only centre of palaeoclimatic attention.

WHAT NEXT – WHEN WILL THE PRESENT INTERGLACIAL END?

Introduction

By now you will have realised that our knowledge of the climate system does not enable us to make accurate predictions about climate change. We can be confident about the astronomical predictions of future variations in solar insolation over Milankovich-sensitive northern high latitudes (Figures 7.11(a)), but we do not understand how these tip the climate from one state to another. None the less, it is worth considering what the palaeoclimatic record can tell us about the length of past glaciations and whether they ended gradually or abruptly, and the possible consequences of the continued build-up of greenhouse gases in the atmosphere.

The Length of Past Interglacials: a Guide to the Future?

Knowing the length of past interglacials can give us some insight into how long we may have to wait until the present one comes to an end. Over the past ~800 thousand years, $\delta^{18}O$ records suggest that interglacials lasted for one half of a precessional cycle – about 11 thousand years. If we assume that the Younger Dryas stadial is the last event in the last glacial, then the duration of the present interglacial is 11 500 years. Taking an earlier warming (the Bølling interstadial) as the start of the present interglacial means that it has lasted 14 500 years. The $\delta^{18}O$ record for the last interglacial indicates a duration of 12 thousand years, although the time interval over which the North Atlantic was almost free of ice rafted debris is 14 000 years. So there is no doubt that the present interglacial has lasted for at least one half of a precessional cycle.

One interglacial was much longer than the 'norm': it occurred ~400 thousand years ago after Termination V, and is part of the 'Stage 11 Problem' referred to on p. 150.

☐ Examine Figure 7.11(a) again. How does the pattern of insolation changes between 450 ka and 350 ka differ from those which occurred before and after this time interval?

■ The difference between insolation maxima and minima are much less than those that occurred before and after this time interval.

☐ Why was this so?

■ The period between 450 ka and 350 ka was a time when orbital eccentricity was at a minimum, so that the precessional driven seasonality cycle was also at a minimum.

☐ Is the insolation pattern between 450 ka and 350 ka different or similar to that for the past 50 thousand years and next 50 thousand years?

■ It is broadly similar.

This suggests that if oxygen isotope Stage 11 is taken as an analogue for the current interglacial, then we could continue to enjoy warm conditions for many more thousands of years.

How Might the Present Interglacial End?

This question can only be addressed by examining proxy records that span the end of the last interglacial, as no high resolution records are available for earlier interglacials. Unfortunately, Greenland ice cores do not extend back as far as the last interglacial (due to ice deformation near the bases of the cores), and so marine and terrestrial pollen records offer the best records. Both show that a long period of stable warmth came to an abrupt end. In the North Atlantic there is a sharp peak in the amount of ice-rafted debris present in sea floor sediments at the end of the last interglacial, followed by several more peaks that indicate that the climate fluctuated between warm and cooler states for some time before it settled into a more continuous cooler state. This pattern is mirrored by peaks in left coiling foram *N.pachyderma*, indicating cooling of surface waters. Pollen records from France show that sharp rises and falls in non-tree pollen occurred at about the same time as the fluctuations in the North Atlantic marine records: the interglacial pollen record is dominated by tree pollen.

It seems then, that if the end of the last interglacial is any guide, the present interglacial may end abruptly, with a long period of warmth and climate stability being replaced by rapid fluctuations between warm and cooler conditions. During the present interglacial the climate has remained locked into a warm operational mode. Marine and ice oxygen isotope compositions have remained largely unchanged, as have sea-levels, yet northern high latitude insolation has declined almost to the level it was at the end of the last glacial (Figure 7.14). Given the fact that palaeoclimatic records indicate that the climate can switch from one state to another very rapidly, we might expect an abrupt rather than a gradual response to changing summer insolation.

Will Greenhouse Gas-Buildup Postpone the Inevitable?

The burning of fossil fuels and other anthropogenically driven effects have increased the content of atmospheric CO_2 from a natural interglacial level of

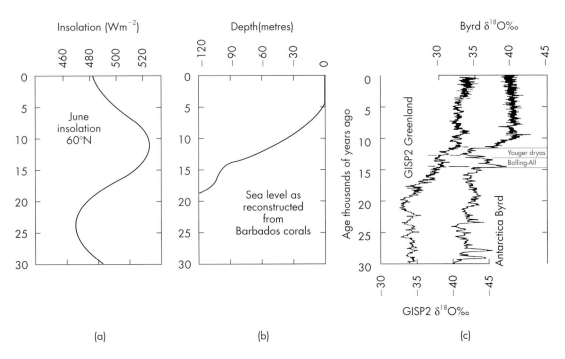

Figure 7.14 Changing high latitude summer insolation (a) and climatic stability during the present interglacial as indicated by sea-levels determined from elevations of coral reefs in Barbados (b) and oxygen isotope records from ice-cores (c).

280 ppm to 365 ppm in less than two centuries. So if human-induced global warming is a reality, are we postponing the end of the current interglacial or pushing the climate system to re-organise prematurely into a cooler state? Global warming will tend to decrease the density of ocean surface water in two ways. As it becomes warmer, its density will decrease, and as a warmer planet delivers more water vapour to high latitudes as the hydrologic cycle intensifies, waters in these areas will become less saline. These effects may not be identical in both hemispheres, and so the current balance between deep water production in the North Atlantic and around Antarctica could be disturbed. Even a diminution in the rate of formation of NADW would have severe consequences for the climate of North West Europe.

The geological record provides relatively high resolution information about climate changes that have occurred over the last few hundred thousand years. This shows that there is a link between global temperatures and the greenhouse gas content of the atmosphere, but as yet the extent to which the latter is the cause of the former is uncertain. Palaeoclimatic records have also shown that the climate system can switch between different states very rapidly, resulting in changes in global average temperatures by as much as several degrees in a matter of decades. Although climate models have yet to successfully simulate changes in the atmosphere–ocean–cryosphere system that result in such rapid shifts, our understanding of the system is sufficient to argue strongly for the precautionary principle to be applied in favour of reducing anthropogenically induced greenhouse gas emissions. Opponents of this approach argue that we should postpone taking effective steps to reduce these until such time as research can convincingly demonstrate the climatic impacts of emissions, and distinguish natural climatic changes from human induced ones. This implies a rather naive faith in science to deliver certitudes. We will only obtain 'proof' as the global experiment we are conducting unfolds. If this experiment triggers a re-organisation of the climate system, proof will come too late for preventive action.

SUMMARY

1 Various explanations of post-Cretaceous global cooling have been proposed. They fall into three categories:
   Magmatic processes A reduction in volcanism from the Late Cretaceous 'high' would have lowered atmospheric CO_2 levels, resulting in global cooling. A decrease in the rate of oceanic ridge and plateaux formation caused a global sea-level fall, increasing the seasonality of climates which would favour ice accumulation at high latitudes. An increase in volcanism in the northern Pacific may have tipped the climate system in northern high latitudes across the 'glacial threshold' at 2.6 Ma.
   Continental drift opened and closed oceanic gateways, reducing poleward heat transport by ocean currents.
   Tectonic uplift particularly of Tibet, changes global atmospheric circulation patterns, and increased weathering rates which led to an increased drawdown of CO_2 from the atmosphere. Around the North Atlantic, uplift elevated coastal areas above the regional snowline/glacial limit and so may have been a glacial triggering mechanism.
2 At about 2.6 Ma, the climate system appears to have passed some sort of threshold which resulted in extensive Northern Hemisphere glaciation. It is likely that a combination of events may have been responsible for this change, but several explanations involve global cooling caused by removal of atmospheric CO_2, plus a final 'push' into northern glaciation. This push could have been caused by tectonic events, or by a period of cooler northern summers resulting in ice sheet growth tipping the climate of the Northern Hemisphere into an icehouse condition from which it has not recovered.
3 Theories of climate change must explain the global nature and synchroneity of glacial/interglacial cycles, the mechanism that amplifies the Northern Hemisphere Milankovich summer insolation signal, and the asymmetry of both the glacial/interglacial cycles and the short term millennial scale events (Dansgaard–Oeschger and Bond cycles).

4 Modelling studies of ice sheet dynamics suggest that isostatic depression of the crust is crucial in determining the timing of abrupt glacial terminations at 100 thousand year intervals. This is because the increase in Northern Hemisphere summer insolation causes the ice sheet to retreat into its crustal depression, resulting in positive feedback effects that rapidly increase ablation rates.

5 The change in the periodicity of glacial–interglacial cycles about one million years ago from 40 thousand years to 100 thousand years (the mid-Pleistocene revolution) may have been caused by:

• the removal by glacial erosion of the mantle of soft weathered material above crystalline rocks beneath Northern Hemisphere ice sheets. When this occurred ice flow rates would drop, resulting in thicker ice sheets (due to higher summit elevation, and deeper underlying crustal depressions due to isostatic subsidence). These thicker ice sheets were able to survive precession related warming episodes.

• postulated long term gradual reduction in atmospheric CO_2 due to tectonic uplift and associated weathering causing global cooling. A threshold was crossed about one million years ago, so that during unusually long periods of lower summer insolation, ice sheets would survive through several successive precessional cycles.

6 Only one model has so far replicated most of the climate changes that occurred since 2.6 Ma. It incorporates three climatic states (interglacial, mild glacial and full glacial) and when driven by Northern Hemisphere high latitude insolation changes and gradual cooling resulting from postulated reduction in atmospheric CO_2, it is able to reproduce the observed ice volume changes, including the frequency change across the mid-Pleistocene revolution, and the changes that occurred across isotope stages 12 and 11 (the Stage 11 problem).

7 The carbon isotopic composition of sediments preserved in the Cariaco Basin off Venezuela suggest that the formation of NADW was shut down for only about 200 years at the beginning of the Younger Dryas. Thereafter the gradual reduction to normal ^{14}C levels suggests that deep water formation elsewhere must have become established. This probably occurred in the Southern Ocean, which would account for the warming of Antarctica during Younger Dryas time. Thus it seems that a thermohaline seesaw may play an important role in driving past climate changes.

8 During the past ∼ 800 thousand years, most interglacials have lasted for one half of a precessional cycle (about 11 thousand years). So perhaps the end of the present interglacial is not far away, but as future changes in northern high latitude insolation are similar to those that occurred during oxygen isotope Stage 11, the palaeoclimate record suggests that the warm conditions of today could continue for thousands of years. Rather than postponing the end of the current interglacial, anthropogenically induced global warming may result in more precipitation over northern high latitudes which could lower the salinity of surface water in the North Atlantic and shut down the oceanic global conveyor, causing rapid cooling.

FURTHER READING

The following articles will provide readers with follow-up reading for this chapter, and references to guide more advanced reading.

Broecker, W.S. 1995. Chaotic climate. *Scientific American*, November, 44–50.

Broecker, W.S. 1997. Thermohaline circulation, the Achilles heel of our climate system: will man-made CO_2 upset the current balance? *Science, 278*, 1582–8.

Broecker, W.S. 1998. The end of the present interglacial: how and when? *Quaternary Science Reviews*, 17, 689–94.

Cane, M.A. 1998. A role for the tropical Pacific? *Science*, 282, 59–61.

Hay, W.W., De Conto, R.M., and Wold, Ch.N. 1997. Climate: is the past the key to the future? *Geologische Rundschau*, 86, 471–91.

Kerr, R.A. 1997. Upstart ice age theory gets attentive but chilly hearing. *Science*, 281, 156–7.

Rahmstorf, S. 1997. Ice-cold in Paris. *New Scientist* 8 February, 26–30. Discusses modelling studies of NADW formation and the likely climatic consequences if this were to shut down.

Raymo, M., 1998. Glacial Puzzles. *Science*, 281, 1467–8.

See also references to figure sources for this chapter (p. 257).

PART 2

ECOLOGICAL CHANGE
AND HUMAN ORIGINS

Signposting

As in Part 1 of this book the last 4 chapters involve some sleuthing. However, the forensic information on which this depends is of an altogether more patchy and imprecise kind. There are no proxy indicators for the evolution of humanity akin to the almost continuous records in oceanic sediments and ice cores. Instead there is just a very sparse and incomplete fossil record of some of the primate species that lay on or close to the human 'bush of descent'. There are gaps in time between hominid fossils that correspond to thousands of generations, and sometimes very little hint of their age at all. Tracking down the true mark of 'human-ness' through evidence for consciousness and sharing things socially has to be imaginative and constructed around archaeological remains of what human ancestors made and did.

We evolved as part of wider ecosystems, together with animals and plants that comprised them. The distribution of these ecosystems fluctuated in response to the climate shifts described and discussed in Part 1. Because this milieu was on land, its record is limited. Chapter 8 outlines how ecosystems shifted, waxed and waned though the Great Ice Age. Chapter 9 examines the record of human evolution preserved as fossils and tools, and documents the gradual expansion through time of the areas in which we find these. It also looks at the heritage of past living people hidden in genetic and linguistic differences in modern populations. It concludes that we emerged, evolved and moved out from the African continent in a number of migratory waves.

Just describing the fragmentary signature and the hypotheses of which ancient creatures may or may not have been on the road to modern people does not take our understanding very far. So Chapter 10 puts the hominid record into the widest possible context. This is changing climate, changing ecosystems and even unique geological upheavals that shaped the details of where we and our tool-using, conscious culture appeared, and equally important where we migrated to. Part of the story is how changing environments and changing habits interweave with the anatomical changes revealed by the fossil record. One of the most dramatic aspects of humanity's rise is that it coincided with the beginnings of major environmental changes and a mass extinction that has the pace, and maybe magnitude of those in the far-off past that stemmed from climatic, geological and occasionally extraterrestrial events. It seems that anthropogenic effects on the world are not modern matters, but date back perhaps a million years or more.

The more detailed record of the last 10 thousand years, the Holocene, including a written history in its latter parts, is the topic of Chapter 11. First, it charts the effects of organised agriculture on the landscape of the 'humanised' world, and how that rebounded on the first tillers of the land. The last two millennia witnessed increasing chemical intervention in the natural world, and the last two centuries a steady growth of impact on climate. That stemmed mainly from a more intense use of energy through burning fossil fuels and releasing the greenhouse gas CO_2 that had been in long term storage in coal and petroleum. We conclude by looking briefly at how the associated climatic warming may interfere with the fundamental, earthly processes that are part of the mechanism behind the ups and downs of climate since 2.6 Ma. This seems likely to present surprising and possibly threatening outcomes that the next generations will have to face.

8

CLIMATE CHANGE AND LIFE ON LAND

INTRODUCTION

In Part 1, we examined the physical evidence that reveals the timing, magnitude and rates of climate change since 2.6 Ma. Now we need to consider how vegetation, animals and our ancestors coped with the rapid climatic oscillations. Terrestrial sediments contain many fossils which can be used to recreate the environment and give us an idea of the changes our ancestors experienced. Indeed, our modern understanding of climate change was founded on palaeontological studies of the kind described in this chapter.

The most spectacular, but unfortunately quite uncommon, fossils are the deposits of large animal bones, like the Trafalgar Square fauna described in Chapter 1 (Figure 1.9), which provide a window onto one scene. But as in the oceans, it is the tiny fossils which are abundant and widespread which give the best evidence of the effects of climate change. Pollen (if you suffer from hay-fever, you will know how abundant pollen is in the air) and insect remains are common almost everywhere and so can give us many glimpses of past vegetation patterns and environmental conditions.

Figure 1.4 shows the current world distribution of vegetation types. It is an idealised picture because so much natural vegetation has been cleared for farming or otherwise altered by thousands of years of human activity (Figure 1.21). The banded distribution shown reflects the Earth's climatic belts (Box 1.1).

Starting in the extreme north, in say Finland, we find a treeless tundra dominated by small herbs and lichens. Here the permafrost and short growing sea-son exclude trees. Passing south the boreal zone of needle-leafed conifer forest gives way to broad-leaved winter-deciduous temperate forest – the natural vegetation of Britain and much of Europe. Further east at this latitude the drier continental interior has treeless grasslands and semi-desert. These merge southwards with the hot sub-tropical deserts of Arabia and the Sahara. Towards the equator moderate rainfall produces savannah and in the wettest parts, tropical rain forest. Savannah is an intriguing vegetation type from which trees are excluded because they are killed when young by frequent fires which stimulate grass growth. Fires are started both by lightning strikes and humans. The balance of natural and human fires and the extent to which human activity has promoted and extended the savannahs during the last million years, or perhaps even longer, is open to debate.

The vegetation patterns shown in Figure 1.4 are a recent phenomenon of only the last 10 thousand years. Similar patterns probably existed during early interglacials, but they persisted only for about one tenth of the time since the onset of the 100 thousand year rhythm of ice ages about 800 thousand years ago. As we saw in Part 1 we live in a world which has undergone 2.6 million years of profound climatic change which is likely to continue (Figure 7.11). At high latitudes advancing ice sheets overwhelmed plant and animal communities. In tropical latitudes, the changes during glacial phases were more subtle but were no less important in affecting current patterns of diversity of living things and the number of species in an area. Examining the fossil record provides us with a picture of vegetation and animal movements in a changing climate. Occurrences of well preserved plants and animal skeletons are rare, but fortunately palynology – the study of plant spores and pollen – has enabled past distributions of individual plant species and biomes to be pieced together.

PLANTS AND CLIMATE CHANGE

Spores and Pollen

Pollen and spores are part of the reproductive system of plants: pollen from flowering plants and conifers, spores from ferns (Figure 8.1), horsetails and mosses. These microscopic grains are very resistant to decay, and often occur well preserved and in abundance in organic rich sediments that accumulated in bogs and lakes. Plants make up for their immobility by producing a profusion of pollen to shorten the odds of successful fertilization. By no means all pollination is by insects, birds or small mammals. Wind pollinated species include all the conifers and many broad-leaved trees such as oaks, elms and hazel, and also grasses that produce dry pollen powder. Pollen aimed at animal transport is often sticky and clumps together. It only occasionally ends up in sediments, usually close to the source plant. Wind-dispersed pollen gets everywhere. Lake beds and bogs provide a column of sediment that record thousands of years of pollen input which can be dated using the radiocarbon method (see Box 8.1).

☐ Given the variation in size and shape of spores and pollen grains shown in Figure 8.1, is it likely that the proportions of them preserved in sediments will directly reflect the relative abundance of the different plants that once surrounded the site of burial?

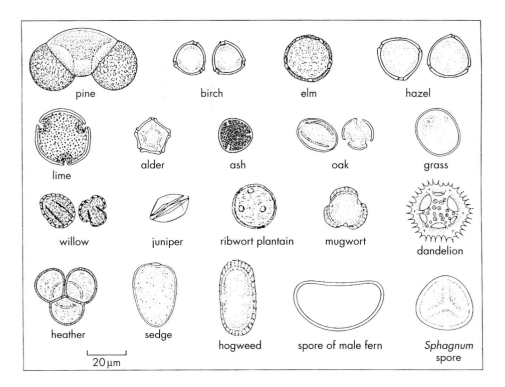

pine   birch   elm   hazel

lime   alder   ash   oak   grass

willow   juniper   ribwort plantain   mugwort   dandelion

heather   sedge   hogweed   spore of male fern   *Sphagnum* spore

20 μm

Figure 8.1 Sketches of a variety of pollen grains and spores.

Box 8.1 Carbon Dating

All living things incorporate the radioactive form of carbon (^{14}C: see Box 7.2) into living tissues: plants directly from atmospheric carbon dioxide, animals by eating the plants. Once the plant is dead, or a structure (like a tree-ring) has been formed, no more ^{14}C is added. All radioactive isotopes decay in a similar way: half the atoms decay in a certain time (the half-life), thus the less radioactivity remaining, the older the sample; ^{14}C has a half-life of 5730 years.

Mass spectrometry measures the ratio of ^{12}C to ^{14}C in very small samples, so precious or small objects can be dated, and some machines can even analyse a single foraminifera. Theoretically this method can date samples back to 60 thousand years ago but at present the limit is 40–50 thousand years ago. Radiocarbon dates are expressed relative to the present (in fact 1950 before ^{14}C entered the atmosphere by nuclear testing).

Counting tree rings gives an independent check of radiocarbon dating. Tree ring patterns from ancient living specimens like the bristlecone pine of western USA and British oaks, matched to those from archaeological and fossil material such as in bog oak and pine, give a time series back to 11 thousand years ago. Radiocarbon dating of rings showed discrepancies between the actual and analysed dates, often more recent than by the tree-ring age, showing that a constant background production of ^{14}C in the atmosphere cannot be assumed. In addition, the ^{14}C composition of the atmosphere and shallow oceanic water is affected by the intensity of deep water circulation (see Box 7.2). For most of the Holocene calibration of radiocarbon dates with tree-rings is well established. Cores from fossil coral reefs help to calibrate radiocarbon dates older than the Holocene, using ^{234}U/^{230}Th and K/Ar dating methods.

Plants which obtain some carbonate from rocks (for example aquatic plants in a lake surrounded by limestone) may produce ^{14}C dates from 200 to over 1000 years older than they should. Contamination can also occur after the plant's death due to movement of mobile humic acids in peats, or sediment mixing by subsequent root growth or disturbance.

■ No, because the size and shape of the grains will influence how far they are transported by wind.

The proportion of different types of spores and pollen preserved in sediments is dependent on a number of factors, including the amounts produced by different plants, how easily they are transported by wind, or whether transport is via animals, and the distance of the sample site from the edge of a former lake or bog (Figure 8.2).

Most pollen grains are only assignable to a genus. Some groups show so little variability they cannot be subdivided. Hazel and bog myrtle pollen are indistinguishable and are plotted together on pollen diagrams. Grasses too are a broad, 'lumped' group, although the larger cereal pollens do give evidence of cultivation.

Interpreting Pollen Diagrams

Each pollen diagram provides a fragment of the overall pattern of vegetation and thus the regional climate, but each also reflects the local peculiarities specific to the particular site. Figure 8.3 is a pollen diagram of a core from Hockham Mere, an infilled lake in Norfolk surrounded by glacial tills and wind blown sands.

Before 9.6 thousand years ago the only tree pollen recorded is birch with some willow and juniper in small amounts. The herb input is high: sedge and grass pollen with a rich assemblage of herbs associated with open ground, including *Artemisia* (mugworts and wormwoods), meadowsweet, docks and meadow rue. Birch pollen probably represents tundra (dwarf and silver birch, *Betula nana, B.*

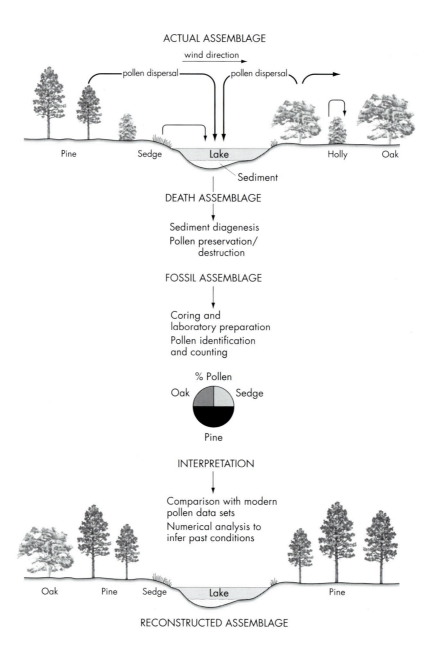

Figure 8.2 Factors affecting the proportion of spores and pollen preserved in sediments. The top sketch shows a lake surrounded by plants, and the dispersal routes of spores and pollen from the assemblage of vegetation. The 'death assemblage' that accumulates in the lake sediment is shown in the pie diagram. The proportions of different pollen present in it are not the same as those in the life assemblage because some plants produce more pollen than others, some types of pollen are more easily dispersed by the wind and have different preservation potentials when buried in the lake sediments. The final reconstructed assemblage, therefore, may be biased towards certain plants. In this example pine trees are more common than they are in the life assemblage, and the presence of holly was not detected.

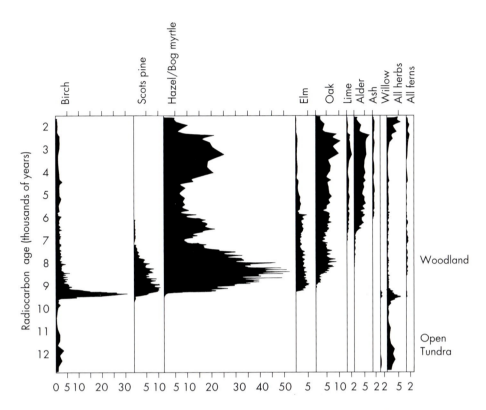

Figure 8.3 Pollen analysis of a core from Hockham Mere in Norfolk. The diagram shows pollen from the main tree types (a full diagram could contain another 60 or more columns!) plotted as number of grains per cubic centimetre of sediment.

pendula). Birches require abundant light to grow. They quickly colonise open ground, but other trees, such as elms and oaks, eventually outgrow them in warmer times. Both silver and dwarf birch are present at Hockham, indicating a landscape of open, grassy tundra that was lusher than the high latitude tundra of today and which contained dwarf birch, crowberry, juniper and scattered stands of birch, willow and rowan. After about 11 thousand years ago, increased sandiness in the lake sediments suggests a windier (also perhaps a colder and drier) landscape, as expected for the one thousand year interval of the Younger Dryas.

The sudden influx around 9.5 thousand years ago of large quantities of birch pollen, closely followed by increases in pine, then hazel/bog myrtle and elm, indicates a transition from open tundra to canopied woodland. Shading restricted the more light-demanding species, so the peak in birch is short-lived, outgrown by large canopy trees. Between 9 ka and 7 ka Scots pine declined around Hockham, while the continued presence of elm, oak and hazel indicates a mixed woodland. Hazel pollen is so abundant it suggests the species was a canopy tree unlike its present habit as an understory shrub. Only after 7 ka do other woodland species such as lime, alder and, finally, ash, appear. The sharp decline in hazel pollen at about 7 ka suggests it was replaced in the canopy by other trees, probably lime. Lime is insect polli-nated and therefore produces little pollen, which tends to clump, and so does not travel far in wind.

☐ What does finding lime pollen in the sediment indicate?

■ Lime must have been abundant in the forest nearby so that its pollen entered Hockham Mere.

From 7 ka to 2.5 ka the Hockham diagram documents a period of forestation that gave Britain a cover of dense deciduous woodland – the 'wildwood' used by our Mesolithic ancestors. The top part of the diagram, from 2.5 ka onwards, records the fate of wildwood. Herb pollen rises and tree species sharply decline. Grasses, plantains (herbs of pasture and trampled ground), heather and bracken all increase. In East Anglia the heather and bracken reflect the formation of the Breckland heaths on the acidic soils of wind-blown sands. In more northern British cores the rise of heathland plants reflects deforestation of uplands where upland heather moor and blanket bog formed. The grass pollen rise, and that of plantains (a weed of pasture lands) indicate more open areas for animal grazing. The grass pollen also includes larger pollen grains of cereals such as wheat and barley.

Most pollen diagrams from sites across Europe show a sudden, sharp decline in elm pollen at about 5 ka. The elm curve for Hockham Mere in (Figure 8.3) decreases sharply at about 6 ka, and then again at 4.5 ka. Although not typical, this illustrates the same Europe-wide loss of elm without much change in other pollen frequencies. Initially human activity was invoked for this decline (elm is used as a fodder tree in some countries), but disease (such as Dutch Elm Disease) is a possibility.

Figure 8.4 shows the lower part of a pollen diagram from Abernethy Forest in the Cairngorms, Scotland. From 12 ka to 10 ka, pollen input is extremely low – typical of tundra vegetation. It rises slightly between 11.8 ka and 11 ka when birch and heathland shrubs (such as crowberry), grasses and sedges increase. At 11.2 ka the dwarf shrubs decline and

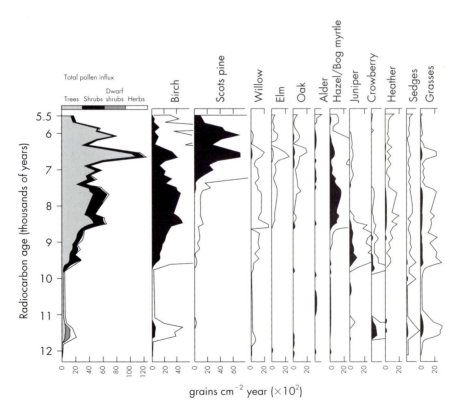

Figure 8.4 The occurrence of the commonest pollen and spore types from Abernethy Forest, Inverness-shire.

glacial tundra vegetation returns. This correlates with the last cold period before the onset of warm Holocene conditions – the Younger Dryas.

☐ When do birch, hazel and pine reach abundant influx levels in Figure 8.4?

■ They reach abundant influx levels at 9.7 ka, 8.7 ka and about 7.4 ka respectively.

☐ Compare these dates with the arrival of the same species at Hockham Mere from Figure 8.3.

■ At Hockham arrival dates are 9.6 ka, 9.4 ka and 9.5 ka. Birch may have been present in scattered copses since before 12 ka, but pine arrives just before hazel at a similar time.

Changes in sediment at Abernethy suggest a change from open vegetation at about 9.7 ka. In Scotland, birch forest survived for 2.5 thousand years in association first with juniper, then hazel, before pine became abundant. Pine, which has buoyant pollen with air sacks (Figure 8.1) that can be blown long distances, did not reach the site until about 6.8 ka. Pine-needles create a deep acidic litter in forest situations and this may be responsible for the subsequent decrease in hazel and rise in acid-loving heather. The Hockham record shows the development of deciduous forest after 9 ka, but it is sparsely represented at Abernethy, for it did not thrive this far north, even during the warmest part of the Holocene, except in sheltered lowland sites.

The rapid rise in birch at both Hockham and Abernethy, at about 9.6 ka to 9.7 ka, indicates that tree birch probably survived in isolated pockets across Britain during the Younger Dryas. However, hazel and pine took some time to reach northern regions. Species that had to spread from southern or eastern survival sites appear at different times in different locations. Even though pine reached Hockham before hazel, it appears to have migrated north much more slowly despite having smaller wind-blown seed.

Spreading Vegetation

Once radiocarbon dating fixed the major events in lake and bog cores from sufficient places, it became apparent that Holocene pollen events across Europe occur progressively later in more northerly locations. Tree species surviving in the south migrated north once the climate improved.

The spread of species can be tracked using precise dating at widespread sites. The first appearance of pollen of a type of vegetation at each site is plotted on a map. Joining sites with similar dates shows the advance (Figure 8.5). Such maps indicate possible sites of refuge of species during glacials and the speed of invasion, as well as colonisation patterns.

☐ From Figure 8.5, what is the average spreading rate (in kilometres per year) of oaks in eastern Europe from 10.5 ka to 7 ka?

■ About 0.4 km per year.

Tree species, despite their quite different seed size, all had similar migration rates across Europe and north America during the early Holocene, ranging from 0.1 to 1 km per year. There is no evidence that the Straights of Dover (which flooded at about 8.3 ka) acted as a barrier to plant migration into Britain. The wider Irish Sea did act as a barrier – only about 80 per cent of the mainland flora occurs in Ireland and several mainland vertebrates are absent.

☐ How does the spreading rate of trees compare to the rates of climate change documented in Part 1?

■ It is very slow.

The ~7°C rise in temperature that heralded the start of the Holocene probably occurred over a few decades. As a 1°C rise in temperature is equivalent to a 100–150 km shift in the spatial distribution of trees, this temperature change would have promoted a 1000 km northward extension of tree ranges.

☐ How long would it have taken for oaks to adjust to the warming into the Holocene?

■ Given a spreading rate of ~0.4 km a year, about 2.5 thousand years.

Herbs are able to spread faster than trees, as they can produce seeds within a year of germinating; annuals may be able to spread faster still.

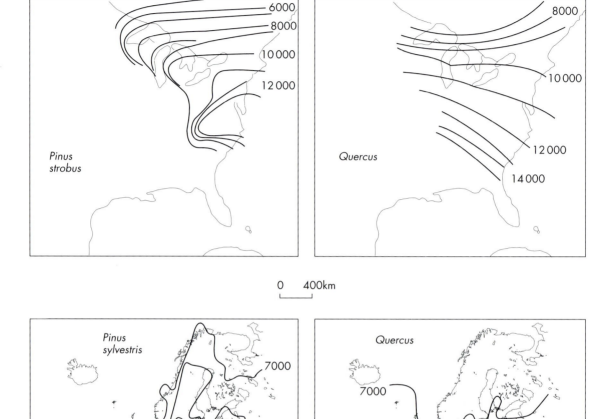

Figure 8.5 The advance of pines and oaks across North America and Europe. The contours show the northern limits of these trees at different times (thousands of years ago).

Despite the relatively slow spreading rates of plants in response to climate change, some pollen records do show rapid variations that are probably related to millennial scale events (Heinrich events, etc.) observed in sediment cores from areas around the North Atlantic (Figure 8.6). This is a puzzling paradox.

Today's Missing Biome: Glacial Tundra

Notwithstanding the fact that pollen diagrams record a combination of climate change, plant spreading rates, local environmental changes and several other factors, they have enabled the changing vegetational

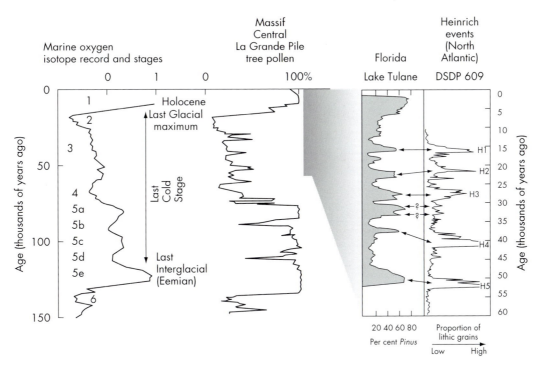

Figure 8.6 Millennial scale variations recorded in pollen diagrams from lakes in Europe: Grande Pile, Massif Central, (France) and North America: Lake Tulane, (Florida) that show probable links with ice rafting episodes (Heinrich events) in the North Atlantic.

cover of the Earth to be documented in some detail through the last two glacial–interglacial cycles. Figure 8.7 shows the enormity of the changes that occurred in Europe between the coldest and warmest climatic extremes. If humans had not colonised the continent, it would still be largely covered in forests today (as it was only 8 thousand years ago). Yet only 20 thousand years ago, these forests did not exist. Trees had retreated to small montane areas – refugia – from which they would spread when the climate ameliorated.

During glacial maxima, much of Europe, Asia and North America was covered in *mid latitude* glacial tundra – a biome that does not exist today. Today the northern *high latitude* tundras are treeless grasslands rich in sedges, dwarf shrubs, mosses and lichens. One of the main reasons tundra is treeless is that the only protection from biting winter winds is under the snow. Any plant above snow cover will be blasted

with ice crystals which strip its protective cuticle. A second reason is the permafrost which prevents the deep root development present in most trees. A third reason is the very short growing season. Many plants are cushion shaped evergreens which use their leaves to photosynthesise as soon as the snow melts. Flowers often develop in the centre of the rosette of leaves which act as a parabola reflecting sunlight and warmth onto the bud.

The low pollen input at the base of pollen diagrams (Figures 8.3 and 8.4) represent these treeless tundra-like grasslands, usually called steppe-tundra or glacial tundra, growing in what are now temperate forest latitudes. At these lower latitudes, the range of day length and extremes of seasonality are more similar to mid latitude alpine conditions than to present day Arctic tundra.

☐ How would day-length and seasonality differ for

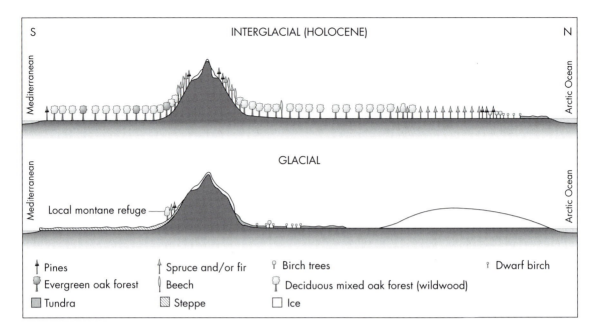

Figure 8.7 Diagrams showing the vegetation cover of Europe during glacial maxima and interglacials.

tundra in southern England compared with tundra within the Arctic circle?

■ English winter day length is about 8 hours long, summer day length about 18 hours. Within the Arctic circle, in winter the sun may not rise above the horizon for several days and in summer not set at all. Thus seasonality is much more extreme at the higher latitude.

The glacial tundra that developed to the south of the huge ice sheets was a mix of plants not found in true Arctic tundra today. This community is now dispersed between a wide range of different habitats: in Arctic tundra, above the tree line on mountains, in heathland, saltmarshes, coastal shorelines, scree slopes and cliffs. Another group are common weeds. The mountain avens, *Dryas octopetela* (Figure 8.8) was widely distributed in the late glacial and its distinctive toothed leaves are common in Danish late-glacial deposits, hence its use to name the stadial. Thrift has retreated to mountain cliffs and rocky coastal habitats.

INSECTS AS CLIMATE INDICATORS

In the last section we looked at the migration patterns of plants during extremes of climate change; but communities also include animals. These too were strongly affected by ice advances and retreats. Most animals require a precise range of environmental conditions to survive and breed, and their presence in a sediment can track climatic conditions. Two groups in particular – beetles and midges – have provided information on past environments independent of the pollen record.

Beetles are abundant. There are perhaps three million species living today. Most are choosy in terms of diet and temperature. In the early 1960s Russell Coope and his co-workers at Birmingham University pioneered the use of beetles to reconstruct climates and showed that these insects provide a more sensitive record of climate change than does pollen.

☐ Why might beetles be more sensitive to climate change than tree pollen records?

Figure 8.8 The mountain avens (*Dryas octopetela*) is a glacial tundra species. These plants give their name to the Younger Dryas stadial. Now they grow in many dispersed sites including the Scottish highlands, the west coast of Ireland, the Alps and Scandinavia.

■ Beetles are mobile (most can fly) and their reproductive cycle is short (usually less than a year) so populations can expand rapidly into suitable habitats. Pollen may travel a long distance, so the presence of a species may not indicate occupation, thereby 'blurring' the record. Beetles, obviously, are physically present or absent.

Many beetle species live in restricted habitats such as animal dung, in acid bog pools, or associated with a particular plant. The climate range for each beetle species may be large, but climate assessment can be refined by considering the overlap in range of several species so constraining the possible climate.

Occurrence of up to at least 200 beetle species at some fossil sites provide a precise climate assessment. Carnivorous and detritivorous (i.e. eating fragments of organic matter) beetles are selected as climate proxies, in order to avoid distributions caused by changes in vegetation. Beetle species appear not to have evolved much during the past two million years. Their metabolic requirements have remained similar too, as recognisable beetle communities are found throughout this time. This contrasts with plant species which to some extent shifted independently of each other, forming different associations such as glacial tundra described on pages 170–2.

☐ How might a rapid response to climate change affect the short term evolution of beetles?

■ They can remain in an optimum environment by migrating rapidly, thus avoiding changes which might select for different characters in the population. However they may show adaptations to migration.

To assess a beetle assemblage requires identifying species from fragments, such as wing cases, and understanding the present geographical and environmental limits of the species. Some insects found in British glacial deposits now live in Siberia or central Asia, so to interpret beetle assemblages requires knowledge of beetle habitats throughout the world.

The sensitivity of beetles makes them ideal for examining the rapid climate change from, for example, stadial to interstadial. Figure 8.9 shows temperature changes for the period since the last glacial maximum in Britain interpreted from fossil beetle assemblages.

There is surprising evidence that winters were less extreme during the glacial maximum compared with the rest of the glacial. This may explain higher snowfall. The transition from Arctic to temperate assemblages is sudden at the onset of the Allerød Interstadial at about 13 ka when the polar front moved rapidly from Portugal to between Greenland

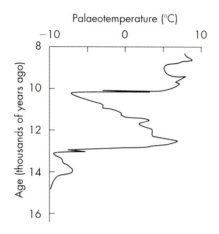

Figure 8.9 The average temperature curve for Britain for the glacial maximum to early Holocene, as assessed from beetle remains.

and Iceland. Summers then were similar to the present day, and the winters became warmer, yet pollen continues to record treeless grassland. Cool-temperate beetle species replaced those preferring warmth at about 12 ka, and remained for another two thousand years accompanied by birch-dominated pollen records. In Yorkshire, beetle records show that temperatures rose dramatically at 13 ka by 12–14 °C for July and 19-21 °C for January, to a summer maximum of 20 °C and winter minimum of 5 °C. Yet juniper scrub did not develop until about 12.5 ka, and birch forest not until 12 ka to 11.5 ka. By then summer temperatures had fallen again. Abernethy Forest (Figure 8.4) pollen shows that the development of a warmer heath and herb flora with birch was delayed until 11.7 ka to 11.2 ka.

The rapid climate deterioration into the Younger Dryas around 11 ka ago is marked by British beetle faunas that are wholly Arctic; finds from Scotland are so rare and impoverished that polar deserts are indicated.

CLIMATE CHANGES AT LOWER LATITUDES

Considerable changes in vegetation occurred at high latitudes between glacial and interglacial scenarios (Figure 8.7). The tropics and sub-tropics now support a range of vegetation (Figure 1.4) from lush tropical rainforest to barren sand-dunes. How did the glacial–interglacial cycles affect these vegetation types?

The Diversity of Tropical Forests

Tropical forests attract attention because of their high biodiversity, and the threats to it from anthropogenic deforestation. The evergreen rainforest of Amazonia is the richest, most diverse community on Earth. One hectare may contain as many as 300–400 species of broad-leaved trees despite the fact that there are only twice as many mature trees present. How this incredible diversity arose and is maintained has taxed botanists for many years.

Most of the trees are pollinated by insects, but bats, opossums and birds are also important. Animals eat the many fruits thus helping to disperse seeds into surrounding soils. Forest regeneration usually occurs in gaps created in the canopy by tropical storms. Some wind pollinated trees colonise larger gaps; growing in groups facilitates pollination. Conifer forests have not developed the high degree of diversity shown by tropical forests because they are wind pollinated. Unlike pollination by animals, wind action is random: unless a tropical species has a particular life-cycle which causes clumping, the chances are low of windblown pollen reaching another tree of the same species some distance away in the forest.

Many biologists have asked why tropical forests are so diverse. One explanation involves the relationships between trees and animals as pollinators and seed distributors. The animals occupy different niches and are adapted to spread seeds and pollen of specific trees, and the trees in turn provide habitats for massive numbers of insects, lianas (climbers) and epiphytes (plants attached to trunks and branches), birds and bacteria. Another explanation is that as far as resistance to the ravaging effects of disease and parasites (pathogens) are concerned, there is safety in diversity. This is because the more thinly scattered a particular susceptible host for pathogens is, the more difficult it is for the pathogens to thrive. The pathogens have evolved to avoid extinction, but the diversity of the potential hosts will check their spread. But how did the relationships first described develop?

There are two main lines of explanations and they are exact opposites! One is that the longevity of the forest (the stasis view) has provided the opportunity for evolution of many forms, the other that repeated upheavals caused by 2.6 million years of climate change produced the diversity.

Stasis and Co-evolving Niches and Species

The stasis view is summarised so: because the world's tropical rain-forests evolved during the Tertiary, they have a long history at low latitudes. Their tremendous diversity is thus due to their stability. Somehow (the argument goes) this stability allowed speciation into an ever-increasing range of ecological niches partitioning to a highly refined degree. The smaller and more refined the niches, the more species there are that "fit" in the ecosystem.

A niche is the space occupied by a species in the ecosystem: a complete description of how it fits into its environment. By definition no two species can have the same niche. If they did, competition should result in one species becoming extinct or evolving to occupy a new niche, thereby removing conflict with the other species.

Early in the evolution of a group, the potential niches will not all be filled. For example, tree habit could only develop once plants evolved sufficient stem strength to grow tall. Once all the niche space is occupied, species diversity can only increase by partitioning niches so that each species occupies a smaller niche. This results in specialisation in some parameter such as feeding habits, mating behaviour, food type, tolerance to toxins, and so on in animals; flowering time, seed dispersal methods, drought tolerance, nutrient requirements and so on in plants.

Diversity from Stress: the Refugia View

Although at the time of writing, some workers still favour the stasis view, many question whether forests have been stable in the lowland tropics for millions of years. Some areas in tropical forests are rich in species but others are less diverse. Documentation of this patchiness in the 1960s seemed to indicate that some areas had been havens or refugia for many species, perhaps during periods of climate change. Drying of the tropics during glacials may have caused the tropical forest to shrink, so that only small, isolated fragments survived. These refugia would have existed for many thousands of years. This too is a plausible background for speciation. Genetic differences arise more quickly in small isolated populations. When the forest expanded again, such populations remained discrete, producing new species. Up to 50 glacial and interglacial shifts in refugia and the population sizes in them could account for rapid speciation in the 2.6 million year span of the glacial epoch.

Evidence comes from the organisms themselves. An excellent example is among the *Heliconius* butterflies (Figure 8.10). The two butterfly species (*H. melpomene, H. erato*) are both poisonous to birds and have developed bright markings as a warning. If both butterflies look similar, then birds learn the warning coloration faster (this is called Müllerian mimicry). In different parts of South America the two *Heliconius* species have both developed similar wing patterns, but each dual-species pattern differs from area to area. With the vast extent of forest under today's interglacial conditions we would expect a continuum between the patterns. The fact that they are sharply divided suggests separate evolution in isolated populations in the past.

Ecologists plotted possible refugia sites in the Amazon using the patterns of species diversity (like the butterflies). High biodiversity and abundant endemics (species which only occur in one small area in the world) suggest a refuge. Plant, butterfly, bird and lizard distributions have been used to pinpoint possible refuges where forest survived climate change (Figure 8.11).

☐ Compare Figures 8.10 and 8.11; do you think they agree?

■ Not a bad match but the butterfly data is only for

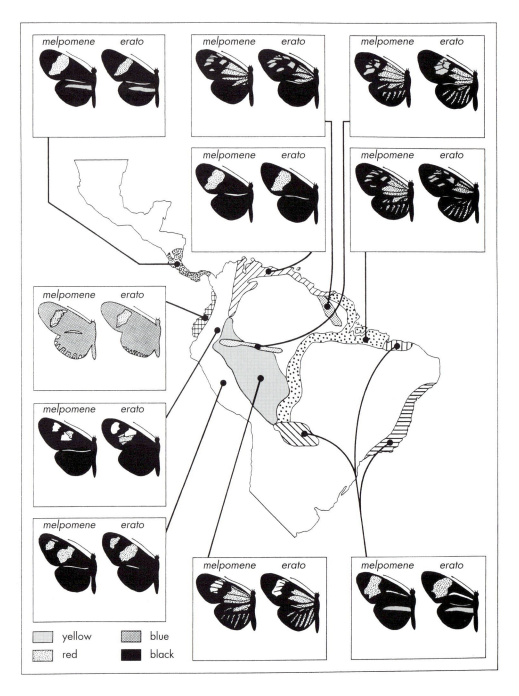

Figure 8.10 Areas in South America where *Heliconius* butterflies show different wing patterns suggesting the existence at some time of several isolated populations.

Figure 8.11 Possible locations of South American tropical forest refuges surrounded by grassland during the last glaciation based on the abundance patterns in plants and animals.

two species. It suggests the butterflies survived in the refuges where each pair co-evolved a different wing pattern. The area where each butterfly pairing is found now is larger than the refuges suggesting they have expanded their range during the Holocene.

One flaw in the refugia theory is that collecting bias in the Amazon (high around towns such as Manaos, low in the more distant impenetrable jungle) could produce false patterns of abundance and endemism. Another is that the patterns of species diversity and endemism are not due to refugia: natural barriers to dispersal in the forest such as rivers or high ground could be responsible. The forests of the Amazon are not uniform. They are lower in the west, where palms form an important part of the canopy and where rainfall is high (2500 to 3000 mm per year) and seasonality less obvious. In central and eastern parts rainfall may be only just sufficient, at 1500 mm per year, to maintain tropical forest. Paul Colinvaux argued forcefully that the Amazon is a vast area the size of Europe with varying rainfall, seasonality and flooding regimes. Disturbance is high from

erosion and storms and even fire, providing a very heterogeneous area with many ecosystems.

To settle the argument independent evidence of change or stability is required, such as pollen analysis. New results from pollen studies of sediments recovered from the Amazon submarine fan suggests that forests were not extensively replaced by savannah vegetation during the last glacial as predicted by the more extreme version of the refugia hypothesis.

Changes in Andean Vegetation Belts

The best areas for pollen records in South America are the upland basins and lakes of the Andean Mountain range: some sites, such as the Bogota Plateau, go back to the Miocene. By the end of the Tertiary, uplift had raised the Andes by about 2000 m. Already, on the Bogota Plateau, the more lowland tropical and sub-Andean palms, legumes and tropical trees had been replaced by oaks and podocarps (strap-leaved gymnosperm trees) and high Andean (paramo) species of ericaceous shrubs, grass and montane meadow herbs. About 800 thousand years ago, vegetation changes reveal longer, stronger climatic oscillations since when ten major glaciations are recorded.

Montane vegetation communities define horizontal bands (Figure 8.12). In the low-latitude Andes tropical forest covers the mountain bases. Above this are cooler forests running up to the tree line. Higher still are various grasslands or paramo (high, treeless grassy areas) similar in many ways to glacial tundra, growing up to the permanent snowline. When the climate was cooler than now, vegetation zones grew at lower altitude, and the tree-line descended the mountains (Figure 8.12(a)), such shifts providing a sensitive measure of climate change.

During the last glacial maximum, the upper Andean forest grew about 1 400 m lower down the mountains than now, to below 2 000 m. This represents a temperature about 8°C lower than the current climate. However, the temperature difference may not have been as much as 8°C. This is because the rate at which plants transpire water is inversely proportional to the atmospheric content of CO_2. In

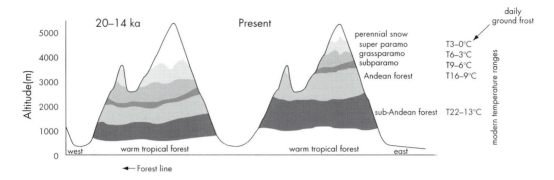

Figure 8.12 Montane vegetation zones and climate in the northern Andes showing (a) the position of the zones during the late glacial maximum suggested from pollen analysis, and (b) the present latitudes and associated temperature range of modern vegetation.

other words, halving the CO_2 content is almost like halving the annual rainfall as far as plants are concerned. So it is not surprising that during the late glacial, when conditions were drier and CO_2 levels lower, that the high altitude dry paramo expanded into lower elevations. The relationship between atmospheric CO_2 and transpiration probably accounts for the discrepancy between temperature estimates for low latitudes during the last glacial maxima based on oceanic proxy records (2–3°C) and the lowering of tree lines (5–8°C).

As in the Northern Hemisphere, some Andean sites record short climate oscillations from 13 ka to 9 ka which caused considerable changes in vegetation. Columbia experienced two cold stages: the first from 13 ka to 12.4 ka is characterised by paramo down to 2 000 m directly in contact with semi-open vegetation and no Andean montane forest between. The following interstadial, during which Andean forest developed between 1 800 and 2 800 m with the increase of oaks and alder indicates a wet, cool–temperate climate. During the second stadial, which appears to correlate with the Younger Dryas from 11 ka to 10 ka, the forest tree line descended about 500 m indicating a temperature drop of about 3°C. The climate was probably drier than during the interstadial, marked by the loss of alders.

Figure 8.13 shows how tree pollen in a core from an Andean site in Columbia varied between 400 ka and 1.2 Ma. The fluctuations show a clear match in the general ups and downs of climate, as recorded by the ocean-floor $\delta^{18}O$ record.

In Peru and Chile (south of the Equator), the record appears contradictory for recent events. Treeless dry-paramo was displaced to higher levels by podocarp woodland from 11 ka to 10 ka, and replacement of scanty beetle faunas by richer tree-dwelling species at 13 ka indicate an increase in wet mountain forest with little change, until drier conditions at about 7 ka. There is no evidence for the Younger Dryas. By 12.8 ka, closed southern beech forest was widespread and shows no sign of later decline. It seems clear that the Younger Dryas was not a strong influence on biota in southerly latitudes, despite the fact that there is evidence for the advance of mountain glaciers.

In the Holocene, from 10 ka onwards, the vegetation zones moved to higher altitudes in the northern Andes, reaching their highest at about 5 ka, when the tree-line may have reached 3 600 m. After this, vegetation zones seem to have retreated slightly reaching their present positions about 3 ka with a tree-line of 3 200 m.

Climate Change in Lowland South America

The humid Amazon basin, with its deep soils and seasonal flooding that rework sediments does not lend itself to the preservation of long-term records of vegetation change. Several pollen analysts have searched recently for evidence of changes in the Amazon dur-

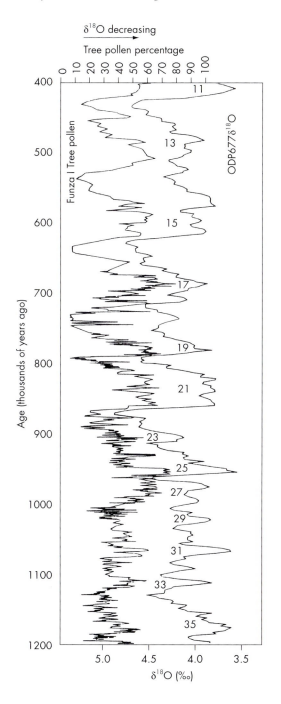

ing the last 100 thousand years. The simple refugia theory that tropical forest was dissected at the beginning of glacial phases only expanding once more in interglacials is seen to be more complex. The temperature and precipitation parameters changed independently, at least in part. The lowlands may have experienced a drop in temperature of up to 6 °C during the coldest period of glaciation (the glacial maximum at 22 ka to 18 ka). However, the main factor affecting the relative abundance of forest and grassland is rainfall. Lake levels, river erosion patterns and the expansion of savannah indicate dry intervals occurred at 65 ka to 60 ka, 40 ka and 20 ka. Pollen assemblages for each period suggest a decrease in rainfall of about 1 000 mm per year. Refugia may have been defined not by diversity in the forest, but by rainfall levels. A regional and uniform fall in Amazonian rainfall by 1 000 mm per year would result in only the areas shaded in Figure 8.14 having the 1 500 mm minimum required to maintain tropical forest. This is a different pattern to that determined by species abundances (Figures 8.11).

☐ Is it reasonable to determine rainfall patterns during glacials from uniformly decreasing current rainfall patterns?

■ It is risky to assume uniform drying or that rainfall patterns during a glacial are similar to interglacial distributions. The pattern obtained in Figure 8.14 does not explain all the refugia identified from species distribution shown in Figure 8.11, suggesting that a more complex climate change may have occurred, or that the species distributions are not all due to refugia.

Colinvaux pointed out that climate models show Amazonian rainfall may only have decreased by

Figure 8.13 Variation in percentage of tree pollens in an Andean lake core from Funza in Columbia (left curve) together with the variations in $\delta^{18}O$ from a Pacific deep sea core (the numbers shown are oxygen isotope stages). The correlation shown between the two curves is tentative, being based on visual comparison between them.

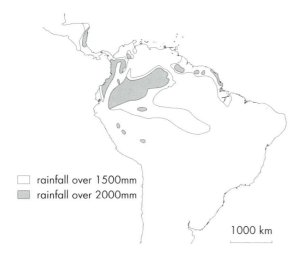

Figure 8.14 Possible location of refugia assuming a uniform drying of the Amazon basin by 1000 mm a year, based on current rainfall patterns.

10–20 per cent, by perhaps 250–600 mm during glacials. This may not be enough to restrict the forests except at the dry margins. Unfortunately, only a handful of lowland palynological sites are known: not enough to show the existence or location of refugia.

Tropical Forest in Africa

The picture for late Quaternary tropical Africa is similar to that for Amazonia. Palynological sites are not abundant, but indicate that glacial temperatures were probably between 3 and 8 °C colder than now: lowland sites indicate a decrease of about 4 °C, while shifts in treelines suggest a fall of up to 8 °C (but remember this is probably an overestimate because of the effect of lowered atmospheric CO_2 levels). High land now covered in tropical forest carried montane conifer forest in central Africa during the glacial. At the last glacial maximum high altitude grassland descended by 1 500 m.

Rainfall decreased by perhaps 30 per cent (about 400 mm) triggering some degree of forest fragmentation. Low land was probably covered by open vegetation with podocarps forming gallery forest (podocarps now grow mainly above 2 000 m). Just as glacial tundra was an eclectic mix of species at high latitudes, so the podocarp–juniper–ericoid–grassland mosaic in the tropical lowlands has no modern equivalent.

Similar patterns of species richness, endemism and distinct distributions are found in the African forest as in South America suggesting the survival of refugia. There were probably fewer refugia for forest in Africa compared with the Amazon.

☐ How might fewer refugia affect diversity?
■ Fewer species would survive periods of contraction and fewer populations in which speciation could occur would also result in fewer species. In fact many related groups do have more species in America than in Africa.

Why Not So Diverse? A Third View

We still do not know just how diverse the tropical forests are: we have not recorded all the tree species yet, nor all the insects, fungi and bacteria. We do not know how diversity is maintained or how it changes over time. Trying, therefore, to understand past change from the scattered pollen records is rather tricky. If we cannot identify the trees, how can we hope to recognise the pollen? Also the few wind pollinated tree species, like *Cecropia*, tend to dominate the record and many animal pollinated species are represented by only one or two grains which are easily overlooked.

Given all these problems, this next piece of information needs treating with caution. The average number of pollen types recognised in a Holocene sample from a tropical lowland flood plain is about 140, which is approximately the number of species found in a 0.1 hectare area. The number of pollen types recognisable in a similar sediment from the lower Miocene floodplain, when conditions were warmer and wetter overall, is 275!

☐ Does this indicate that the Miocene tropical forest was twice as diverse as the Holocene forest?
■ Perhaps. The patchy diversity of today may have existed in the Miocene, so we may just have an

extreme example. However, patchy diversity is supposed to be due to the refugia.

If Miocene tropical forests were more diverse, then the stasis view of forest evolution in the Tertiary was correct all along. In some ways this evidence is supported by palaeontologists who explain some mass extinction events as due to periods of oscillating climate. In other words glacial cycles may reduce diversity, not create it. There is evidence that this happened in NW Europe, where the number of tree species recorded in interglacials has diminished from 47 over two million years ago to 15 in the last interglacial.

CHANGES IN DESERTS

Introduction

At high latitudes today, temperate, boreal and Arctic vegetation are more influenced by temperature fluctuations than by changes in rainfall. The short growing season, deep winter cold and low summer temperatures tend to limit plant growth. At tropical and subtropical latitudes, temperatures are higher (excluding mountain situations of course), so the effects of temperature-restricted growing seasons are not important. Because temperatures are high, however, so is evaporation. The availability of water as rainfall or in underground aquifers becomes paramount.

As the name suggests, tropical rainforest needs abundant water, but in many tropical areas evaporation exceeds precipitation. This is because there are extensive rain-starved areas due to rain shadows on the lee sides of mountains, and the deflection of global wind patterns by huge uplifted areas such as the Tibetan plateau. Lack of water restricts tree-growth so that only scrub or scattered small trees live in a grassland vegetation (the savannah in Figures 1.3, 1.4). Because rainfall is often seasonal, there may be a distinct growing season. Extremely low rainfall may result in no vegetation cover at all (desert) except where water collects due to run-off or emerges from underground aquifers to produce oases.

With such a scenario, it might be expected that vegetation changes at high latitude will reflect temperature changes more than precipitation, while low latitudes will reflect changes in humidity and rainfall. As you have seen so far, this is indeed the case. In the following section we can look at the areas most sensitive to changes in rainfall: the desert/grassland transition. In particular, the region of north Africa dominated by the Sahara today illustrates these changes.

A number of techniques have been used to determine the history of vegetation in dry low latitudes including pollen analysis, lake levels and salinity changes recorded in sediments, pollen input to off-shore sites (indicating wind directions and strength as well as regional vegetation), distribution of animal remains and human artefacts. Figure 8.15 shows some of this evidence for the Sahara: the occurrence of animal bones and the human record of the living herds depicted as rock art.

☐ What does Figure 8.15 indicate?
■ That areas of the Sahara that are now unable to support a variety of large grazers such as elephants did so in the past.

The Sahara

The Sahara today lies at the centre of a series of vegetation zones. In the north there is a Mediterranean climate with winter-dominated annual rainfall of 400–1 000 mm per year resulting in oak woodland along the North African coast. South of this, at about 35–33°N, more patchy rainfall creates a steppe vegetation of *Artemisia* and alfalfa grass. This grades into desert proper, with less than 100 mm of rain a year and drier air producing inhospitable conditions for plant growth in the Sahara desert. South of the desert lie the Sahel grasslands with scattered acacia trees, which grades into wooded savannah and finally rainforest. Today the Sahel lies in a band between 15 and 18°N, while the Sahara covers 18 to 30°N.

Terrestrial sedimentary sequences from Africa are rather patchy and incomplete so most of the evidence for African climate and vegetation changes come

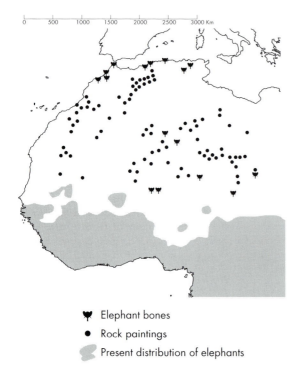

♆ Elephant bones

● Rock paintings

▨ Present distribution of elephants

Figure 8.15 Records of elephant remains and rock art depicting several species of megafauna across the Sahara.

from offshore sediments despite the loss of detail due to long distance transport of pollen, and therefore over-representation of wind-pollinated species. Some of the longest records of North African vegetation change come from offshore palynological records. These records go back nearly a million years and reveal a number of oscillations in the boundary between the Sahel and Sahara from about 14°N during periods of desert expansion to above 23°N in humid phases with few areas of open sandy desert. Interglacials appear to have been wetter before 280 thousand years ago. Since then the Sahel has not encroached above 23°N, and the glacial arid phases appear more extreme (Figure 8.16).

During the last two glacial cycles from 185 thousand years ago there were significant changes in the vegetation zones and the extent of desert. In the north the *Artemisia* grasslands, with scattered pine woods, expanded right up to the Pyrenees during glacial

maxima. Broad-leaved trees probably only survived in refugia in the Atlas Mountains. However, the major changes occurred during the last and present interglacials when oak woodland became abundant along the northern fringe of the Sahara. The expansion of forest indicates changes in humidity and precipitation in the north of Africa, possibly due to weakening of the north east trade winds in interglacial periods which would have allowed sub-tropical depressions to reach further north (Figures 1.19, 1.20).

To the south the Sahel savannah made large inroads into the desert during interglacials. The arid–humid cycle (also reflected by periods of high lake levels – see Figure 5.20) does not correlate exactly with glacial–interglacial patterns, but maximum aridity seems to correlate with glacial maxima, followed by a period of high humidity, as in the early Holocene. As you can see from Figure 8.16, the northern and southern boundaries of the Sahara move independently. Evidence from Egypt indicates the penultimate interglacial was wet. Rising water tables produced lakes in windblown depressions where many remains of reptiles, birds, fish and mammals are found. These indicate large lakes surrounded by savannah and scattered trees. Rodents now living only in central and southern Africa occurred at least 250 km further south, indicating an early Holocene rainfall of 500 mm, whereas now it is about 100 mm. Clearly, for low latitude arid to semi-arid lands, interglacials may have been similar in temperature, but not in terms of humidity.

During the last glacial, increasing aridity caused gradual expansion of the Sahara to peak during the glacial maximum from 21 ka to 15 ka. After about 13 ka climate began to change but the pattern is complex: different sources of data record different stories. The final post-glacial shift from arid to warm and wet started at low latitude and spread north while the vegetation change from scattered pine and steppe to Mediterranean oak forest spread south from Spain. Yet again the vegetation seems to lag behind rapid changes in climate.

Most Saharan data identify the Younger Dryas as a dry phase. Sediments in the Sahel record a short

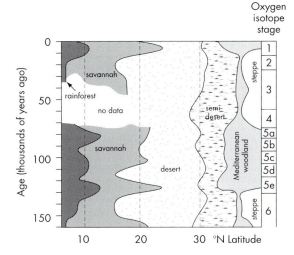

Figure 8.16 Changes in latitudinal extent of the Sahara and Mediterranean vegetation since 150 ka.

intense wet phase during the Younger Dryas, while others show no clear sign of the stadial.

The wettest part of the Holocene was from 8.5 ka to 6.8 ka when rainfall was probably 100 to 400 mm higher than now (Figure 1.20). The monsoon belt encroached further north and most of the Sahara may have been wet enough (200 to 400 mm per year) for grassland to support grazing animals. The boundary of the desert and Sahel was 500 km further north than at present and 1 000 km further north than it had been during the glacial maximum. The Sahel vegetation began to be replaced by desert species about at 6 ka and by 4 ka aeolian sand dunes started to form. The extent of desert at 3 ka indicates a similar monsoon limit to that of today.

CONSEQUENCES OF CLIMATE CHANGE

The Tropical Latitudes

In the tropics – if the refugia hypothesis is correct – a drier climate during glacials probably resulted in expansion of grassland which dissected forest into refugia that may have been linked by narrow strips of forest growing along waterways. Vast expanses of forest were probably eliminated wherever rainfall decreased below about 1 500 mm per year. Under these circumstances very little migration of tropical forest species could have occurred. The refugia areas were already occupied by fully developed forest so reducing the chance of invasion by other species. Inevitably this must have led to some extinctions: perhaps half the Tertiary tropical species. Cooling caused vegetation belts to descend mountains resulting in an increase of high-level grasslands.

During interglacials, forest species expanded from refugia to re-colonise the savannah. In Africa, even during interglacials, forest is confined to the west and grassland is now quite extensive in central and eastern areas. In South America natural tropical grassland covers only a small area of the western Amazon. In mountain regions the vegetation belts ascended during interglacials, constricting the upper vegetation belts because there is less area at the top of cone-shaped mountains. During interglacials the mountains become refugia for many species which were more widespread in glacial times.

Higher Latitudes

At higher latitudes the glacial–interglacial cycles produced different evolutionary pressures. The vegetation was not dissected into a mosaic, but almost completely changed (Figure 8.7). At the onset of glacial conditions temperate and boreal tree species were eliminated except in refuges far to the south in southern Europe and America. As in the tropical forest, habitats at low latitude were already occupied, so most of the northern tree populations may have died and been replaced by northward spreading lower-latitude species. This pattern was complicated by the position of mountain belts. In China and North America, large north–south trending mountain ranges provided a diversity of habitats without preventing migration into lower latitudes. In Europe, the Alps and Mediterranean acted as barriers to southward population migration. Far more Tertiary relic trees such as hickory, sweetgum, tulip tree and magnolia survive in China and America than they do in Europe.

Glacial tundra expanded in huge belts across Eurasia and Laurasia – it was dominant in these areas for 80 per cent of the time since 2.6 Ma. South of the ice, glacial tundra was far more widespread than Arctic tundra is today and the many herbs, dwarf shrub species and accompanying faunas expanded their range and population sizes. During interglacials tundra vegetation is restricted to high latitudes, high altitudes and patchy lowland refugia.

As we have seen, the end of the glacial was very rapid, producing favourable conditions for plant growth over vast areas of relatively bare land. Selection pressure at the amelioration of glaciation would be strong for characteristics which improved migration rates. There is some evidence that species evolved to grow faster, reproduce at a younger age and produce smaller, more widely dispersible seeds.

Polyploidy in the Northern Hemisphere

One characteristic of higher latitude plant species today is hidden in their genes. They show high levels of polyploidy, a phenomenon described in Box 8.2. In the tropical angiosperm flora about 25–40 per cent of species are polyploid, in Britain it is 53 per cent and in higher latitudes 65–86 per cent (Figure 8.17).

A single new polyploid plant will have no other polyploids to cross-breed with, but this may not be as serious as it sounds for the new species. Plants can reproduce asexually by vegetative means. The single plant may be able to self fertilise to produce seed. Often polyploids can produce seed directly without requiring pollination: the offspring are identical clones of the parent. If other polyploids have formed in the same area, then outcrossing becomes possible. Polyploids seem to form when something is going wrong with normal breeding: for example if two related species come into contact and produce sterile hybrids. A hybrid swarm forms, some of which may misdivide and create polyploids.

This may explain why the high-latitude plant communities have more polyploids than those at low

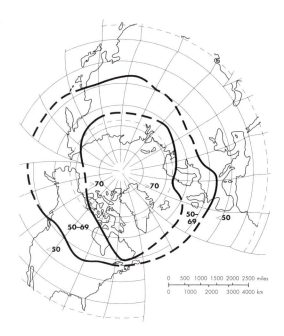

Figure 8.17 Trend of increasing polyploidy of floras with increasing latitude in the Northern Hemisphere.

latitudes. During the Quaternary, migrating plant species had a free-for-all, when community structure vanished. Inevitably related species would have met and hybrids would have formed. Perhaps this caused an increase in the polyploids.

☐ What evidence do we have from the tropics which fails to support the explanation for high latitude polyploids?

■ The refugia theory predicts species in refuges expand and mix after glacials. Surely here too many hybrids would form, yet polyploidy is low.

☐ Polyploidy in West Africa is 26 per cent, on African mountains it is 45–50 per cent. What does this suggest?

■ Polyploidy is associated with both high latitude and altitude. Although the effect is less obvious on African mountains, vegetation belts did change altitude during the Quaternary so the hybridisation explanation could be the same as for latitude.

Box 8.2 Polyploidy

All plant and animal cells have a nucleus containing DNA. These extremely long chain molecules are the templates for life: they record, in code form, the structure of all the enzymes and other proteins. Each length of DNA which codes for a particular substance is called a gene.

When a cell divides, the DNA folds into a very compact structure visible with a light microscope as a dark sausage-shaped chromosome. Each cell has several chromosomes. Most cells in an organism have a double set: each with a 'mirror' copy of all the genes. These are diploid cells. Sex cells form from a special division which leaves each cell with only one set of chromosomes: these are haploid cells. When a sperm fertilises an egg each provides a haploid set of chromosomes, so the resulting embryo has diploid cells.

Some organisms have cells containing four sets of chromosomes, or six, or even more; these are polyploids. A polyploid arises if, during the first cell division after fertilisation, the two new cells (each with two sets of chromosomes) merge together, leaving four sets of chromosomes in the cell. If this cell then divides to form an embryo, the organism will be polyploid and have twice as many chromosomes as either of its parents. It cannot reproduce with the original diploid populations so the polyploid becomes a new species, evolving independently.

Many plants are found to be polyploid, but only a few animals. If a polyploid animal embryo forms, it usually dies very quickly. This is probably because the developmental processes in animals are different to those in plants. Too many chromosomes has a fatal effect. Polyploids are often vigorous, larger than their diploid ancestors and sometimes have new characteristics. For example, stinging nettles are polyploid, but there are rare non-stinging species that are diploid.

An alternative theory is that the vigour and genetic richness of polyploids (having so many sets of genes) gives them an advantage in post-glacial conditions. Polyploids could survive the extreme environments during climatic upheaval as they are able to adapt rapidly to change. A single vegetatively reproducing or self fertile plant could found a new population in an isolated area.

Unfortunately our inability to determine from the fossil record whether high latitude floras of the Tertiary were also polyploid, or if polyploidy increased at particular times and places in the Quaternary, means we cannot determine between these possible causes. Either way, it seems that the glacial cycle may have selected for polyploids in severe conditions at high latitudes and high altitudes.

As we have seen in the last section, there is compelling evidence for evolutionary change in response to the continual and periodically sudden changes in climatic conditions. It is best documented from plants, but inferred for animals. But much the most important evolutionary change in the Great Ice Age was our own evolution, which resulted in us becoming the only conscious agent of environmental change.

SUMMARY

1 Evidence for the effect of Quaternary glaciation on global biodiversity can be gathered in a number of ways. Fossil remains of pollen, plants, beetles and vertebrates in sediments all show the changes between the glacials and interglacials. The current distribution of species may indicate the existence of refugia in the past, especially in tropical forests, and even the genetic make-up of plants can suggest modes of speciation.

2 In low latitudes, increasing aridity during glacials seems to have caused the tropical forest to be invaded and dissected by grassland. Trees and their accompanying animals survived only in lowland refugia until post-glacial wetter climates allowed them to re-invade the grasslands. On mountains a decrease in temperature of a few degrees caused montane vegetation belts to descend. Reduced atmospheric CO_2 levels would also have affected the distribution of plant species by causing an increase in transpiration rates, the effect of which is equivalent to reducing rainfall. Once temperature and CO_2 levels rose again vegetation belts rose and montane species became isolated in refugia.

3 At higher latitudes cooling during glacials allowed glacial tundra to expand at the expense of boreal and deciduous broad-leaved forest. Some Tertiary trees became extinct, but many survived by migrating south. Individuals of short lived species, such as beetles and annual plants, were probably not directly affected by climate change, being expanded into new areas when favourable climate improved. The longer lived trees (and perhaps large mammals) may have been killed during cooling phases.

4 There would be a strong pressure to evolve improved migration and colonisation characteristics such as reproduction early in life, long distance dispersal and ways of reproducing in isolation to found advanced colonies. The mixing of species from many different communities may have produced hybridisation and an increase of polyploidy which was further selected for by preferential survival of polyploids in extreme conditions.

FURTHER READING

Bell, M. and Walker, M.J.C. 1992. *Late Quaternary Environmental Change: Physical and Human Perspectives.* Longman, Harlow Essex.

Moore, P. D., Webb, J.A. and Collinson, M.E. 1991. *Pollen Analysis* (2nd edition). Blackwell Scientific, Oxford.

Roberts, N. 1989. *The Holocene: an Environmental History.* Basil Blackwell, Oxford.

The three books listed above plus chapters in the following books listed earlier, provide useful follow-up reading: see Lowe and Walker (1997), Chapter 4 and Williams *et al.* (1993), in reading list for Chapter 1 (p. 29).

Wills, C. 1996. Safety in diversity. *New Scientist,* 23 March, 39–42. A summary of how the biological diversity of tropical rainforests may be a mechanism for keeping parasites at bay. This subject is also treated in:

Givnish, T.J. 1999. On the causes of gradients in tropical tree diversity. *Journal of Ecology*, 87, 193–210.

9

THE RECORD OF HUMANITY

WHAT IS HUMANITY?

What is it to be human? A great deal of metaphysics has been written about what makes humans tick. The problem lies in trying to place limits on (the precise meaning of the verb 'to define') humanity, which is continuously changing, and in fact consciously changes itself all the time. Today is the only useful starting point, as in all the Earth and life sciences, when one species bestrides the planet. It occasionally leaves its terrestrial confines, and is perpetually consumed by gathering the information that teems throughout the rest of nature, and on which our survival is totally dependent. This dependence is inextricably bound up with means of passing on and extending that knowledge; with foresight and means of communicating with the future, as well as with other living members of our species.

The computer and the word-processing software with which I wrote this chapter are artefacts, extensions of my body and brain, and so too is the language that I use to communicate my ideas. As it happens, the hardware and software were invented and made by other humans of whom I will likely never know. But in essence they are no different from a simple cutting edge that anyone might make by breaking a hard, fine-grained rock. Each is a tool made to make life easier and more assured. Moreover, I used them in a context that extends beyond myself and the present. Like anyone else, you and I have obligations of all sorts to the society that surrounds us, and depend in one way or another on billions of other people interacting socially with the natural world. (Of course, all of us can pick and choose which obligations we fulfil or neglect at any moment in time, but there is no real escape!) The Pentium 150 chip at the core of my computer, which brings in the meat and potatoes, was made in Malaysia from silicon, which is the second most common element in the Earth's crust. Without tools of any kind and without society you or I would surely die of starvation or thirst within 30 days at most. To survive in even Britain's balmy June we would need tools, in the most general sense that includes shelter, fire and clothing as well as a rabbit snare and a digging stick. Our jaws, fleetness and sureness of foot, and digestive system are not up to the performance of our nearest biological relatives among the primates.

The hallmarks of the human are social use of tools, and making them. Sure, many other animals have social graces, and some even pick up natural objects and use them for one function or another. But not one other transforms natural objects, and none use them within a social structure. It is quite common to see bits and pieces of what we do among this or that species, be it bird or mammal, but outside of our own all these attributes are never assembled together. Tools are part of the natural world, taken from it and transformed, to be used as means of intervening in natural processes. How such materials are transformed to become tools is passed on as the central element in culture. The information is passed on as part of a universal sharing of labour and goods, whatever economic form this takes, and within whatever social group we belong. Today there is a global economy and much of the sharing goes on without any individual being fully aware of it.

Humans consciously create new conditions for sustenance and survival around them; a 'second nature' taken from the rest of the natural world, yet used within it. In the sense of evolutionary biology, fitness is conferred by what humans create adding to physical attributes that stem from genetic inheritance. That is our uniqueness. In this may lie a fundamental break from Darwin's concept of natural selection. Billions of individuals in the human line survived to reproduce, not because they were necessarily fit in a Darwinian sense, but because they were cushioned from 'nature, red in tooth and claw' by their own conscious actions and those of other people through tools. Human evolution is inextricably linked with consciousness and social being, and they have provided unique opportunities for physiological development of many kinds. We pick up this theme in Chapter 10.

A FRAGMENTARY RECORD

Earlier chapters in this book describe how the period in which human evolution took place was one of complex and often dramatic change in the environment. So the human story is the outcome of an interplay between climate change, ecological forces and the wholly new 'window of opportunity' provided by consciousness, social sharing and creation of 'second nature'. Not surprisingly, changes in climate, terrestrial ecosystems and sea-level, mean that direct evidence of hominid evolution in the form of fossils is extremely rare. As always, marine biota are best represented in the fossil record because of the better chance that remains are buried and preserved. The Great Ice Age and the ups and downs of sea-level meant that deposition of continental sediments that might potentially bear fossil evidence of our evolution was greatly outweighed by erosion in all parts of the world. Potential preservation sites, such as lake deposits or those formed in peaty swamps, suffered episodes of uplift relative to sea-level, so that they were cut into by rivers and streams again and again. Periodic episodes of drying repeatedly oxidised and dispersed the anoxic environments in which terrestrial preservation is best achieved.

The skeletons of vertebrates are loosely connected by gristly tissue and so break up easily. On land they are often eroded or eaten by scavengers. In the primate fossil record teeth outweigh other remains by far – they are virtually indestructible once they are out of reach of things of which our dentist warns us. Parts of the head, such as jaws and skull fragments, come a long way second, followed by major limb bones. The crucial details of feet and hands are as rare as hens' teeth. From the scanty remains, therefore, we can most often make some observations from wear and tear about what a particular set of dentures bit, but rarely the nature of the biter and how he or she obtained food. Jaw fragments inspire comparative anatomists to model lower face shape, while bits of cranium produce some evidence for brain shape and size, and whether a primate was 'high-' or 'low-brow'. Where the spinal cord enters the skull depends on whether the head balances precariously on top of an upright spine or sticks out in front of a quadripedal creature. That is information of the highest value, but extremely scanty. Likewise, how the spine joins the pelvis is a key to whether a creature was bipedal, quadripedal or habitually switched between the two forms of locomotion; again a rare find, as are the small bones of the foot that provide similar clues. Without pelvic remains, telling male from female among primates is exceedingly difficult. However, firm identification of a female pelvis tells us about the size of a new-born's head, and, together with other clues to body mass and head size from adult individuals of the same species, suggests the pace at which infants developed in both stature and brain capacity. This information is of enormous importance.

For an upright gait there is a limit to the width of the female pelvic girdle, beyond which walking becomes difficult, if not impossible. This limits the size of the birth canal and thus the size of an infant's head if it is to be born intact and without fatal risk to its mother. Human mothers experience enormous pain during extremely complex labour. Women walk with a distinctive gait compared with men, which is due to their pelvises being around the size limit for bipedalism. Because of this limit and the far greater brain capacity in relation to body size of human adults compared with other primates, one conclusion from comparative anatomy of modern primates is that human infants emerge at an earlier stage in foetal development. They are entirely dependent on adults for longer than other young primates, and experience much more rapid brain growth in relation to body growth before attaining reproductive age. This is one of the burdens of being capable of developing consciousness, and from it stems the need for protracted infant care, and arguably the development of a unique social structure, including permanent male–female bonding among humans. The advantages of the last outweigh the risks associated with the first, and so increasing neoteny, as early foetal delivery is called, has been selected for during human evolution. Context prevents our taking this aspect any further until Chapter 10.

THE FIRST TOOLS AND THEIR USERS

One approach to charting human evolution is to seek tangible clues in the geological record. You might well think that the search goes like this: first the search is for primate remains, then to see if they provide evidence for upright gait and third to find out if a fossil is associated with tools. Basing themselves on the only definition of humanity that is of any use, palaeoanthropologists ought primarily to seek evidence for the appearance of consciousness – in other words: tools. A fossil primate preserved clutching an axe (a highly unlikely find as the fossil record is so poor) shouts to us, 'Human!', whether or not the individual brachiated in the upper canopy of a forest, possessed a brain the size of an aniseed ball, or had the face and table manners of a baboon. Detailed anatomy does not matter a jot in the face of clear evidence for consciousness. Consider for a moment an astronaut's finding a pocket calculator behind a rock on Mars. Must the undoubtedly conscious alien that lost it there have had free hands? Not necessarily; it may just as likely have had several prehensile noses or nimble tentacles. It may be hard to imagine an Earthly creature habitually scampering around on all

fours either to have the manipulative ability to make and use tools, or, for that matter, the need for them. But we should not rule that out.

Of course, research cannot proceed fruitfully by basing itself exclusively on ideal discoveries of tools with their users, because of poor preservation. Any palaeoanthropologist will leap with glee on the tiniest fossil scrap that bears any resemblance or relationship to a higher primate, milk it dry of information content and endlessly compare it with equally rare remains from elsewhere and other times. New finds and the insights that they provide generate re-evaluations of earlier finds. Not surprisingly, because ideas about our origins sell books, newspapers and magazines, the human story gleaned from fossils changes almost by the month – research funds are often generous in comparison with those for other sciences. Never has so much argument, so much speculation rested on such meagre evidence.

The earliest recognisable tools, simple cutting edges produced by deliberately breaking pebbles of flinty rock (Figure 9.1), appear in the geological record of the Hadar area of eastern Ethiopia, at about 2.4 to 2.5 Ma. They may not look particularly impressive, indeed a sceptic might claim that similar broken bits could be found by any river or even in the garden. Closer examination reveals that the sharp edges are not the result of a single clean break, but are jagged from repeated smaller breaks superimposed on the main one. Used as cutting tools they all fit snugly in the hand, without annoying edges that cut the fingers (interestingly most fit best in the right hand). The variety of shapes suggests a tool kit, including scrapers and borers (with possible connotations for the use of skins for clothing), although some scientists dispute this. Crucially, several of the rock types used in the tools only occur 10 to 20 km from the site. There is no evidence for natural means of their transportation. They were deliberately carried after careful selection – none of the objects are made from rocks that will not take and retain an edge. There can be no doubt that they are tools. So far, there are no definite fossil signs of whoever made and used them at Hadar. But humans were definitely around. I use the term human for the obvious reason that no other

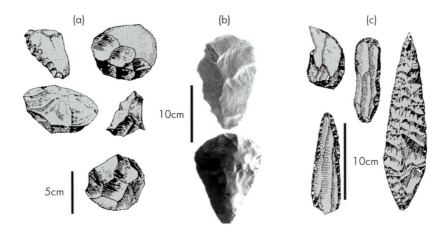

Figure 9.1 (a) An Oldowan tool kit, (b) bi-face hand axes, (c) stone tool kit from 30 thousand years ago.

genus makes tools, and toolmaking demands consciousness, however rudimentary. In fact, without being shown how to craft one of these tools, you or I would need days of trial end error (and some nasty wounds) to become proficient. Only at around 2 million years ago is there a clear association between tools and the humans who used them.

Olduvai (formerly Oldoway) Gorge in western Tanzania has been a 'gold mine' for studies of human origins and evolution, thanks to decades of work there, mainly by the Leakey family, following the discovery of human-like remains there in 1960. The earliest tools are known as Oldowan, since the first examples were found there, although they are now known to be much younger than those at Hadar. At Olduvai, the tools occur within the same deposit as human-like fossils, indeed so close to them that the fossils must be of tool makers. These first human remains are scattered and fragmentary, either because rotted corpses soon break up in flowing water, or scavengers run off with the bits. None the less, a picture of the tool makers of 2 million years ago can be pieced together. Their association with tools lends them their name, *Homo habilis* or 'Handy Man' (Figure 9.2). Your average adult habiline (that is an easy way to remember them) was a cut above apes as regards brain capacity, at around 0.60 to 0.75 litres. That is about half the modern human average of 1.40 litres and significantly larger than that for

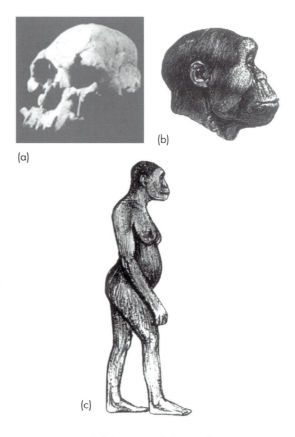

Figure 9.2 Homo habilis: a typical skull and reconstruction.

apes (between 0.35 to 0.50 litres). Scanty remains of limb and torso bones means reconstruction is difficult, but they walked upright, as their skulls have a *foramen magnum* more or less where ours is. Adult skulls come in two distinct size groups, which some workers assign to males being 20 per cent larger than females, others to two separate species. Microscopic examination of their teeth shows wear patterns likely to have arisen from a mixed vegetable and meat diet. The Oldowan cutting tools show clear signs of wear due to cutting into bone. Coupled with this is similar microscopic evidence of marks on the bones of other animals in the same deposit. They could only have been produced by using stone tools for butchering. These signs raise the question of *H. habilis'* relationship to other animals and the high-quality protein source that they represent. Did habilines hunt or did they scavenge?

☐ What sort of evidence from animal food remains might cast light on this issue?

■ Finding the remains of a whole animal, with dismembered bones, each of which had cut marks, would suggest hunting. If butchered bones were mainly of large, meaty limbs this too might suggest hunting and the carrying to safety after butchery of food from a kill site.

Sadly for those who have fantasies of 'Man the Great Hunter', neither piece of evidence is associated with habilines. The food remains are statistically low in limb bones and most would have carried poor reward in terms of meat. Lots of the bones have been smashed for marrow. Some of them show clear signs of cut marks superimposed on predators' teeth marks. As regards their meat-acquiring activities, habilines were almost certainly opportunist scavengers. Sharp tools gave them the chance to lop off a gnawed limb from a predated animal and then beat a hasty retreat. But the food bones at Olduvai and other habiline sites show signs of being in concentrations, which suggest that food was brought to a common site, perhaps for sharing among members of a group. The well-studied Olduvai deposits have produced another exciting indicator of the habilines' life style. At the 1.8 Ma level there are crude rings of large stones, that

suggest the building of shelters, perhaps from skins held up by a wooden pole (highly unlikely to have been preserved) and kept in place by stone weights.

From about 1.9 Ma onwards, Oldowan tools turn up over large areas of Africa, though rarely with fossil humans. So the culture spread; it had been remembered and carried by roaming habilines. The picture of humanity that stems from later fossils is by no means so simple or clear cut. The remains, still sparse because they are from so long ago, display a bewildering variety of subtle differences in cranial capacity, face shape and dentition, plus variance in stature and the robustness of long limb bones. There are two main views of this diversity in form. One accounts for it as a reflection of several distinct human species that cohabited in Africa, if not specific habitats. As well as *H. habilis*, palaeontologists of this school recognise three others in sequence of the age of their defining fossils, *H. rudolfensis* from Kenya, *H. ergaster* ('Action Man'!) from a variety of sites, and *H. erectus*, of which more shortly. The other school compares this ancient diversity with that so characteristic of modern humanity, which is far more varied in form and facial features than is the case for species of other animals. The alternative view accounts for the different forms by genetic drift of features that have no bearing on fitness in natural selection, more accurately that the possession of tools and other aspects of humanity's 'second nature' (shelter, fire perhaps, social living and ingenuity itself) removed many of the pressures of 'normal' natural selection. This is called polymorphism and requires no speciation. The scarcity of fossils really makes it impossible to decide between these two possibilities.

Tool-using humans, whether in several species or as one diverse in physiognomy, were not the only upright apes around in eastern Africa from 2 Ma. There are fossils of another group with very different characteristics which are never directly associated with tools, and so they cannot be regarded as human. Skulls of these creatures are very different from those of the tools users. Their dentition is massive, with large flat teeth, often heavily worn, set in large, square lower jaws. Some have bony crests on the top of their crania, which are easily interpreted as being sites where powerful jaw muscles were attached (Figure. 9.3). Wear patterns on teeth indicate an exclusively vegetable diet, probably hard fibrous plants, seeds and nuts. The first fossil of this type that was found at Olduvai by the Leakey team was dubbed 'Nutcracker Man' (*Zinjanthropus boisei*), dated at around 1.9 Ma. Three other species similar in form survived for about a million years in the same areas as humans. Their parallel existence with humans and their non-human characteristics demanded a more precise name in common. The four species are now placed in the genus *Paranthropus* (near to man). Adult paranthropoids from any site or date come in two varieties, large and small. Specialists believe that they were sexually dimorphic. Among living great apes the clearest parallel is with leaf-, fibre- and fruit-eating gorillas, where a single large male cohabits with several females less than half his size. So perhaps the paranthropoids were upright 'gorillas'. Like gorillas, the paranthropoids' low-grade vegetarian diet would have meant large guts to ensure digestion; for-

midable creatures by comparison with the lighter early humans. The abundance of their fossils and their co-existence with early humans for a million years shows that they were equally as successful in the East African habitats. The two groups, vegetarian and tool-using omnivores, could not have been in direct competition for resources; one or the other would then have disappeared.

Anatomically, paranthropoids and early humans shared the common feature of walking on two legs and both had hands freed to manipulate the world. The human group applied this last feature in making and using tools to become conscious beings. They had 'growth potential' to exploit any opportunity presented by the environment and by their production of 'second nature'. Their subsequent evolution reflects this context. The paranthropoids evolved to exploit plant foodstuffs, the most assured and simple resources. We cannot be certain that paranthropoids lacked tools – wooden digging sticks, or other artefacts made from plant materials would never be

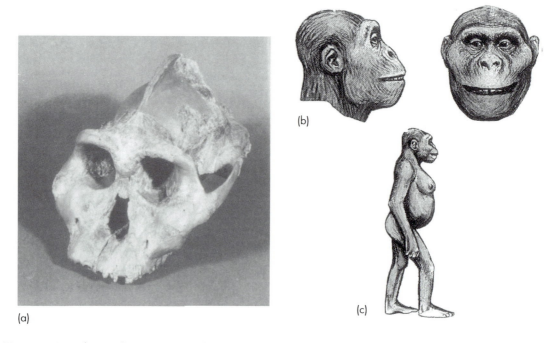

(a)

(b)

(c)

Figure 9.3 Paranthropus robustus: (a) typical skull, (b) facial reconstruction, (c) body reconstruction.

preserved – but stone tool use by them is unlikely. Here we leave the paranthropoids and return to the human story later. Coexistence of two anatomically similar groups implies divergence from an earlier common ancestor. Today our closest anatomical relatives are the great apes, the gorillas and chimpanzees of Africa and the orang-utans of Southeast Asia. Some time earlier than the common ancestor of early humans and paranthropoids there must have been divergences that formed the sources of the four modern groups of higher primates. To which are we most closely related today?

THE ROOTS OF THE HUMAN EVOLUTIONARY BUSH

None of the living great apes, other than ourselves, moves around habitually on two legs. The others occasionally adopt an upright gait, chimps more frequently and successfully than the other two. All three have some grasping and manipulative abilities in hands on their forelimbs, but equally their hind feet are capable of grasping, as indeed they must in tree-living species. No individual would survive for long in a forest canopy up to a hundred metres above ground without being able to grip branches securely with as many appendages as possible. Infants must cling to their mother's fur too. Life in trees demands other physical characteristics, such as good balance, binocular vision and the depth perception which that confers, plus excellent hand-eye co-ordination and reflexes. It is the grasping capability of both hands and feet shared by the other three modern ape species that points to a common arboreal ancestor, even though chimpanzees and particularly gorillas spend plenty of time at ground level. Human descent from an ape-like ancestor who lived in trees has been universally agreed among biologists since the late nineteenth century, if not by some religious groups. One obvious question is: with which living ape species do we share this last common ancestor? Less obvious, but equally important is the question of when this divergence may have taken place. Its importance lies in the dramatic change in lifestyle and potential represented by the adoption of a permanent upright gait that we know characterised both early humans and paranthropoids at around 2 Ma. The change implies that something happened to the ancestor's habitat to which the bipedal divergence adapted both in gait and in diet. There are now two ways of looking for clues. One is the traditional search for human-like (technically hominid) fossils further and further back in time. When this was the only available approach it was dominated by the quest for the 'missing link', of which there have been a number of candidates, notoriously the Piltdown forgery of a modern skull with an ape jaw.

Such is the nature of the divergence of species, the extinction of some and the eventual splitting of those that survive, that modern ideas of evolutionary descent see highly complex bushes of relationships (Figure 9.4). Set such complexity against the very low chances for any individual's preservation as a fossil, let alone that of all species that did exist, and you can easily visualise the problem. Missing links, the founders of new species and new genera are like needles in a haystack. All that fossil hunting can do is to link often fragmentary finds on the basis of physical similarity. If dated, the finds can be assembled in some sort of sequential order, but any bush of relatedness is open to many interpretations as the true one must depend on genetic relatedness at the level of molecules in the cells. The relatively new science of comparative molecular biology offers some help in rationalising the fossils as they are found.

Sequences of amino acids in proteins, such as the haemoglobin in red blood cells, or of nucleotides in RNA and DNA are either determined by genes or direct evidence of genetic structure. Comparisons of such sequences among and between living apes and ourselves are pretty extraordinary. You and I differ by up to five million nucleotide positions in our full DNA sequence, but most involve 'introns' or genetic junk. In terms of functional genes, humans differ across the planet by 0.15 per cent at most. Comparing the two very similar species of chimpanzee, the common and pigmy chimps or bonobos, they differ in this respect by 0.7 per cent. Comparing chimps with us reveals a difference of about 50 million nucleotide

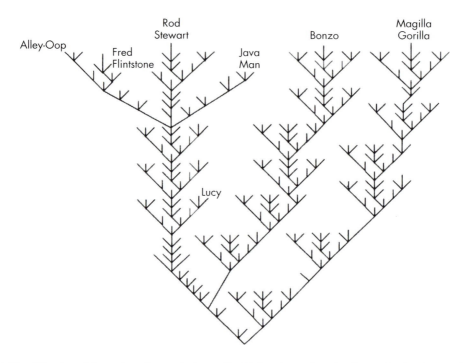

Figure 9.4 Simplification of the notion of an evolutionary 'bush' that shows the relatedness of individuals and species. The present is at the top, so that with increasing geological time the direct line of ascent of some modern individuals can barely be picked out from those of species, genera and families that have become extinct.

positions, or about 1.5 per cent. Chimpanzees and ourselves share the same sequence of amino acids in haemoglobin. It turns out that humans are genetically closer to chimpanzees than they are to either orang-utans or gorillas; there is less difference than that between a mouse and a rat, or among those common song birds the warblers. This might support a view that humans are the third species of chimpanzee, were it not for the immense differences in behaviour.

Genetic differences arise through random mutation on which natural selection operates at the level of the individual. So, differences in genetic relatedness express in some way the length of time since the evolutionary divergence of the subjects being compared. Potentially such comparisons provide a molecular 'clock' with which to date the important branchings in the evolutionary bush. Such branchings in the past led, in increasing order of antiquity, to surviving local groups, regional populations, global species, genera

and the whole set of hierarchies in the anatomical classification of modern lifeforms. There are many difficulties remaining, which we must gloss over here, before the dates from such clocks become universally accepted. Figure 9.5 shows the timing of separation for the branches that led to modern apes and ourselves. Of course we cannot tell what other important branches there might have been if they led eventually to extinction. All we can find are fossils over which debate may rage interminably, unless, that is, some means of extracting minute amounts of genetic material from bones is developed, and that such material has been preserved. The important point to note is that the divergence of the routes to living humans and living chimps falls around 5 to 7 million years ago. Since we are quite sure that the common ancestor of both lived in trees, there is a long period of 3 to 5 million years before the first humans in which to search for fossil evidence of the earliest

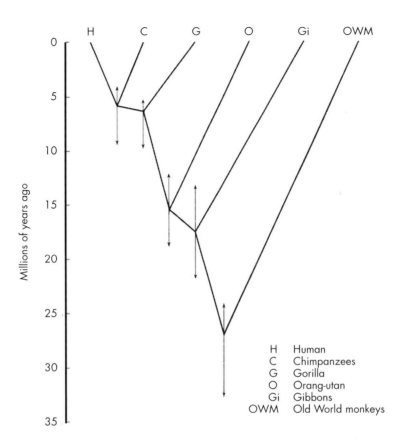

Figure 9.5 Timing of divergences between the principal groups of Old World (Europe, Africa and Asia) primates based on genetic studies of modern groups. The bars represent uncertainties in dating. Other families, genera and species that have become extinct would fill the diagram to make it resemble Figure 9.4, if we knew of them, when they first appeared, and if they had preserved genetic material.

appearance of upright apes and perhaps set up other tentative and lesser evolutionary branching points.

Research on human origins has progressed in such a way that field workers have found pre-human hominid remains in progressively older strata. The period around 2 Ma is particularly rich in such fossils, if rich can be used to describe the source of at most the remains of a few dozen individuals. The first was described by Raymond Dart in 1924 from a site in South Africa.

☐ What features would bear the stamp of bipedalism in a fossil primate?

■ As well as similarities to the angle of connection between the spine and pelvis and the position of the spinal chord's entry to the skull (the *foramen magnum*) in humans, foot bones must show signs of being capable of bearing all the creature's weight.

As upper body mass bears on the feet through the connection between the upper leg bones and the pelvis, the angle involved in that linkage is quite different between humans and other apes (Figure. 9.6). Dart's find and subsequent ones with similar basic morphology going back to 2.5 Ma all show clear evidence for permanent upright gait in the critical characteristics. The stature and body shape of these hominids that just precede tool-using humans, and

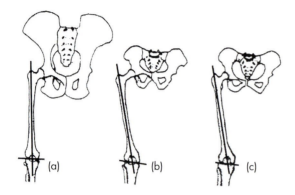

Figure 9.6 Comparisons of pelvis and knee joints in (a) ape, (b) *A. afarensis*, (c) modern human.

which may well overlap them in time, fall into two distinct groups. Both were much the same size, but one had powerful jaws and skull architecture buttressed to carry large chewing muscles, the other having a lighter skull and smaller dentition. Like chimpanzees both had prominent brow ridges connecting fully across the lower forehead, yet their dentition was not so protuberant, so both had flatter faces. For a while controversy raged about whether these 'robusts' and 'graciles' were merely male and female of the same species. Examination of the teeth shows conclusively that the first had the same diet as the paranthropoids, indeed the first discovered is now regarded as the earliest of the genus *Paranthropus*. Members of the gracile group have teeth that indicate a different diet, needing less chewing power, probably dominated by softer, higher grade plants and fruits and possibly some meat. They belong to the famous *Australopithecus africanus* ('Southern Ape of Africa'), which for many years was regarded as the 'missing link'. Yet many are younger than the first tools discovered by later workers in Ethiopia, so they must have been a species separate from those unknown tool makers that lived at 2.5 Ma. The evidence from the period since then suggests at least three lines of upright apes. The earliest paranthropoid (*P. aethiopicus*) occurs in deposits near Lake Turkana in northern Kenya that give a date of 2.6 Ma. Earlier hominids are all of a more gracile type.

Because all previous hominid finds were associated

with ancient lake and river deposits in the southern part of the African Rift, attention spread to the northern Rift in the 1970s where such lakes had formerly existed. At Hadar in the northernmost rift basin of Afar in Ethiopia, a team of American and French palaeontologists led by Don Johanson and Yves Coppen made a discovery that not only pushed back evidence for bipedalism by a million years, but by its near-complete preservation gave a unique insight into many aspects of early hominid form and function. The young researchers, being fans of the Beatles, were blasting out 'Lucy in the Sky with Diamonds' as they painstakingly excavated their 3.5 million years old site. They were soon convinced by the pelvic architecture of the unearthed skeleton that it was a female. 'Lucy' (Figure. 9.7) became a celebrity as well as the first known member of the australopithecine species *A. afarensis*. She retained 40 per cent of her bones, and since information from one side can be used to infer structure on the other, 'Lucy' represented an 80 per cent complete find. Her pelvis-to-leg connection is very close to that in modern humans, her dentition is intermediate between that of chimpanzees and ours, her hands are close to ours but with smaller thumbs relative to other digits. The only major difference that bears on a commitment to life on the ground should lie in her feet: the trouble is, 'Lucy' is missing the critical small bones of her big toe!

Fortunately *A. afarensis* specimens have been unearthed elsewhere in Africa. Most spectacular of all was Mary Leakey's discovery at Laetoli in Tanzania of a track of footprints preserved in volcanic ash reliably dated as being 3.5 million years old. These are of an adult and a young australopithecine walking side by side, clearly upright, and perhaps trudging out of an area left useless by the volcanic deposit (Figure 9.8). Certain features of the tracks possibly show that two adults were present, one following exactly in the footsteps of the other through the glutinous mud. Each print shows weight transmission at the ball and heel of the sole barely different from that of our own. The story does not end there however. Finally ankle and big toe bones of an *A. afarensis*, or at least a creature with geologically the same age, turned up in South

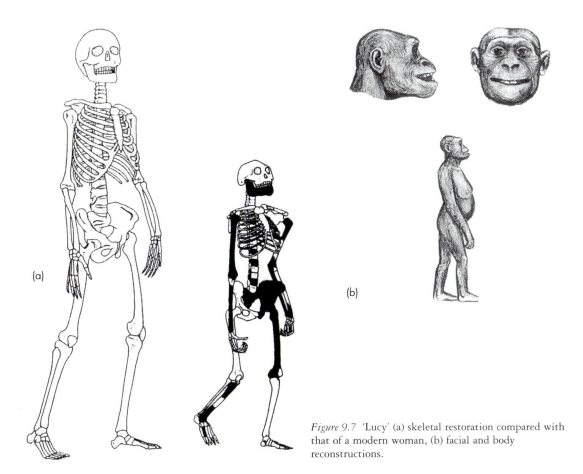

Figure 9.7 'Lucy' (a) skeletal restoration compared with that of a modern woman, (b) facial and body reconstructions.

Africa. These are perfectly articulated, coming from the same individual, and reveal what appears to be the wide separation of big toe from other toes that characterises modern apes' ability to grip branches with their feet. Inevitably, the interpretation that australopithecines also climbed trees as well as walking upright is controversial, but there seems little intuitively unlikely about such a combination – perhaps they sought trees for safety, particularly for sleeping. Whatever, *A. afarensis* walked rather than scampered. Not only did the species migrate down the Rift valley to southern Africa, but remains crop up in Chad, 2 000 km west of the Rift. So *A. afarensis* had an extremely wide geographic range compared with modern great apes. Such a range helps confirm that the species was not confined to a narrow

ecosystem as chimps and gorillas are today. It must have exploited a considerable range of foodstuffs.

The relative wealth of *A. afarensis* fossils, now dated between 3.9 and 2.9 Ma, especially the remains of more than 10 individuals from the same site at Hadar might seem good cause for celebration. That it is more one for a battle royal among specialists results from there being two clearly different sizes of such australopithecines. 'Lucy' and others, which are certainly adult from their dentition, are midgets at barely 1 metre stature, whereas others reached heights of 1.5 metres. One school (as it happens 'Lucy's' discoverers) holds that this represents sexual dimorphism ('Lucy' supposedly being the better, if more diminutive half!) and that *A. afarensis* is on the direct line of descent to humans. The other is

(a)

(b)

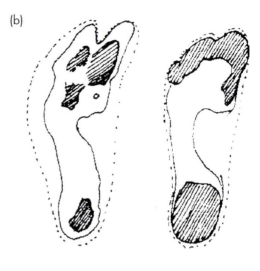

Figure 9.8 (a) Footprints dated to 3.5 million years, probably of *A. afarensis*, discovered in a volcanic ash bed at Laetoli, Tanzania by Mary Leakey during 1978–9, (b) comparison of a Laetoli footprint with that of a modern human.

convinced that the height differences are too large, and that a pair of species is represented. Comparative studies of the smaller pelvises in 1995 swung the argument to a pair of species, for one of them is probably male, and it is 'Lucy's'! With heated disputes in the public arena over a topic of great interest to many lay people (and buyers of popularised science magazines) a flood of funds poured into palaeoanthropology in the 1980s and 1990s. Not surprisingly older and yet more enigmatic finds were made.

'Weighing in' at an age of 4.2 to 3.9 million years old are teeth, a partial jaw and a leg bone from Lake Turkana in northern Kenya that Maeve Leakey and her co-workers claim to be from an upright australopithecine called *A. anamensis*. They were beaten to press by a year by Tim White's team, who worked in southern Ethiopia. Their first report in the prestigious scientific journal *Nature* claimed 'the long

sought-for potential root species' for the human family in the 4.4 million years old bones that they called *A. ramidus*. Their teeth are like those of apes but their foramen magnum is more forward; an indication of uprightness. However the material is so different from that of other australopithecines that White and company arbitrarily corrected their naming of the beast to *Ardipithecus ramidus* (the ground ape root) seven months later; a new genus! At the time of writing all is not such a welter of confusion as you might surmise. There are two basic models for the descent of humans after divergence of the lines ending with modern chimps and us. One is a simple linear model linking all the early finds directly to each other, with a late divergence of australopithecines into paranthropoids and humans. The other is best regarded as fragments of an evolutionary bush, in reality with complex branchings, most of which are not represented by known fossils (Figure 9.9). In this model every species found so far lies on separate branches, maybe with the real human pathway having sneaked through time without detection! It is quite possible that soon after the chimp–bipedal ape divergence the latter entered completely new African habitats. Novel opportunities, and perhaps threats, drove rapid radiation of new, but short-lived species into a variety of 'experimental' ecological niches. We return to this aspect of evolution, the role of the environment, in Chapter 10. Now is the point at which to examine another kind of explosion among early humans; that to new lands.

CULTURE AND HUMAN WAVES

Evidence from Asia

Archaeologists from China, the US, Canada and Holland excavated a cave floor at Longgupo in central China during the early 1990s. It had previously

yielded old human remains. On reaching the deepest levels of accumulated debris they began to find tools of the most primitive kind; Oldowan tools in fact. In the same place scientists also excavated the left side of a human mandible and one upper incisor. Meagre pickings for most fossil studies, but painstaking taxonomic work on the teeth reveals similarities to those of African *H. erectus*, with the caveat that they are more primitive. They are most like those of 'Action Man', *H. ergaster*. Dating such remains and the strata that contain them is no easy matter, and usually depends on several methods. Contradictory results generally raise doubts about the usefulness of any of the dates. In this case, all methods support one another. The most convincing is that based on the magnetic polarity preserved in the strata at the time of their deposition. In the deepest levels at Longgupo the remanent magnetism is the same as that today. Although there have been several such global periods of normal magnetic field in the last few million years, the reversed field that characterises the overlying strata here is likely to be the one that occurred

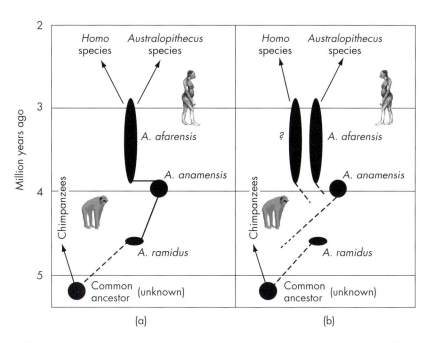

Figure 9.9 Models for pre-human hominid evolution (a) linear or 'ladder' model, (b) 'bush' model.

between 1.68 and 0.93 Ma. Fossils of other animals within them are too primitive to link to the last magnetic reversal between 0.87 and 0.69 Ma (Figure 5.10). On this logic the human remains at Longgupo occur in the same period of normal magnetisation that affected the primitive tool-bearing sediments in Tanzania, which spanned the period between 1.85 and 1.68 Ma. That itself was precisely determined from studies of the abundant volcanic rocks in the Rift there, and is called the Olduvai period.

Early humans carrying the oldest tool-using culture reached central China at least 1.68 million years ago, and perhaps at the same time as the date of their oldest remains in Africa, around 1.9 million years ago. You should remember that such dates have errors of the order of tens to hundreds of thousand years, so coincidences do not mean instantaneous shifts in population. Nevertheless, Longgupo represents a migration of over 10 000 km in at most 200 thousand years. This amounts to an average of only 50 metres per year, and the fastest movement that might remotely be supported by the evidence would be only one kilometre per year. This was not some purposeful shift of population as might be encountered with nomads of recent times or the herd followers that crossed from Asia to North America around 20 thousand years ago. It is best regarded as a sort of steady diffusion, but the first out of Africa. It requires little imagination to envisage the changes in climate, terrain, vegetation and animal life that these early humans encountered during their wanderings, corresponding to around 40 degrees of latitude.

The 'colonisation' of east Asia seems to have been permanent. Fascination with early human remains in Asia began in 1891, when Eugene Dubois discovered a single skull cap at Trinil in Java. Its heavy brow ridges, yet large cranial capacity provided the first convincing evidence of brainy beings that predated modern humans. The earlier discovery (1856) in Europe of what we now know as Neanderthals was greeted with scepticism, and many naturalists regarded them as remains of Russian soldiers who chased Napoleon's army from the gates of Moscow in 1805! Dubois called his discovery *Pithecanthropus erectus*, the 'Erect Apeman', also known as 'Java Man'.

Within 20 years similar remains were found in the Zhoukoudian cave near Beijing – 'Peking Man', and now there is a wealth of such bones from China and Indonesia. All have been reclassified as human subspecies of *H. erectus*, the earliest of whom are now known to have lived in East Africa from about 1.9 million years ago.

The erects, as it is convenient to call them, have as large a fossil record as all preceding human and hominid species. Consequently, we can be much more confident in assessing their build and features. They had brains twice the size of modern apes (around a full litre) and two thirds that of ours. They reached almost two metres in height and weighed up to 70 kg (11 stone, or light middleweight), with a slender build. Apart from having an extra lumbar vertebra, an erect's skeleton is hard to distinguish from our own without seeing the skull. (Incidentally, the canal that carried their spinal nerves is narrower than ours in the region of the rib cage, a feature that some authorities have used to imply that erects did not have the fine control over breathing that we need for fully-formed speech.) Erects' skulls are very different from ours (Figure 9.10a). They have thick walls reinforced by top, back and side bony ridges. Their eye sockets are dominated by glowering brow ridges (Figure 9.10b), above which there is a barely noticeable forehead. While their teeth are quite similar to ours, their jaws jut forward, and the heavy lower jaw has no chin. Their faces (Figure 9.10b) would worry most of us, if we encountered an erect, even in broad daylight.

Several fascinating insights emerge from a century of work on the record of erects from Asia. With the Longgupo results, occupation by erects spanned the period from 1.9 million to 100 thousand years ago: 1.8 million years! Cohabiting the region were huge herbivorous relatives of modern orang-utans, the *Gigantopithecus* whose commonly found teeth figure in Chinese medicine. (Rumours that such creatures still inhabit remote temperate forest in China spurred a group of Chinese gourmets in 1996 to mount a major expedition in search of yet another menu item!) How odd that a direct parallel existed with the lengthy human–paranthropoid coexistence in Africa. Nowhere

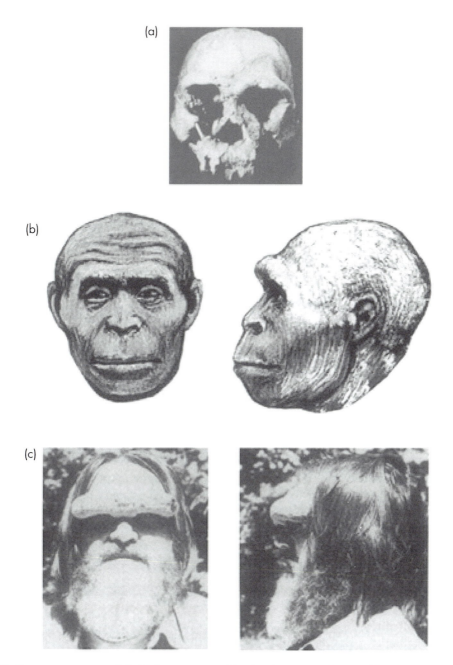

Figure 9.10 Homo erectus: (a) typical skull, (b) reconstructed head and (c) Grover Krantz of Washington State University wearing his *H. erectus* brow ridges to demonstrate their combined effect as anti-glare devices and to keep the face free of hair!

in east Asia have archaeologists discovered stone tools other than those of Oldowan type from this long period. Because African culture soon made a decisive and easily recognised shift after this first intercontinental diffusion, Asia was not a destination of choice for Africans thereafter. That there was not some return by Asian erects is by no means so certain. There is not a single shred of evidence that Europe was colonised by any human or hominid before 1.5 Ma, despite its geographically closer proximity to East Africa. Remains at Dmanisi in Georgia show that erects were at the gates of Europe by at least 1.6 Ma, but the shin bone and single tooth found at Boxgrove in Sussex, the earliest human remains in Europe, appear to be only half a million years old. So called 'Boxgrove Man' was not an erect but a more advanced human.

An intriguing possibility emerges from East Africa. Remains bearing a strong resemblance to the Asian erects enter the fossil record at between 1.9 and 1.8 Ma; they form part of the range of morphologies separated by some into erects, habilines and 'Action Man'. After this time erect-like fossils are the only human remains in Africa for the next million years. Were they the descendants of 'returnees from Asia'? This is one of the great enigmas of human prehistory and palaeoanthropologists class all humans from this period as subspecies of *H. erectus*. Whatever the answer, the African population's greatest invention – the hand-axe – proves that none returned to east Asia, for its distinctive form is never found there.

Acheulean Culture

At around 1.6 Ma the human tool kit received an additional item in Tanzania. It is a pear-shaped, deftly worked hand axe with cutting edges along both edges of its sharp end (Figure 9.11). Fitting perfectly in the hand, the bi-face Acheulean hand axe (after the village of St Acheul in France, where examples were first found in very much younger strata) was struck from the core of large pieces of suitable rock. The waste flakes from flint, volcanic glass or fine-grained lavas and quartzite are razor sharp, and would have found many other uses as makeshift knives, scrapers and

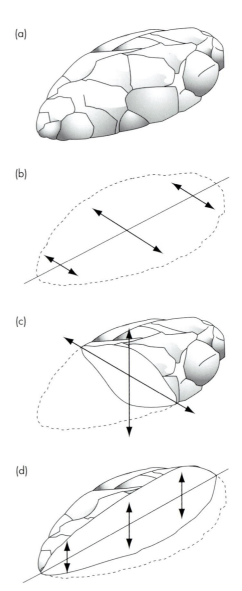

Figure 9.11 Basic design of an Acheulean bi-face or hand axe, showing its threefold symmetry.

borers. Whoever first made such an axe must have been capable of visualising the tool *within* the raw material, in much the same way as a modern sculptor imagines a subject to be crafted. Neither you nor I have the skill to make such a beautiful object, for it takes many years of patient practice to learn the method unaided by modern technology. It involves striking flakes more or less symmetrically from either side of a block of stone, perhaps using a bone or wooden baton, and delicate retouching to maintain its cutting edges. We can imagine a whole range of uses for the axe, a true multi-purpose tool of great value to its possessor. Many bear the wear marks of cutting wood, and there can be little doubt that one of its uses was to make other tools. It was the first machine-tool, albeit very different from a lathe. Another thing seems equally as likely. Unless its owner leapt onto the backs of fleeing prey, it could hardly have been a hunting weapon. A product of perhaps weeks of labour would not be idly thrown; or would it? William Calvin of the University of Washington has engaged the services of discus throwers to see what happens with a thrown bi-face axe. Many turn out to be aerodynamic, like a discus or frisbee. A heavy razor-edged frisbee would be a formidable hunting weapon.

The bi-face axe marks a leap in human consciousness to an appreciation of the hidden world of potential uses within natural materials; to conscious abstraction. Its uses must have produced a leap in its maker's ability to make use of the rest of the natural world – wooden tools and weapons, skins for clothing and shelter. While the first undoubted evidence for controlled fire comes from a charcoal-bearing stone hearth in a cave at Ascale in southern France, dated at about 1 Ma, being so common as a natural phenomenon it seems hardly likely that people capable of making hand axes would not have harnessed fire at a much earlier time. Despite the erects' dauntingly primitive appearance to us, they had a basic culture no different from the bare necessities of modern gatherer-hunters nor from those of our own direct ancestors a mere 100 thousand years ago. The Acheulean culture was the mainstay of human life for a million and a half years. As you will learn, it was a culture passed on to what many palaeoanthropologists regard as several succeeding human species, including our own. In archaeological terms, the bi-face axe is the 'signature' of the Lower Stone Age or Palaeolithic. You might well note that this term was defined in terms of the tools, but was originally

applied only to fully modern humans. It came as a great surprise to *Homo sapiens sapiens* oriented archaeologists when the progress of research revealed it as a thread linking us to those people regarded previously as little better than beasts.

Acheulean culture did not spread like proverbial lightning. Not until 1.4 Ma is there evidence to show that it superseded the Oldowan throughout Africa. Of course, this might reflect the great intrinsic value of the bi-face axe; few would be discarded and most finds could be assumed to have been mislaid. Oldowan tools are easily made, and the possessor of a magnificent hand axe could casually make use of them in the same way as a disposable razor! For a million years the African erects and their culture thrived, diffusing outwards to Palestine, the Indian sub-continent (but never to east Asia) and to Europe. Despite the fact that the world's most avid collectors of curios and the seat of anthropology and archaeology's beginnings reside in Europe, Acheulean tools and fossils of erects dated definitely before about 500 ka are unknown, although fires of undoubted human origin date to 1 Ma in Europe. The most spectacular evidence for the first European culture, though sadly without human fossils, comes from sites around 400 thousand years old on the Castillian Plateau of Spain and the Mediterranean coast of south-west France.

In the steep-sided headwaters of the River Ebro, at Torralba, is a charnel house with the remains of at least seventy elephants. Nearby are remains of twenty living sites with circular heaps of long bones, doubtless involved in some kind of shelters. A sharpened tusk may have acted as a support for a skin tent. As well as elephants there are remains of substantial numbers of rhinoceroses, deer, horses and extinct oxen known as aurochs. The site is swampy, and large numbers of charred branches and twigs throughout the killing ground suggest that large, herding animals were driven by burning brands into the swamp to become bogged down. The excavators found no pointed stone weapons, but some suggested that the unusual numbers of round, weighty stones at the site may have been used to bludgeon the panicked animals to death. The sheer size of the deposit, together with the large number of dwellings suggest either seasonal use of the unique terrain, or a single orgy of feasting by more than a hundred erects. Elephant skulls do not occur there, with the obvious implication that they were carried off elsewhere, either for their tasty brains to be eaten at leisure or as trophies. The coastal site in France is rather more homely, with middens of both large mammals and shellfish. Near Nice is a group of post holes, hearths and aligned stones marking out oval huts up to five by seven metres in size. Imported flat blocks of limestone may well have served as seats. Among the remains are a wooden bowl and pieces of red ochre trimmed into pencil-like points. Other Spanish sites of the same period yield cleavers, bi-face axes, saw-like blades and stone carving tools, bone and ivory spatulas, scoops, blades and scrapers. Most significant are hacked, hollowed and polished pieces of wood, some of which are likely to be remains of spears. Evidence such as this begins to show that their makers were not dissimilar from modern humans lacking modern technology in their relationship to nature. Native American sites less than 20 thousand years old show similar characteristics. The pre-*H. sapiens sapiens* developments have some time to go, however.

THE LAST 500 THOUSAND YEARS

Fossils of late erects from Africa show increasing signs of polymorphism, but still with the characteristic heavy-boned skull, full brow ridge, forward jutting jaw line and lack of a chin. Around 500 to 300 thousand years ago the geological record witnesses a marked change in human skulls. Still robust by present standards, they show a higher cranium and a jump in brain capacity, less protruding face, smaller teeth and chins! Heavy brows there are, but distinctly separated and more rounded (Figure 9.12). Because polymorphism is abundantly clear, and there are still insufficient remains to decisively state that here is a new human species, there is little agreement other than a change took place over a protracted period. Some authorities call these fossils 'late' erects, others favour 'archaic' *H. sapiens*, but a more acceptable designation is perhaps some kind of transition. Two fos-

sils are famous. Heidelberg in Germany provided a jaw of this type in the early days of human palaeontology, together with the first African representative from Broken Hill or Kabwe in modern Zambia. Today the term Heidelberg or Kabwean 'grade' is favoured by the Kenyan ecologist turned palaeoanthropologist Jonathan Kingdon. To avoid the confusion engendered by raging disputes among more formal scientists we can use Kingdon's last term until the issue is settled. Kabweans occur in Europe and Asia too. The famous Swanscombe skull and the Boxgrove remains in Sussex are likely Kabweans, and they represent yet another north and westward migration from the African heartland. In the period up to 100 ka Kabweans (locally called Mapas) finally brought Acheulean culture to East Asia, where they cohabited with the long-isolated erects who still survived over large areas. It was the first discovery of Kabweans with erects in Asia, together with polymorphism among all erects that prompted Milford Wolpoff to propose several lines of descent for fully modern humans – his multi-regional hypothesis for the origin of human 'races'. More on this later. At 100 ka the erects finally disappeared from their last stronghold in east Asia.

Kabweans were inventive. Beginning around 250 ka, their remains in Africa occur with a new type of tool made in an entirely new way. Instead of the massive bi-face axe fashioned from large pieces of rock, Kabweans turned their attention to developing flakes broken from these cores. The product is a set of tools based on large, blade-like flakes prised from larger cores. They are light, razor sharp and easily trimmed into a multitude of shapes, including points that can be attached to wooden hafts as spears. The first appearance of flake and blade industries mark the beginnings of the archaeologists' Middle and Upper Palaeolithic. More importantly they opened up the possibility of projectile weapons and true hunting, instead of trapping and herding prey over cliffs and into bogs. Following this cultural breakthrough, fossil humans in Africa show a steady reduction in the bone mass of the skull, a general increase in skeletal lightness, steadily smaller teeth and flatter faces with emphatically jutting chins. By now you should have

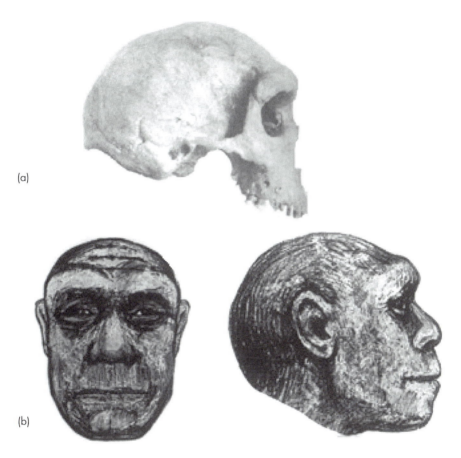

Figure 9.12 Kabweans: (a) The 'Broken Hill' skull, found at Kabwe, Zambia in 1921, (b) facial reconstruction of (a).

realised that the term 'chinless wonder' is quite misplaced as a pejorative for the dim and genetically disadvantaged; fully modern humans had arrived on the African scene.

The human picture in Europe, Central Asia as far east as Tadzhikistan and the Middle East (Figure 9.13) for the period after 300 ka is different. Neanderthals were the only occupants, their undisputed remains appearing in the fossil record at about 200 ka and possible Kabwean ancestors at 300 ka. In the period following scientists' recognition in the early twentieth century that their remains were indeed much older and different from those of modern humans, Neanderthals suffered a 'bad press'. Partly this stemmed from one of the best preserved speci-

mens of the time being an old man with badly deformed spine and limbs, partly from the low brow, prognathous face and small chin common to all Neanderthals. Until very recently our immediate predecessors in Europe were regarded as shambling brutes, the epitome of the 'caveman'. Indeed, Neanderthal's were very different from us in several respects. Their skulls retain the robustness of the Kabweans with rounded brows but extremely enlarged nasal passages; Neanderthals had enormous noses. Crania are flat and long (Figure 9.14), yet brain capacity is often well in excess of that for modern humans. Much of the brain expansion from their presumed Kabwean ancestors occupied the rear of the cerebral cortex, the parietal and occipital lobes. In all

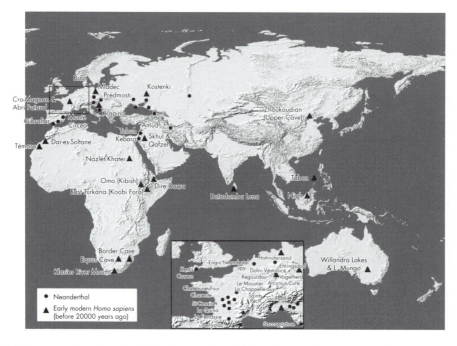

Figure 9.13 The range of the Neanderthals in Europe after 300 ka, showing the main archaeological sites.

higher mammals visual processing goes on in the occipital lobe. The parietal lobe is associated in us with information storage, language, learning and memory. It is a key region for intelligence. There seems no reason whatever to regard Neanderthals as dim; quite the contrary. Although bodies were within the same range of stature as our own, the robustness of Neanderthal's limb bones with massive sites for muscle attachment indicate enormous bodily strength. Muscles seem to have been so powerful that their use produced several kinds of bone injury and distortion. Many adult specimens show signs of arthritic deterioration.

Neanderthals were well equipped with all manner of late Palaeolithic tools and probably hafted weapons. They undoubtedly wore clothing, made shelters, hunted, used fire and famously lived in caves, where available. Arguably, finds of deliberate burial of their dead, in some cases with remains of flowers and personal ornaments, indicate some form of ritual and belief system. Their occupancy of Europe and west and central Asia went unchallenged for 200

thousand years, so they were extremely successful, but it ended suddenly. After 36 ka there is not a single trace of the Neanderthals anywhere on the planet. Palaeoanthropologists are at a loss to explain this disappearance.

Anatomically modern humans and Neanderthals overlapped in their ranges in the Middle East at around 100 ka, but the poor precision of dating the remains permits back and forth movements of both, perhaps with no direct contact. From 40 ka, there was definitely co-occupation by modern humans and Neanderthals in France. The earliest modern humans did not gain a foothold by virtue of a more advanced technology. Their tools in Europe are functionally little more advanced than those developed by the late Neanderthals, although some scientists have suggested that the latter were copied or traded from fully modern sources. An overlap of at least four thousand years is certain, so the disappearance of the Neanderthals is highly unlikely to have been genocidal. Neither is there evidence for interfertility between the two groups and genetic 'swallowing', no 'hybrid'

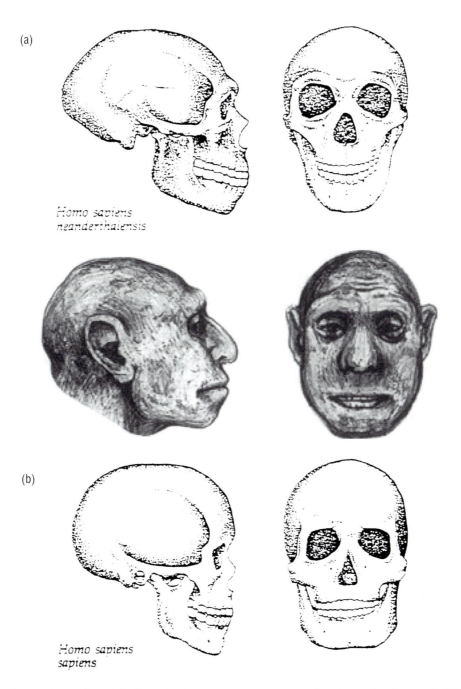

(a)

Homo sapiens
neanderthalensis

(b)

Homo sapiens
sapiens

Figure 9.14 Typical Neanderthal skull and facial reconstruction (a) compared with a modern human skull (b).

fossils having been found. Distribution of modern and Neanderthal sites gives a vague indication that the latter lived in upland areas whereas moderns occupied lowlands, perhaps restricting contact to the extent of the rural–urban division that characterised Europe in the Middle Ages and even in some areas of the world into the twentieth century.

The relationship between anatomically modern humans and earlier migrants to the Far East is undocumented, yet the record is extremely clear as regards the way moderns spread their influence. From a foothold in mainland Asia at around 60 ka they progressively colonised the Archipelagos of the West Pacific, including New Guinea at least 60 ka. By 40 ka at the latest, moderns entered Australia, and by 20 ka they had crossed the Bering Straits to the Americas to reach South America within a few thousand years. The humanisation of the remainder of the Pacific is well-known to most of us, becoming complete within the last 300 to 500 years.

Despite the clear anatomical differences between modern humans and Neanderthals, we have no conclusive evidence of significant cultural differences between the two groups during the period of their cohabitation of southern Europe. Following the disappearance of the Neanderthals modern human culture in Europe exploded. The more or less common stone tool kit expanded to include delicately crafted blades and points, some of which could be for arrows or harpoons, and a host of so-called microliths including chips for sickles, wood- and bone-working tools. The Eurocentric view of modern humans' cultural revolution took a plunge in 1996, when excavations in Zaire of sites more than 70 thousand years old unearthed equally sophisticated tools in abundance. Their appearance 40 thousand years later in Europe may well signify import of technology from Africa.

Within a thousand generations at most, human culture had domesticated animals, begun cultivation of a wide range of crops, started the smelting of metals in Africa, Europe and Asia, and built the first city states within which society was for the first time divided into wealth owners and wealth producers. In the Bronze Age of the Tigris–Euphrates valley history begins with the first writing on clay tablets. Much

the most astonishing outcome of the human cultural explosion is the appearance almost simultaneously in Africa, Europe and Australia at 30 to 40 thousand years ago of art; arguably the first representative art yet of the most exquisite quality and accuracy (Figure 9.15a). By 12 ka we see for the first time actual portraits of humans made by themselves (Figure 9.15b). How this sudden burst of 'talent' emerged will be difficult to judge, but where its executers came from can now be tied down with a satisfying level of certainty. The source of this confidence stems from the fact that living human beings carry the evidence in their cells as their gene sequences.

OUR AFRICAN ANCESTRY

Molecular biologists have revolutionised knowledge of the source of all living people through examining differences in gene sequences between individuals from widely separated populations. The most important approach has been to compare nucleotide sequences in the DNA of cell mitochondria (mtDNA) which are passed on exclusively in the female line. The global variability of these sequences is no more than one tenth of 1 per cent. Of this, the greatest proportion is that between different African populations and between them and non-Africans. It is possible to calibrate the rate of genetic change by comparing sequences from populations whose physical separation is well known from accurately dated geological evidence. Good examples are those provided by the geographic separation of the first Australians from people in New Guinea at about 50 ka, and over the last 20 thousand years by the successive spread of people throughout the Pacific. Such a calibration allows the timing of other genetic differences to be dated by a sort of molecular 'clock', and living populations to be linked by a genealogical tree (Figure 9.16). Interestingly, comparing the relatedness of vocabularies and grammars between different languages in different parts of the world produces a 'tree' that bears close resemblance to that provided by mtDNA. One of the outcomes of the mtDNA method is that the 'root' of the tree, the source of the

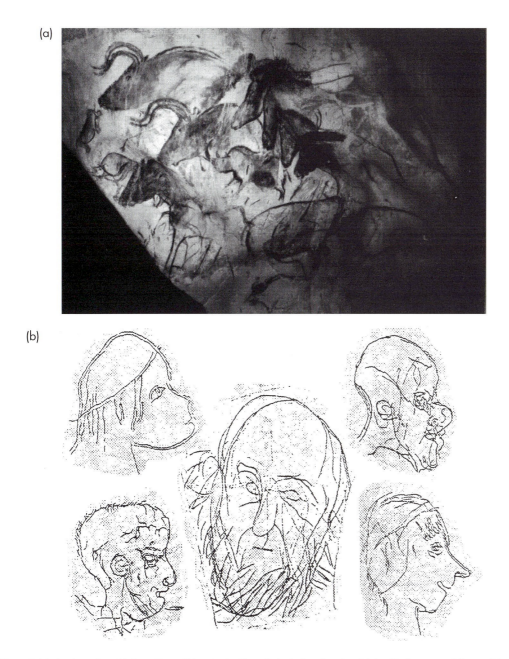

Figure 9.15 Ancient Art: (a) from Pont d'Arc in the French Ardeche, dated at 31 ka (b) faces traced from a 'sketchbook' on the walls of the La Marche cave in France.

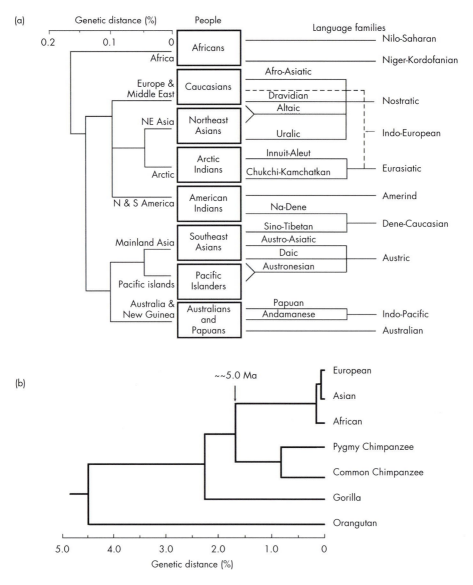

Figure 9.16 (a) Relatedness of modern people based on mtDNA analyses (left), matched to that of associated language groups (right), (b) genetic differences based on rRNA among primates.

genetic divergence among all living people, began at between 150 to 250 ka, starting with a single African female ancestor. This does not mean that this ancestral mother was a solitary 'African Eve', the first modern human, but that all other women living at the same time did not begin a line of descendants that survived to the present. It provides a minimum date for the appearance of the truly modern humans.

Several corollaries of mtDNA studies help resolve some major disputes. First, any interfertility between moderns and Kabweans or Neanderthals is ruled out, since the surviving genetic imprint of such liaisons with more ancient lines would produce clear anomalies in the data. None have emerged so far. Second, by the same logic, had modern humans in Asia, Europe and Africa evolved separately from the earlier inhabitants of those regions, either erects or Kabweans, the genetic patterns in each area would show clearly separate groupings. The fact that there is a continuum among the living individuals who were tested is powerful evidence against Milford Wolpoff's 'multiregional evolution' hypothesis and any attempt to assign different 'racial' characters, such as intelligence, which some people miguidedly read into Wolpoff's hypothesis. The third strong implication, supported by linguistic studies, is that whenever migrants left Africa, they carried language. Before the molecular evidence, many palaeoanthropologists believed that language appeared at the same time as art, only 30 thousand years ago, another distinctly Eurocentric view!

The first modern humans now seem likely to have been speaking as they moved out from their African source. This opens up the distinct possibility that language ability is extremely ancient. Parts of the brain used in language, Broca's and Wernicke's areas, leave discernible imprints on the inner skull wall. These have been seen on the skulls of *Homo erectus* and *H. habilis*. Their presence is not absolute proof, for similar but unused features occur in some living apes. Moreover, one may have the wit but not the means for speech, which is dependent upon the structure of the larynx and the tongue. Such is the case among chimps, though whether they have either the wit, or anything astonishing to say outside of 'Give Koko orange, orange Koko give, Koko orange give' etc., that they sign, seems increasingly unlikely after early grandiose claims by a few earnest primatologists. The search is on for small throat bones and subtle structures in earlier human fossils. Some specialists claim from detailed anatomical studies that Neanderthals must have had problems with vowels. The respected psycholinguist, Steven Pinker, has retorted, 'E lenguege weth e smell nember ef vewels cen remeen quete expresseve'!

Issues such as the origin of language and art are entirely beyond the context of this book, as are those relating to recorded history. Having covered what may seem to be a detailed account of human evolution in terms of shifts and drifts in past forms, functions and 'second nature' (in fact extremely abbreviated and necessarily biased), we now examine the links with the drastic fluctuations in the environment within which they took place.

SUMMARY

1 The essence of what makes humans is their use of tools, and all that means in terms of consciousness. More than that, tool use is bound up with a social form of living; not gregariousness as in some other primate species, but a continuous sharing of both labour and its products. Society and its culture, of which tools are the quintessential part, create a unique 'second nature' from the rest of the natural world. This acts not only as a buffer that enabled humans to populate all corners of the globe, but a transformation of environments within which human biological evolution takes place. Fitness, or the ability to survive and reproduce fit progeny within this context is the only valid way to approach human evolution and the various physiological changes that we find in the fossil record of our forbears.

2 Two criteria help point to the human line in the fossil record of primates, the presence of tools and evidence in skeletal remains of an upright gait. The first confirms humanity's presence, whether tools occur with bones or not, the second only that such and such a fossil primate is human-like or hominid. Any fossil primate that we can demonstrate to have made and used tools, and to have passed on the requisite skills can only be regarded as human, however 'primitive' it might seem. The earliest definite tools, simple faceted objects of rock capable of taking a sharp, durable edge, date from 2.5 Ma in the Hadar area of eastern Ethiopia, but there are no associated fossils. The tool–user link only appears at about 1.9 Ma with remains of *Homo habilis*/*H. ergaster*.

3 Molecular evidence supports the idea that the

human line emerged from a 'bush' of evolutionary diversification, which rooted in a division from another sequence of descent leading to modern chimpanzees. Assuming a molecular clock gives an estimate of 5 to 7 Ma for this divergence. The earliest evidence for at least an occasionally bipedal primate, an unusually forward position for the *foramen magnum* in a fossil cranium, dates back to 4.4 Ma in *Ardipithecus ramidus* from southern Ethiopia. In the succeeding 2 million years a variety of upright hominids scatter the record, several species seeming to have co-existed in the East African environment at any one time. Tracing the branches of the evolutionary bush is difficult, so scanty are the remains. But two main trends stand out, one adapted to diverse, probably opportunistic dietary habits, the other more specialised to survival exclusively on vegetable foods – the australopithecines and the paranthropoids. The first has a more delicate head anatomy than the last, for which the term 'robust' is no overstatement. Provisionally, until further finds are made, most workers assume that the tool users diverged from the 'gracile' australopithecines sometime before 2.5 Ma.

4 The first 'handy' humans shared their environment with closely related paranthropoids, each having a quite separate ecological niche. At that stage, there is no question that one or the other was more successful, for paranthropoids survived for another million years, although only in Africa. Shortly after the tool–human link can be pinned down, human fossils and the earliest stone tool kit turn up in China, in cave deposits at least 1.6 million years old, and possibly 1.85 million years old. 'Second nature' had permitted a species adapted to tropical grasslands to migrate across two continents and survive in much cooler climes with a totally different range of foodstuffs. This migrated population seems not to have been followed until 200 ka, and evolved successfully for more than 1.5 million years as what became known as *H. erectus* from finds made in Java in the nineteenth century. Its stone tools did not evolve, remaining as primitive as when first invented, though we have no record of the uses of less durable materials.

5 Anatomically similar to the Asian erects, those of East Africa made a tremendous breakthrough in culture at about 1.6 Ma. Someone visualised a more useful, indeed multi-purpose tool *within* raw stone that could take an edge. The bi-face Acheulean tool, usually referred to as a hand axe but whose purpose we have not yet divined, with a wide range of specialised flakes formed the core of a tool-kit that served humans until 250 ka. Like the earlier, Oldowan tools, the Acheulean tool kit passed from one species to the next and to our own, and signified at least a hardware aspect of culture that stagnated for immensely long periods. Only early *Homo sapiens*, in the form of the so-called Kabwean morphological group, made a decisive advance at 250 ka. Instead of exhuming useful stoneware visualised within rock, the Kabweans and their successors focused on developing delicate, sharp flakes prised from the outsides of such cores. Development of considerable skill and ingenuity with flake tools led in rapid succession, geologically speaking, to the multiplicity of specialised, hunting, fishing and leather working implements with which we are most familiar in museum displays. Far from the old Europe-centred view of human industrial development, each phase of stone-tool innovation emerged first in Africa. Those from fully modern human sites in Europe after 30 thousand years ago, that were long regarded as evidence for an explosive development in culture there, are matched by recent finds from Zaire that are 70 thousand years old.

6 Despite evidence for stagnation in part of human culture, the evidence from fossils clearly shows that physiological development was not in stasis. Through the crude sequence, erects to Kabweans to fully modern humans, we see little change in overall body plan. Changes concentrated on the head. Skulls become increasingly less robust, lighter and with larger cranial capacity. Jaws follow the same trend to lightness, robustness of skull being closely related to jaw musculature, which must seat on the upper head. Clearly chewing power became less favoured by selection than growing braininess, for which skull flexibility and

therefore lightness is *de rigeur*. Brow ridges went out, chins and foreheads came in!

7 One unique characteristic of human remains, despite their comparative rarity, is the great variation between specimens, particularly as regards skulls. This is as noticeable in collections from single or closely situated sites as it is in the global population for any age. It causes great confusion, where some use morphological differences to split species and subspecies, whereas others regard the diversity as polymorphism among very close genetic relations within species. An extreme case of the first approach is the notion of multiregional human evolution from diverse types of erects across the Old World, that must date back more than a million years. The molecular evidence from all living human populations rejects that view decisively, for we are all so closely related that any evolution-ary divergence must have started only 150 to 200 thousand years ago, almost certainly in Africa. Polymorphism and rapid evolution are characteristic of two basic types of population: small ones in isolation, as on islands; those somehow sheltered from the full force of environmental selection pressure. In the last, otherwise deleterious genetic traits that might be extinguished by conferring poor fitness are allowed to pass on. As well as encouraging intra-species diversity, it preserves features that later evolution might build on with decisive advantage.

FURTHER READING

See list for Chapter 10 on p. 235.
See also references to figure sources for Chapters 9 and 10 (pp. 258–9).

10

CHANGING ENVIRONMENTS AND CHANGING HABITS

AFRICA'S GREAT RIFT

The story of human origins summarised in the last chapter is clearly that of 'Out of Africa'. Why was that continent humanity's nursery?

Africa today has a more varied range of ecosystems than any other continent (Figure 1.4). Spanning 70 degrees of latitude, 35 either side of the Equator, it encloses a more or less symmetrical set of climatic zones and their associations of natural vegetation. At least, that is the impression given by small-scale ecological maps in school atlases. In truth it is the most ecologically varied on *any* scale of all continents. Figure 10.1 reveals one reason why this is so. Despite much of it being a low dome rising imperceptibly from its coastline to regional swells 1 to 1.5 km high, sudden changes in topography interrupt Africa's general shape. Superimposed upon it are several large upland areas and intervening basins. The most striking feature of all is a huge cleft with high flanks that runs in a curved and branching form from Mozambique to the Red Sea and terminates at the Gulf of Suez. This East African Rift system sharply divides the continent into two regional climate systems.

West Africa's climatic zonation is predictable from the global pattern of air movements conditioned by the tropical convection system and the Coriolis Effect (Figure 1.2) Mediterranean–desert–savannah–tropical rain forest–savannah–desert–Mediterranean along its north–south axis. East Africa, in contrast, has strongly seasonal climates, particularly as regards precipitation. In the northern part there is a monsoon

system related to that of south Asia. It has its own dynamics brought about by the Ethiopian Plateau and the roughly north–south line of topographic highs along the Rift. Its vegetation zones are far more complex than those of West Africa. Moreover, they are conditioned by the superimposed effects of sharp and irregular variations in elevation. As well as creating cool and even frigid climatic 'islands' in the tropics, such as the glacially capped Mounts Kenya and Kilimanjaro that top 5 km, the mountains encourage rainfall by their orographic effects. Water flowing

Figure 10.1 Shaded relief map of Africa.

from these elevated 'condensers' feeds semi-permanent rivers and fills a necklace of large and small lakes along the Rift. Lower elevations have a cover of mixed grasses, thorn scrub and isolated large trees to give steppes, scrubland and savannah. Lacing through them are riverine and lake-shore forests, with swamps and occasional more arid areas. Each mountain has an upward zonation through dense grassland, mountain forest and peculiar mist forests with strange plants such as giant lobelias (compare with Figure 8.12b). On the highest ground is tundra, not greatly different from that encountered at lower elevations in upland Britain.

East Africa is an unparalleled mosaic of natural ecosystems. Many of them form islands of vegetation diversity separated from others of similar kind by hundreds of kilometres of the more pervasive steppe, scrub and savannah. Although the tropical rain forests of the Amazon and Congo basins form the greatest repositories of plant, insect and small vertebrate life on the planet, East Africa hosts by far the greatest number of large land-vertebrate species. The tropical rain forests of Africa and South America have probably existed for over 50 million years, but the ecological mosaic of East Africa is geologically very young.

Throughout tropical Africa occur areas dominated by brilliant red soils. Today such iron-rich soils, or laterites, are still forming beneath the rain forests of the vast, low-lying Congo river basin. But similar soils exist at very much higher elevations too, where they are now being eroded away. Such laterites formed in the geological past; they are palaeosols. The youngest of them formed between 30 and 60 million years ago in the early Tertiary. Laterites are notoriously difficult to date as they contain neither fossils nor radio-isotope-bearing minerals that formed when they did. Their inferred age span is simply that between the youngest rock on which they sit and the oldest which rests upon them. Besides their dramatic contribution to scenery, the palaeosols provide us with a starting point for examining the relationship between human evolution and that of ecosystems and the environment in general. By analogy with similar soils that are forming now, the old laterites represent

highly corrosive chemical processes beneath tropical rain forests that existed during the early Tertiary. Various sedimentary features within them, such as relics of river channels, show that they formed on a huge plain where basins were separated from local watersheds by only a few metres. This was a plain that covered all of northern Africa at least, and which sloped gently to the surrounding oceans. In Arabia these laterites pass into marine sediments of around 50 to 60 million years old.

So the red palaeosols are a proxy record of dense forests that, from their now isolated but widespread occurrences, stretched unbroken from the modern Congo basin to Arabia. The range of elevations of this huge jungle was probably no greater than 500 m. Today the distinct, once horizontally continuous red rocks occur as high as 3 km above sea level in Arabia, Eritrea and Ethiopia. In that area they are covered by hundreds of metres of basaltic lavas, erupted between 30 and 18 million years ago, which themselves rise to as high as 4.6 km. They represent the youngest major episode of continental flood magmatism. Wherever erosion has exposed the base of these basalts the laterites are always present.

☐ What can you conclude from this observation that is relevant to tectonic processes?

■ Whatever contributed to their high elevations of the laterites must have occurred less than 18 million years ago – insufficient relief was present before the flood basalts for the laterites to be eroded.

In the area flanking the Red Sea sedimentary rocks were not deposited again until around 13 Ma. The lowest of them contain cobbles of the flood basalts, which indicate that the area at last had become sufficiently rugged for high-energy erosional processes to begin. A similar, but not so clear picture emerges from other areas in East Africa. The onset of forces responsible for the unique and ecologically influential topography of eastern Africa therefore began between 18 and 13 Ma. The source of these driving influences is the Earth's last superplume event. Locked more or less stationary after its collision with Europe to form the Alps at about 70 Ma, the thick continental crust of the African plate seems to have formed a 'lid' over such an internal phenomenon, inducing it slowly to evolve and release its power.

The most profound expression of the Rift is the Red Sea, where slow sea-floor spreading now drives Arabia and Africa apart from their former union. Various studies show that this spreading and at least part of the 2 to 3 km of uplift on the Red Sea's flanks began at the start of the Pliocene, at 5 Ma. Where the Rift is within the African continent, although volcanoes were active before the Pliocene, the faulting and flank uplift that makes the Rift so prominent began at 5 Ma as well. The East African Rift is now the most volcanically and seismically active area in any continental interior. Although disturbed by earlier Tertiary tectonics and volcanism, the monotonous and simple zonation of the original African ecosystems broke up decisively during the Pliocene to the present, so forming the rich mosaic that hundreds of thousands of tourists (and a few dozen palaeoanthropologists) visit each year.

- ☐ What effect do you think these events had on Africa's climate?
- ■ Without the north–south barrier formed by the Rift the south-westerlies that dominate tropical West Africa's climate would continue to the east. (Figure 1.19)

Formation of the Rift allowed air masses from the Indian Ocean to replace this Atlantic influence. The high volcanic and tectonic massif of Ethiopia, although much smaller than that of Tibet, now exerts a similar influence in modifying the Hadley Cell and controlling local monsoons. So, here we have a linkage between the Earth's internal processes and those of climate and the biosphere. We have no means of assessing the global influence of these events, since they are overwhelmed by those, such as CO_2 drawdown, for which a link with the Tibetan Plateau is widely accepted. They were, however, events with an intimate touch on our own origins.

OUT OF THE TREES

The initiation of the East African Rift at 5 Ma coincides with the youngest estimate from rRNA comparisons of the divergence that led to chimpanzees and ourselves (Figure 9.5). Pure coincidence is a possibility, but the ecological changes wrought on Africa by rifting do suggest a relationship linking geological processes to our own evolution.

There is some evidence that arboreal apes, which were ancestral to modern higher primates, spread widely across northern Africa before rifting. At Fayum, west of the Nile in Egypt, the 30 to 18 million year old basalts sit upon riverine sediments that contain a fossil forest flora and fauna. Among the fossils are early, tail-less ape-like creatures, together with arboreal ancestral monkeys. Similar finds, though rare, occur at a number of sites of early Miocene age. Until the appearance of the earliest australopithecines more than 4 million years ago, Africa presents a blank sheet as regards ape evolution. What happened must be inferred from geological and ecological evidence outlined above. We can justify this because evolution of particular species and bushes of descent is not merely *in* a particular habitat or ecosystem, but as an integral *part* of them and all their associated changes.

The source of evolutionary change is random mutation among individuals within a population. Natural selection pressures test the degree to which such mutations in the genotype confer fitness on the phenotype. Those genes surviving in the progeny of fit individuals spread through the population. In large connected populations or those in stable ecosystems, genes that produce either no immediate benefit or advantages that are only selected under changed more stressful conditions, dissolve into the large gene pool of succeeding populations. They have a slow, sometimes minimal consequence on a species as a whole. In small interbreeding populations nonharmful and advantageous mutations become more common in the gene pool, and show in the phenotype more rapidly. A large population may be reduced either by physical decimation through some sudden event, or parts can become isolated into localised populations with no physical contact with the main one. Before 13 Ma, Africa's climate and vegetation zones were continuous across a topographically monotonous continent. Any large, mobile vertebrate species would have maintained large interbreeding populations over millions of square kilometres. Since then, geological processes and climatic change (some caused by tectonics, others by the Milankovich pacemaker) have broken up these broad ecosystems, and with them the physical connection between groups of the various original animal populations. West of the Rift, Africa's broad zonation remained more or less intact with interconnected populations of animals, through which genetic changes could spread and dissolve in the large gene pool.

The separation of dwindling and rising 'islands' of the original forest by the spread of seasonally dry plains covered with scrub, grasses and widely spaced savannah trees would have isolated species with arboreal habits in several ways. For tree-dwelling creatures, digestible foodstuffs of the forest became separated by wide tracts of the inedible. Into the plains spread large numbers of rapidly evolving animals adapted to that habitat – browsers and grazers together with predators that ate these herd animals, plus various scavengers and opportunists that survived at the fringes of the main trophic pyramid. Studies of early Miocene pigs, elephants and rhinos show a sudden change in their teeth, making their chewing apparatus more suited to grasses than to leaves. These new inhabitants presented competition and threats to the original fauna. For animals adapted to moving through forest, often in the canopy, though occasionally across open patches, the wide tracts of lightly wooded ground presented a formidable barrier to movement between the 'islands'. Any apes or monkeys attempting to migrate would have faced starvation, dehydration or being eaten. Possibilities for the more rapid accumulation of genetic mutations in small, isolated communities were accompanied by new selection pressures and new opportunities for changed habits. Not only were non-forest environments growing but forest stands became restricted to a narrow range of elevations on the rising massifs, further decreasing their area. Indi-

viduals able to exploit new resources beyond the forests and to move safely within the new habitats would increase their fitness by such diversification. For many types of animal, and also for plants, the fragmentation of ecosystems and isolation of populations prepared the ground for an almost universal adaptive radiation. The geological processes involved were not 'one-off' events, but continuously rejuvenated the landscape and do so to this day.

Early apes and monkeys already possessed some useful attributes for plains life. To a greater or lesser degree all species can eat fruits, nuts, leaves and small animals, such as grubs, eggs and fledglings. Primate diets are rarely highly specialised. Scrub and savannah offer excellent, varied menus, but they are more widely scattered than in forests, and their abundance fluctuates between dry and wet seasons. So the same diet would not always be available. Arboreal life favours keen binocular vision in colour, an ability to find foodstuffs visually, excellent balance and agility, and a tendency to move in bands of up to 50 individuals. All four limbs in primates are adapted to grasping, and any visitor to a wildlife park will have been struck by monkeys' human-like hands as they patter on the windscreen and tear off the car radio aerial. Two evolutionary trends to life in more open habitats characterised the response of primates to mosaic ecosystems. Monkeys, having tails, are not physiologically predisposed to upright walking. Various species of baboon adopted quadrapedal life in bands with powerful and aggressive males maintaining guard on the smaller females and offspring. Apes have no tails and only minor changes in form are needed to adopt a permanent bipedal gait. Freed hands offer means of defence using staves of wood or stones, still commonly used by chimps under threat. Whether moving on two legs is energetically more efficient than using all four is debatable, but freed hands obviously mean that food can be carried easily. This provides a greater measure of security than having to eat every last scrap at a food source or protect it from competitors. Compared with the bizarre trends taken by evolution to fit other animals to tightly defined ecological niches – think of the aardvark or the naked mole rat – developing upright posture among omnivorous, opportunistic apes requires no great transformation.

About half a million years after the Rift began to form we see the products of such a simple bodily re-organisation in the upright *Ardipithecus ramidus* and *Australopithecus anamensis* from the Kenyan–Ethiopian border area. Both are apes with no particular increase in the relative proportion of brain to body sizes, but neither of these have long, ape-like canine teeth. With *Australopithecus afarensis* and the later *A. africanus*, teeth had changed to involve thick enamel and low, blunt cusps. Their jaws are less elongate and closer to the almost semi-circular arrangement of ours. Wear patterns show that both meat and vegetable matter contributed to diet, they were omnivorous diners. Robust australopithecines and paranthropoids have very different dentition and extremely powerful jaws. Their evolution was towards an exclusively vegetarian diet, probably comprising leaves, fruits and woody stems. There is absolutely no possibility that primate digestion could have coped with grass eating, for that requires digestive systems that evolved in tune with this geologically quite recent type of vegetation from which arboreal primates were totally isolated. The emergence of grasses and their peculiar metabolic pathway is linked to the decreasing CO_2 content of the atmosphere during the mid-to-late-Tertiary (see Chapter 7, pp. 146–7). It presented a dietary opening followed by animal groups such as cattle, sheep, antelopes and horses, that separated from the primates more than 50 million years ago.

The preponderant grassy biomass of the open plains has never been directly available to primates. To exploit its potential riches meant intervention in the dominant trophic pyramid involving herding grass-eaters and large predators. Unaided, an australopithecine would not only have been incapable of killing an antelope or horse, but could never have bitten through its hide. By comparison with such fleet prey animals it would also have been an easy victim for early feline, hyaenid or dog-like predators. Even to come upon abandoned carcasses, dried by the sun, would present no opportunities for sustenance. Australopithecines either ate fresh meat exposed by the teeth of predators, or none at all. Early human ances-

tors must have evolved as bystanding observers of the main East African drama between large predators and large prey; opportunistic omnivores, little different in diet from others, such as porcupines or pigs. Opportunists must be capable of recognising and remembering a wide diversity of foodstuffs, seasonal in nature and variable in their location. The dry season presents problems, because the only nutritious vegetable foods are either buried as roots and bulbs, or as fruits on different tree and shrub species. Foraging offers no clear-cut advantages and demands continual movement. Primates carry little in the way of adaptations to this lifestyle, such as the powerful snouts and tusks of pigs or the claws of badgers. Unlike other foragers, primates do not have a keen sense of smell, and rely mainly on vision. They are not so agile as small predators. Physically they are more or less defenceless against large predators. They do however, have the wit to observe, mimic and remember, as anyone who has observed the famous macaque monkeys of Hokkaido in Japan will be aware. That is their speciality – watch pigs foraging and you will find a ready source of food, when they depart.

Lowly as this picture of our origins might seem to be, for creatures whose most developed sense was that of sight, with smell coming a long way down the list, to survive demanded an encyclopaedic memory for every potentially fruitful nook or cranny of their surroundings. That in itself constitutes a rudimentary form of culture. To have intervened habitually in the dominant food chain meant two developments; that it became essential for survival at a time of great stress, and that large-animal flesh was rendered edible to the australopithecine dentition. The last demands cutting tools. Before meeting that challenge we need briefly to consider a specific aspect of upright gait that does not bear directly on getting food, but which none the less is of crucial importance in human evolution.

BIRTH PAINS

To walk upright means that all upper body weight is supported by the legs, and the load on the spinal column is transmitted through the pelvis. This restricts any opening in the pelvis if the body-support archi-tecture is to function efficiently. The most important opening in any mammal's pelvis is that for the birth canal. Primates have bigger brains and heads propor-tionate to their bodies than any other mammal. This presents little problem for birthing among apes and monkeys, but it does for those that always walk upright. In human birth the emerging foetus must rotate through 90 degrees in order to be delivered head-first. The birth canal dilates to an alarming degree so that the head can pass, entailing enormous difficulty and pain for several hours, together with risk to both mother and child. The head and brain of an infant human are one third of their adult sizes. Those of apes are about one half adult size. Were human mothers to give birth to babies with ape-like proportions they would need pelvic dimensions that would render them incapable of walking upright. As it is, the pelvis of female humans is close to the limit for mobility, which accounts for their distinctly dif-ferent gait from that of males.

To avoid the physiological disadvantages of giving birth to babies with ape-like proportions, humans come into the world at a much earlier stage of foetal development than do apes – a tendency that is tech-nically known as neoteny. Children's development to self-sufficient maturity takes 10 to 13 years compared with about 5 years for chimps. The early period of child development is one of complete dependency that demands the care of both parents. Moreover, the brain grows extremely quickly with much larger demands on high-protein and high-energy food resources than among apes. Many anthropologists infer that the onset of neoteny marked the beginning of social living and the sharing of food and joint infant care that are its hall marks, in contrast to gre-gariousness which is common to many primates. Pelvises of female *A. afarensis* from Hadar indicate that with a foetal head size half that of adults, if they retained the ape proportions, giving birth would have been difficult for these early hominids. Alternatively, *A. afarensis* infants were born neotenously. The possi-bility that the Laetoli footprints (Figure 9.8) involved two adults and a juvenile australopithecine could mean that they are the tracks of a family unit. Infer-ence can be pushed too far, but the existence of social behaviour more than 3 million years ago is a distinct possibility.

TIMES OF STRESS; TOOLS AND NEW OPPORTUNITIES

Continental ice sheets began to form in the Northern Hemisphere at 2.5 Ma. This marked the beginning of the astronomically forced fluctuation of climate and sea level that has so far seen 50 warm–cool cycles. Studies in Africa of pollen and lake-levels for more recent full glacial epochs in the Pleistocene show that rainfall declined and the tropical rain forest of West Africa may have fragmented to a small number of relic areas or refugia (Figure 1.12b). Grassland and open savannah increasingly dominated low tropical latitudes and deserts spread at higher latitudes (Fig-ure 8.16). Although the expansion of grassland dur-ing glacial epochs would have increased the number of herbivores and top predators, this would have held few, if any advantages for the australop-ithecines. Indeed it would have posed great stress on the availability of foodstuffs for such foraging species and their competitors. One evolutionary option was to focus on using the protein and starch in grass seeds, which for primates means grinding seeds to a pulp so that they can be digested. This seems to have been the route exploited by the evo-lution of paranthropoids. The other was to use high-grade proteins already processed from the widespread grasses by grazing animals. The first appearance of tools, though not their makers and users, at 2.5 Ma surely marks a response to extremely hard times by one group among the aus-tralopithecines. Curiously, it was probably the wet season that posed the greatest hardships. That is when grass grows quickly, but when fruits and tubers are at a premium.

So how did stone cutting tools of the Oldowan type come to be invented? The notion that sharp edges were somehow discovered by trial and error seems absurd. Imagine starving australopithecines coming upon abandoned kills, frantically poking at the hide and meat with every conceivable object. Finally, after hundreds of generations, in a Stanley Kubrick-like dawning of realisation, one succeeded in administering a cut with a sharp stone! Slow wit of that kind in hard times is the knell of doom. Imagine instead yourself walking barefoot across a landscape strewn with all manner of lavas and volcanic glass. The concept of sharpness would be an everyday experience for an upright creature with fleshy feet. At the very least an immediate response would be to examine the offending shard. Applying sharpness of broken rock to ripe-smelling carrion, about which vultures and small carnivores leaped in delight, would dawn on the dimmest of hungry beings with hands freed to exploit this natural phenomenon. The positive selection pressure attending the use of such rich protein sources in an otherwise inedible environment would have been immense – more individuals would survive to reproduce, and infants would have had dietary benefits previously unavailable. Remember too that our ancestors were apes. To creatures that habitually hurl rocks and use them for pounding, the leap to *producing* cutting edges themselves, and thereby serviceable tools, would not be spectacular in itself. Its consequences would have been revolutionary in many different ways. Primarily it removed by far the greatest problem for opportunistic foragers, that of the next meal!

As any Girl Guide will tell you, a cutting edge opens wide cultural horizons. Wooden and bone tools, and the use of skins need sharp instruments. Digging for roots would be easier, and both the means of production and the fruits of labour could be carried around in bags and pouches. By 2 Ma we see the outcome of this 'second nature' as humans in skeletal form, albeit with physical characteristics that show limitations on brain size. *Homo habilis* had a cranial capacity of between 0.55 to 0.85 litres, compared with the average for modern adults of 1.4 litres. Though heavily boned, particularly around the brows, their skulls are considerably lighter than those of australopithecines. So far as we can judge from scanty remains, the rest of their frame had proportions between those of australopithecines and our own.

CHANGING BRAINS AND BELLIES

Much of human physiological evolution that succeeded *H. habilis* focused on the skull and an increase in brain capacity. Lightness of both cranium and jaws are the main characteristics of our skulls. Both are more easily damaged by impacts than in our ancient forebears – surely a feature that carries disadvantages in a purely evolutionary sense. But what if heavy-boned skulls hinder growth of the brain? Foraging and tool making demand brain power. With such a lifestyle, increasing brain size carries a positive advantage for an individual's survival and successful reproduction. Meat-cutting tools and new access to a high-protein, scavenged diet removed the advantages of having heavy, multipurpose jaws.

☐ Cup your upper head between both hands and go through the motions of chewing. What can you feel?

■ In the region of the temples, to the sides of your eyebrows and about halfway to the back of your head you will feel the muscles of the lower jaw working.

These muscles are quite small, and attach to almost imperceptible ridges on the cranium. Lower down are attachment sites on the cheek bones. For heavy chewers, such as the paranthropoids and to a lesser extent australopithecines and less still for early humans, these attachments were more robust than are yours. They had to be, because the mechanics of a muzzle-like face with large lower jaws demand large muscles to drive their lever-like function. Declining muzzle and expanding cranium go hand in hand, as the reduced need for chewing power renders heavy cheeks and brow ridges redundant. Humans succeeding the habilines became flatter in the face. Curiously, the recession of the arch of lower teeth exceeded that of the lower jaw, and this resulted in the chin! The only primates with chins are humans, and anyone with a little chin is sometimes spoken of unkindly. One avenue for the evolution of increased brain capacity was therefore anatomical redundancy due to a changed diet. Changing environment, different habits and a new diet would not only affect chewing apparatus and the bones that support it. More easily grasped than the gradual change in skull morphology is a tendency for a change in our predecessors' guts. We can surmise this from comparing the bowels of living humans and apes.

Primates evolved powerful digestive systems to cope with exclusively vegetable foods, low in nutrition in proportion to bulk. Take a look at a gorilla, particularly a large male, and the immediate impression is, 'My goodness, what a belly!'. Though there are sights almost equally as worrying in any public bar, a physically fit human is diminutive in that department. A high-protein, high-energy diet made the original primate gut largely superfluous, and so the human gut is the only energy-demanding part that is strikingly small relative to body size, compared with those of other mammals. Relative belly proportion has its reflection in overall body shape. The chests of apes and australopithecines are shaped like inverted pyramids, getting wider upwards. We, on the other hand are built with a barrel-like chest and a narrow waist, and we see lithe frames in all humans from the erects onwards.

You can easily imagine the opportunities presented by this transformation in shape, even if it would not take our earliest ancestors to the front page of Vogue. There is more room for lungs, thereby increasing stamina, and with a decreased bulk comes improved agility and running speed. Large lungs with sophisticated muscle control also open up the route to speech. But by far the most important new avenue stemmed from the decrease in energy demand by digestion with the 2 million year old model gut and a high-protein diet. Our brains and the crania that hold them are very much larger than those in other mammals of our dimensions. Big brains are not only expensive to use and maintain, but building one is the most energy intensive aspect of all biological activity particularly human child development. To bear and then to breast feed an infant means that women's energy intake has to rocket during pregnancy and early motherhood. The tripling in size of the human brain in the period up to puberty places a similar load on children's energy intake. Even the

brain of the most dedicated adult couch potato today consumes more than 20 per cent of our energy intake yet makes up only 2 per cent of our mass. Decreased energy needs of shrinking bowels, plus a more concentrated food input, freed more for the brain. Of course, not all meat-eaters are clever, but primates always had to be smart to live in trees. Given that starting point, all the environmental changes surrounding human beginnings and the selected-for physical characteristics and habits that stemmed from those changes, it is not surprising that human evolution focused predominantly on the head and what lay between the ears. Freed of literally breathtaking bellies, it opened up the physical requirements for becoming predators – speed and the endurance that allows any athlete eventually to outrun a horse.

Modern human skulls are light. By comparison with apes and extinct humans they are more like those of juveniles than adults' skulls. During maturation our faces change much less than do those of apes. We fail to become fully robust. Unlike all other primates, the skulls of human infants are almost disarticulated. Babies' fontanelles are one example of this. Our infantile skulls are plastic and thereby capable of expansion and growth to accommodate the trebling in brain size during child development. The other physiological factor contributing to evolution of the human brain therefore stemmed from neotenous birth, itself a direct legacy of the constraints imposed by upright posture as a selected-for response to an open environment. We can say that these anatomical considerations also provided opportunities for brain growth, but why grow a bigger brain in the first place? That presupposes a need and an advantage set against the rest of the world.

More or less every 50 thousand years over the last 2.5 million, or about every 2 thousand generations, Earth's climate has gone through a warm–cold, and for Africa a damp–dry, cycle. Such has been the pace of ecological shifts to which all participants in an ecosystem had to respond. We see no more examples of human anatomical and cultural changes that coincide with climatic and geological changes after 2 million years ago. Oldowan tools served their makers well for about 25 climate cycles. Likewise the succeeding Acheulean culture. The physiognomic changes, probably polymorphic, that in Chapter 9 we suggested may have been allowed by cultural 'protection' from Darwinian selection, have no rhyme or reason in the context of repeated environmental change.

EARLY HUMAN MIGRATIONS: PATHWAYS AND BARRIERS

The links to wider change among early humans involved migrations. While Africa was well-watered there would be little impetus to the movement of populations of humans who increasingly depended on a meaty diet. Growing aridity would have seen diffusion of herd animals as the vegetation belts shifted. To survive, opportunistic humans would have had to follow, thereby opening new possibilities in new lands.

While reading Chapter 9 you may well have wondered why the first diffusion out of Africa at around 1.8 Ma reached east Asia and not Europe and the Middle East. Tectonics provides an interesting framework. The first requirement for land animals and plants is a reliable source of water. The bulk of rainfall in East Africa either sinks into the sands of fringing deserts on its way to the Indian Ocean, or becomes trapped or channelled within the great African Rift. Lakes, with a chain of vegetated patches around them, have always been present along this system of topographic lows. Diffusion following game and plant resources during periods of climatic change, particularly when conditions became more arid, would inevitably have been channelled north and south. Until about 1.5 Ma the Red Sea was closed across its southern end at the site of the present Straits of Bab el Mandab, and periodically open to the Mediterranean to the north. Early migrants following the Rift would therefore have been channelled to the south-eastern fringes of Arabia and around the coast to the Persian Gulf. Proceeding west from there even during a warm–moist episode would have been barred by the deserts of modern Kuwait, Saudi Arabia, Iraq and Syria. Going east was the only

option that could be survived. Movement must have been conditioned by vegetated land, a northern route to China being impassable by virtue of the Alps–Caucasus–Himalayan mountain chain. The opening of the Straits of Bab el Mandab around 1.5 Ma would thereafter have halted diffusion through Arabia. This helps explain why the *H. erectus* population of east Asia remained isolated until only 100 ka (see Figure 10.3).

Later diffusion from Africa may have had to follow the west flank of the Red Sea, through Palestine. Even today, this is an arduous route blighted by extreme aridity from Port Sudan to Suez. Nevertheless, diffusion permitted the Middle East and eventually Europe to become colonised, but no one has found remains of the migrants along the Red Sea coast. There is another possibility – following the White Nile from its source in the western branch of the Rift. Winding through the most arid area on Earth, the Nile would have formed a narrow, watered and vegetated corridor leading to the shores of the Mediterranean and eventually Palestine. During cold–dry periods the deserts of the Middle East would again present insurmountable barriers, this time to eastward movement, because of their lack of water and food. Equally daunting are the Zagros, Kurdistan and Taurus mountains that rim the Tigris–Euphrates plain to the north. To us, it seems a short step (about 150 km) from the eastern shore of the Mediterranean to reach the fertile strip along the Euphrates leading through Syria and Iraq to the Persian Gulf and the road east followed by the first migrants out of Africa. The Euphrates is fed by the snow fields in the mountains of east Turkey, and would have flowed during glacial periods too. However, the modern route through Aleppo in northern Syria crosses arid mountains and plains. It is impassable without access to water.

☐ Name a barrier to direct westward diffusion from the Middle East.

■ The outflow from the Black Sea to the Mediterranean through the Bosporus and Dardanelles today bars entry without boats to Europe from Asia.

Reaching Europe demands a route around the Black Sea as far as 46°N. Remains of *H. erectus* from between the Black and Caspian Seas in modern Georgia date back to 1.6 Ma, showing that this option may indeed have been followed.

☐ What important process that characterised glacial periods might have removed the barrier presented by the Bosporus?

■ When ice sheets accumulate on the continents sea-level falls, thereby drying out some sea barriers.

The Bosporus is shallow enough (about 50 metres) for it to have been passable at the height of any of the glacial periods. Though channelled by barriers that would have been impassable by virtue of lack of water and food, the diffusion of early humans had no physical limit within Europe–Asia–Africa. Without fire and clothing, the extent to which this diffusion was possible had only the limitation of survivable climate.

CLIMATE AND HUMANS

Until 1 million years ago there is no direct evidence of the use of fire. Consequently we have to assume that diffusion across lines of latitude was probably back and forth as conditions cooled and warmed before that time. Apart from the clear differences between east Asian and Afro-European *H. erectus* heads and physiognomy, it is impossible to judge if there were significant differences between regional populations of erects in Africa and Europe because there are few remains.

☐ How many generations of 15 years would have arisen during each roughly 10 thousand year long interglacial period, when diffusion out of and back to Africa would have been unnecessary?

■ About 650.

The 650 or so generations would probably not have allowed distinct trends in physiology to have arisen. We can be quite certain of this for even 3000 generations of separation between modern human

groups such as the separate Australian and New Guinean populations, has produced no significant skeletal changes and only a tiny shift in genes. So, as regards the Afro-European erects we can consider them to have been part of essentially the same interbreeding population. The change that did arise was the appearance of the distinct Kabwean facial characteristics, albeit a polymorphic mish-mash. Kabwean features appeared in Africa at between 300–500 ka, a period athwart three cold–warm, dry–moist climatic cycles. There is no clear correlation between environmental change and this step in human evolution. It seems that it emerged out of a population with enormous advantages over other animals by virtue of their use of 'second nature' and the consciousness indissolubly linked to it. Increasingly cushioned from 'natural' selection, genes that might otherwise have been extinguished, such as the tendency for neotenous birth and long childhood, and for light skulls without great crunching jaws, overwhelmed the population. The advantages of increased brain size permitted by these tendencies, and the more sophisticated use of such an addition of information-processing power would have allowed those individuals so endowed to outcompete other, less brainy individuals.

Kabweans definitely were pretty clever, having around the same cranial capacity as our own, and sometimes considerably more. Their most famous branch, the Neanderthals, entered Europe at around 200–250 ka during an extended warm period, armed with much the same tool-kit as their erect predecessors. They proved capable of surviving full glacial conditions and were the permanent, sole population of southern Europe until about 50 ka. As we have seen in Chapter 9, the Neanderthals did become distinct anatomically from other human groups, with enormous strength and equally enormous noses. The latter is a clear adaptation to frigid conditions, without which sub-zero air enters the lungs and causes pneumonia and ultimately death. Another line of anatomical evidence permits a more rigorous appraisal of the climatic adaptation of both modern and ancient humans.

All mammals and other warm-blooded animals

evolve sophisticated means of regulating their body temperature, to keep it constant. The bodily modifications depend on whether heat needs to be lost or conserved – elephants have huge ears to help their bodies lose heat, seals have thick blubber to insulate them against heat loss. Much the most basic adaptation is to adjust the body's surface area, from which heat is lost by radiation and in our case by evaporation of sweat, relative to volume, which governs how much heat is generated by metabolism. The greater the surface area of a body in proportion to volume, the more heat is lost, and *vice versa*. It is a matter of geometry. Spheres have the lowest area:volume ratio of any object, so an ideal adaptation to frigid conditions is to be spherical! Conversely, to lose heat in hotter conditions is best served by increasing departure from a spherical shape, perhaps to that of a cylinder. Lapps and Inuit today are short and pretty round, while the inhabitants of hot plains, such as the Masai, are tall and slender, with the rest of us falling somewhere in between these two extremes. Erik Trinkaus of the University of New Mexico used a simple anatomical comparison to assess human 'roundness', which he used in comparison with average temperatures of the areas where different groups live. The best measure is the ratio of shin length to that of thighs (in skeletons that of tibia to femur lengths). Figure 10.2(a) shows a plot of tibia:femur ratio against mean annual environmental temperature for a wide range of modern humans. Although not a perfect correlation, there is a clear trend of increasing ratio with increasing temperature, showing that people adapted to warmer climates are increasingly less 'round'.

☐ Compare the positions on Figure 10.2(a) of white and black South Africans and Americans. What can you conclude from them?

■ White people group with Europeans, showing their recent origins by colonisation. Black people plot at the high-temperature end, naturally for those from South Africa and showing their origin as slaves from tropical Africa in the case of American blacks. This is a genetic matter, not one of individual adaptation.

Trinkaus measured tibia:femur ratios for Neanderthals and African *Homo erectus* skeletons, and these clearly show adaptation of the former to extreme, even Arctic temperatures, and the latter to life in hot tropical plains (see Figure 10.2b). Neanderthals were genetically habituated to glacial conditions.

NEANDERTHALS AND MODERN HUMANS

Further human evolution, both in form and in culture, emerged in Africa with fully modern humans at the height of the second-last cold–dry episode (~150 thousand years ago). The interface between the first Europeans and the latest was in Palestine. Archaeological evidence since 100 ka indicates both groups in the area. Modern human skeletons there are of tall, slender people, while the Neanderthals are short and stocky. However, it is not possible to judge whether or not these markedly contrasted groups ever came into contact; perhaps one group migrated in as the other moved out. Whatever, from 50 ka fully modern humans and Neanderthals cohabited Europe for 20 thousand years. How relations were we do not know, except insofar as they could not have been so intimate that Neanderthal genes entered the human genotype! Such are the differences in anatomy and therefore physiognomy that Steve Jones of University College, London has made the following observation. If an unwashed modern human from 50 thousand years ago was dressed in a twin set and rode the London Underground, most people would prefer to change seats if such a person sat next to them, but they would probably change trains if a besuited Neanderthal did the same thing! Both groups had much the same tools, and the Neanderthals definitely had the edge in terms of experience. However, the full humans entered Europe at a very bad time, at the depth of the last Ice Age. They and their genes survived, the Neanderthals did not.

All kinds of theories, including genocide, have been put forward to explain the 20 thousand year

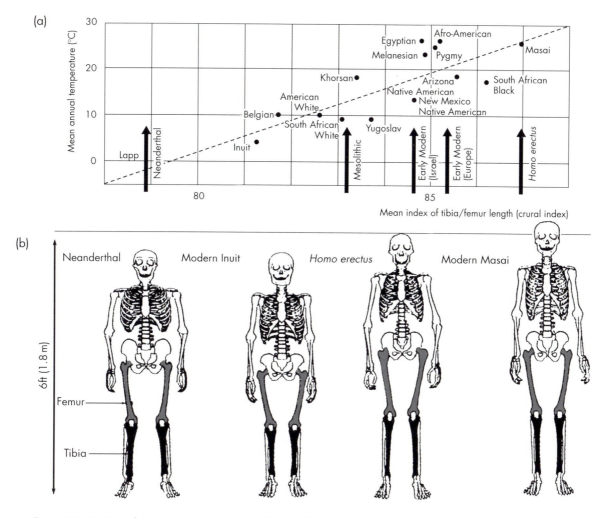

Figure 10.2 (a) Plot of the ratio between human tibias and femurs against local mean annual temperature today. (b) Comparative builds of Neanderthals, modern Inuit, *H. erectus* and modern Masai. The tibia: femur ratios of *H. erectus*, Neanderthals, Middle Eastern and European early modern humans, and post-Neanderthal (Mesolithic) European humans are plotted on (a).

decline and eventual demise of the Neanderthals. Research conducted by a team from Harvard University seems to provide the final answer. They studied the remains of food animals associated with modern human and Neanderthal sites, particularly the teeth of ruminants, which continually grow to replace wear and tear. In doing so they acquire a layered structure (incidentally used by horse traders when assessing the age of a potential purchase), dark layers being laid down during winter when nutrition is low and light layers when the grazing is good. The research focused on the last layer formed before the animal was slain and eaten. For Neanderthal sites both light and dark layers are always mixed. Modern human sites have a preponderance of either dark or light final layers, with little mixture.

□ What can you conclude from this difference?

■ Neanderthals seem to have occupied sites permanently, whereas modern humans probably had a nomadic lifestyle with shifting summer and winter hunting areas.

Imagine the situation. Permanent occupancy means obtaining food from a more or less fixed landscape, with the ever present danger of your depleting local resources. Nomads move from place to place, as local resources wax and wane. Either can achieve a balance between populations of humans and food items, but not both together. Nomads entering the hunting and foraging grounds of a fixed population mean that the same natural resources temporarily have to support maybe twice the usual human numbers. Foodstocks drop rapidly. The nomads move on to pastures new, leaving the locals to face starvation. Managing a game reserve and eking out an existence demands very hard physical work at the best of times, whereas roaming to find rich pickings is less demanding. Perhaps the former explains the enormously muscular physiques of Neanderthals and the frequent

occurrence of healed fractures, osteoarthritis and deformity among their skeletal remains. Their cohabitees show far less evidence of crippling labour. The only surprise is that the Neanderthals lasted so long in the face of such inevitable competition for sustenance. Population densities were probably low, and maybe some Neanderthals did shift camp. Evidence from Central France suggests that late Neanderthal sites are more concentrated at higher and therefore colder elevations than those of fully modern humans. Whatever, the last Neanderthals seem to have died out 30 thousand years ago in the rocky fastness of Zafarraya near Malaga in southern Spain, where fully modern humans had previously been conspicuous only by their absence.

MODERN HUMANS GO EAST

Lest we descend into Eurocentricity, much more dramatic events were taking place elsewhere during the last Ice Age (Figure 10.3). Fully modern humans arrived in east Asia at least 60 thousand years ago and

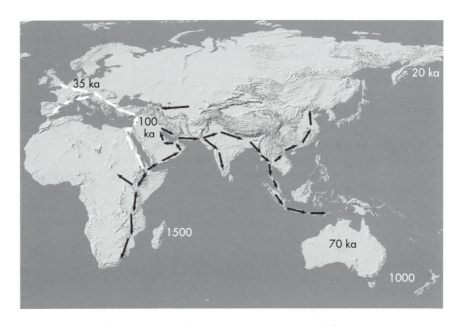

Figure 10.3 Possible human diffusion paths. Black arrows: erects to east Asia before 1.5 Ma; white arrows: post-1.5 Ma routes. Dates give arrival times of fully modern humans.

Box 10.1 Population Dynamics

Both theory and study of natural ecosystems indicate a simple pattern of population growth among non-human species. A population founded by a small number of individuals (either a newly emerged species or a small population entering an ecosystem for the first time) increases slowly at first but then undergoes a rapid growth in numbers (Figure 10.4(a)). This is characteristic of exponential growth where new offspring exceed deaths at an annual percentage rate. Numbers double approximately every (70/annual percentage rate) years in the same way as savings grow by compound interest. Various environmental factors ultimately limit the species numbers to a maximum level; the ecosystem's *carrying capacity*. Rapidly growing numbers can temporarily overshoot this limit. In such cases lack of resources or effects of overcrowding increase mortality to decrease numbers, and eventually they reach equilibrium at the carrying capacity provided there are no changes in the rest of the ecosystem. Where large numbers are involved, it is usual to plot population on a logarithmic scale (Figure 10.4(b)) so that a period of exponential growth is represented by a straight sloping line and carrying capacity by a plateau. Figure 10.4(c) is such a graph for total human population for the 10 thousand years of the Holocene.

☐ By comparing Figures 10.4(b) and (c) how would you characterise human population growth over this period?

■ There is an increase in the rate of exponential growth from about 9 to 5 ka ago, then a reduced rate up to 2 ka ago, when estimates are more constrained by records. Since then the slope of the graph increases towards the present, which shows that the rate of population growth is increasing.

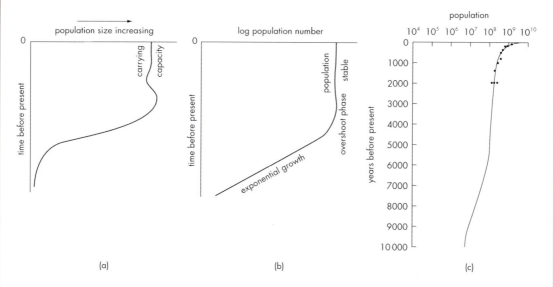

Figure 10.4 Population dynamics: (a) Typical curve of population growth for a species living in an environment with limited resources. (b) Same data with population plotted on a logarithmic (powers of ten) scale. (c) Changes in world human population during the Holocene, with population plotted on a logarithmic scale. Data are estimates for the period up to 2 ka.

spread quickly throughout what is now the western Indonesian archipelago. Through falls in sea-level as glaciers grew at high northern latitudes, Malaysia, Sumatra and Java were connected by dry land. At around 50 ka, people arrived in New Guinea and shortly afterwards in Australia. (At the time of writing evidence had been published to suggest that the first colonists of Australia, together with primitive art, reached there about 176 ka at the height of the glacial before last. If confirmed, this poses a host of new problems, for at that time they could not have been fully modern humans.) To reach New Guinea, people must have travelled in boats or on rafts, 'hopping' from island to island in the Indonesian chain. With sea-level depressed by more than 75 metres because of continental ice build-up, Australia could be reached from there simply by walking. At the depth of the last glacial maximum the Bering Straits were dry land connecting Asia to the Americas. By 20 ka, and maybe earlier, north-east Asian populations discovered the Americas a thousand generations before Icelanders and Christopher Columbus. Within a hundred generations they had occupied both continents, previously virgin as regards humans and their culture. The much stormier conditions of glacial epochs delayed colonisation of the ocean hemisphere of the Pacific until the Holocene, but in its 10 thousand year duration before Europeans entered the Pacific, virtually every island had been visited if not colonised by intrepid Melanesian and Polynesian voyagers. Their exploits put the Apollo missions to the Moon into sharp perspective.

THE HUMAN IMPACT ON ENVIRONMENTS

The geological record before the Holocene bears witness to how conscious humans, themselves a product of naturally changing environments, rebounded dramatically on the rest of life. The last 100 thousand years mark a mass extinction event comparable in scale and pace with any previous one, yet neither an impacting comet nor a geological catastrophe is in any way involved. With today's widespread concerns about environmental degradation it is easy to view the impact of humanity as being modern and somehow related to human population growth. Both offer only a narrow and misleading view.

Humanity is presently in a period of super-exponential growth, and has yet to reach the carrying capacity of the planet. This reflects modern humans' unique ability to control their environment to a large degree, and population theory based on observation of other species and restricted ecosystems is probably inappropriate for us. The world average population of humans from 2.5 Ma until the Holocene has been estimated at between 200 000 (around the combined attendance at British Premier League soccer matches when a winter Saturday comes) and 2 million. That is, between 1 person per 1000 km^2 to 100 km^2, which are the averages for foraging gatherer-hunters in the Arctic or arid central Australia, and in tropical rain forest respectively. Figure 10.5 shows estimates of global population and how they might have changed over the last 1 million years. There are three sharp shifts in the graph. The first starts at a very low population (100 000) of stone tool users that may have continued since 2.5 Ma, building to about 1 million by the end of the Pleistocene, as humans occupied larger areas of the planet and their numbers reached the carrying capacity for foragers in various habitats. The second reflects the leap in carrying capacity presented by agriculture at the outset of the Holocene. The third, for which real figures are available, is the rapid jump in numbers after the beginning of the Industrial Revolution. Our concern here is with the first phase.

Use of the environment by a tiny number of humans foraging for food and clothing, shelter and fuel might suggest that its impact would be immeasurably small. However there is evidence that the appearance of humans in previously uninhabited areas had a dramatic effect on the environment. The clearest cases come from the colonisation of the Americas and Australia, beginning at around 20 and 50 thousand years ago respectively. We come to them shortly. First we need some sort of yardstick.

When animal genera are considered world-wide over the Tertiary Era, the number of new additions to

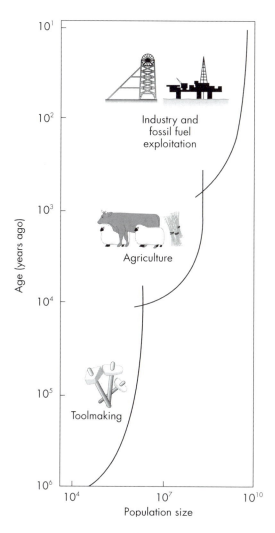

Figure 10.5 A hypothetical logarithmic plot of human population variation (*H. erectus* and *H. sapiens*) from 1 Ma.

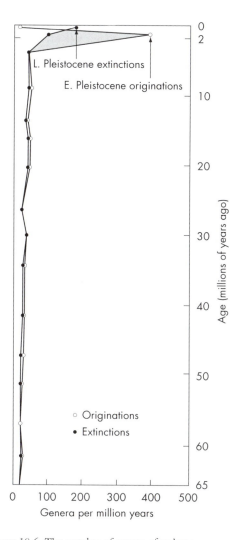

Figure 10.6 The number of genera of rodents, one group of grazers and terrestrial carnivores that became extinct each million years during the Tertiary and Pleistocene.

the fossil record is roughly matched by corresponding disappearances, presumably by extinction. Figure 10.6 shows the record for three terrestrial mammal groups with a good fossil record that has been studied enthusiastically: rodents, one group of grazers and carnivores. The gentle rise in the graph towards the younger Tertiary is mainly due to increasingly good preservation. Suddenly, at the end of the Pliocene the appearance of new genera jumps by ten times, and the number of extinctions increases by about four times.

Since Pliocene and Miocene strata are preserved just as well as those of the slightly younger Pleistocene, these sudden changes are unlikely to be an artefact of the fossil record.

☐ Can you offer an explanation for this change in the record?

☐ Climate and ecosystems underwent a decisive change when continental ice sheets began to wax and wane in the Northern Hemisphere after 2.6 Ma. Climatic cycling and novel diversification of vegetation zones both opened up new opportunities and placed stresses on all terrestrial animal life.

The jumps in new genera and extinctions shown on Figure 10.6 are probably evolutionary responses to sudden environmental change. The new diversity of habitats resulted in a pulse of adaptive radiation to fill new niches. The large number of extinctions must be dominated by the demise of some of these new genera rather than that of faunas that appeared in the earlier Pliocene – there simply were not enough of the latter. Perhaps diversification overshot the eventual carrying capacity, so that some new genera were outcompeted. So how can we judge any human influence? Simply by looking specifically at animals that might conceivably have provided prey items for human hunters or for whom humans represented competition. By focusing on large mammals weighing more than 45 kg it is possible to get an idea.

Africa has had the greatest diversity of large mammals anywhere since the start of the Pliocene. The earliest Pleistocene (2.0–0.7 Ma) saw the loss of 21 genera there, at a time when *H. habilis* and *H. erectus* had no tools designed for hunting. In the Middle and Late Pleistocene (0.7–0.1 Ma and <0.1 Ma), when hunting was more likely, the number of African extinctions is only 9 and 7 genera. The Early Pleistocene extinctions seem likely to have been due to natural causes through being outcompeted by other mammals. Now look at the extinctions for the same periods in other areas on Figure 10.7, together with the dates of colonisation by Kabweans and modern humans.

□ In what way do the records from elsewhere differ from that for Africa? Is there any pattern relating extinctions and human colonisation?

■ The largest extinctions in Europe are in the Middle and Late Pleistocene. For Australia and the Americas the peaks of extinction occurred in the last 100 ka. There does seem to be a coarse pattern following human diffusion.

Europe was colonised by Neanderthals in the late-Middle Pleistocene and was continuously occupied thereafter, whereas Australia and the Americas only received human migrants in the Late Pleistocene. The extinctions outside of Africa are emphasised when expressed in percentages. In Australia the Late Pleistocene witnessed the demise of 86 per cent of all large mammals, including a wombat that weighed more than 2 tonnes. For the same period South America lost 80 per cent, and the North American fauna suffered a 73 per cent loss.

Clearly, there is a significant difference for the Middle and Late Pleistocene between Africa and previously unpopulated continents – a much smaller proportion of large mammals became extinct in Africa than in the latter. Yet Africa had both more genera and a more widespread and possibly denser colonisation by resourceful humans than either the Americas or Australia. The natural question to ask is whether the declines were due to human factors or some other combination of circumstances. There is no doubt that humans with modest hunting equipment have devastated large animal populations in historic times. The record from Madagascar and New Zealand is a good yardstick. Madagascar was colonised at about 1.5 ka. Within 600 years ten genera of large mammals, including 7 lemurs, together with the strange flightless elephant birds were eaten to extinction. In AD 1000 Polynesians reached New Zealand. Within 700 years 34 bird species, including the giant flightless moas, had disappeared. Bones of all these groups in both areas show clear evidence that they were hunted and eaten. In both cases the new prey, although in some cases formidable, were defenceless, mainly because they had never encountered humans before; they were naive. Although assuming naivity may seem frivolous, there is a large body of evidence that animals pass on experience of predators from generation to generation – they adapt their behaviour. This is easily observed when we encounter familiar species in wilderness areas. They are often surprisingly 'tame' compared with those in our back gardens.

Although both the Americas and Australia had fearsome mammalian predators before humans

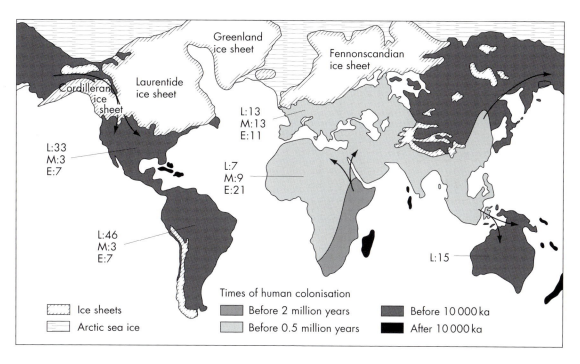

Figure 10.7 Losses of large mammal (>45 kg) genera during the Pleistocene in different continents, relative to times of human colonisation.
Key: E = early (2 Ma to 700 ka); M = mid (700 ka to 100 ka), L = late (after 100 ka).

entered the arena, there were none with two legs and spears. Many authorities couple the disappearance of large mammals with the arrival of humans there as a consequence of indigenous naivity. Although small in number at first, colonising bands may literally have eaten their way through three continents in a matter of only a few thousand years (Figure 10.8). There is abundant evidence that such extinct large mammals as American elephants and Australian giant wombats and kangaroos were food items. However, there are other possibilities. Humans probably followed game, and therefore other predators across the dry Bering Straits during the last Ice Age. Perhaps these other incomers presented insuperable competition to the native fauna. That might be a justifiable conclusion for high latitudes in North America, where mammals would have been adapted to much the same conditions as were those crossing from Asia. However, it cannot be invoked either for isolated Australia or for tropical latitudes in South America, as non-human

mammals would not wander outside their normal climatic range and suitable feeding grounds. The second main objection is that other animals could have crossed the Bering Straits up to 50 times during the Pleistocene – every time sea level fell. Yet there is no evidence for pre-human decimation of comparable magnitude. For the Americas human-induced mass extinction seems inescapable. Australia's extinctions seem to have been delayed relative to human entry by about 20 thousand years. Some have put this down to the effect of a period of increased aridity between 25 and 15 ka. However, Australia is not well provided with water supplies in its interior, and perhaps the early colonists lived along the coast, only venturing into the Red Centre when population pressures enforced it.

In the areas in which human induced extinctions are well supported, large predators were also decimated to the same extent as potential prey. Examining the record of smaller animals reveals no such

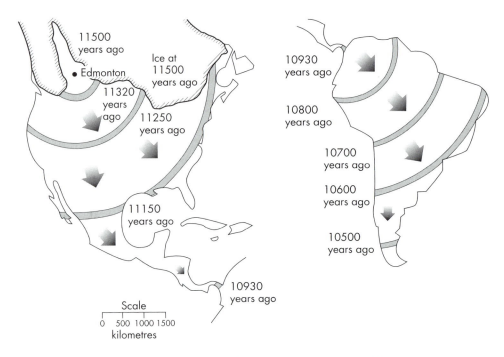

Figure 10.8 Possible rate of human colonisation of the Americas during the Younger Dryas.

dramatic events. No doubt they were hunted and eaten (if edible), but their larger populations would have protected them as individual taxa. Moreover they reproduce extremely rapidly, whereas larger mammals have inter-birth periods up to several years long, have small numbers of offspring and are therefore more precarious. The last mammoth or giant deer would have been a more valuable target for human predation than its equivalent weight in voles. Meat is not the only inducement for hunters; in treeless steppes early native Americans and Europeans built shelters from elephant bones and probably skins, and very few would have worn vole-skin clothes (except perhaps as intimate undergarments).

Africa, in contrast, has many large mammalian prey animals that have adapted to all indigenous predators, including humans, and have devised a variety of defensive strategies. Neither a Cape Buffalo nor an African Elephant can by any stretch of the imagination be regarded as naive. Despite the fears of

a mainly urban, educated and well-fed section of modern humanity that it is today's economic system that threatens mass extinction, the process began 50 thousand years ago. It is also possible that similar changes were inflicted by early humans on vegetation.

In East Africa and Australia tree and woody shrub communities are dominated by fire-tolerant species and those whose seed germination depends on fire. To some extent this may stem from the effects of natural fires lit by lightning in seasonally arid areas. But fire has been part of human culture for at least a million years. As well as for warmth and cooking, it can be used in two other important ways by gatherer-hunters. Fire can be used to panic game into traps and ambushes, a particularly effective strategy for hunters lacking projectile weapons. It can also be used to clear impenetrable scrub, giving less cover to prey animals. Fire-tolerant flora have been endemic in East Africa for more than a million years, possibly as a result of erects' and later humans' strategies. In Australia they became dominant in the last 30 thousand years at the same time as increased aridity set in. In this case, either the aridity encouraged fires and the adaptation of vegetation, or widespread human deforestation through burning was a causative factor in climatic drying. Transpiration by trees has a major effect on the humidity of regional airmasses, while the resins and other dusts emitted during transpiration act as important 'seeds' for the condensation of clouds and rain. Whatever the possibilities of a direct human influence over vegetation from a very early date in our evolution, a proved connection remains elusive and other explanations are possible. It is only when we examine geological records that can be dated very precisely in relation to well-established sequences of climatic events that such proof of the human touch

becomes available. Sadly, such investigation is limited to the last 30–40 thousand years when radiometric dating reaches the decadal level of precision. Evidence for forest clearance and both pasture lands and cereal cropping in Britain comes from the pollen records of well-dated lake and bog sequences (Figure 8.6). The first proof of human occupation, outside of older skeletal and archaeological evidence, comes from the sudden decline in elm pollen and the onset of increasing grass, weed and cereal pollens at around 5 ka, which occurs at the same time throughout southern Britain. Land was being cleared for agriculture and the great deciduous forest of Europe began progressively to be cut and burned down. What passes for most European landscape today is of no great antiquity.

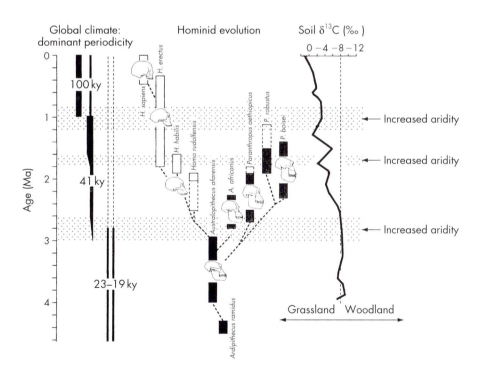

Figure 10.9 Branching of hominid descent into human and australopithecine/paranthropoid lines (bars show time ranges of fossil remains) matched to evidence for changes in the dominant periodicity of global climate change (left) and for steps in the increase of aridity in Africa (dotted zones). The carbon isotope data from soils show a trend of decreasing ^{13}C that indicates the expansion in East Africa of grasses compared with woody plants – the cell metabolism of grasses takes up less ^{13}C than does that of woody plants (from de Menocal, 1995).

SUMMARY

1 The environment of human origins and evolution, East Africa, has been one of great geological, geographic and climatic change since 5 Ma. Figure 10.9 summarises the rough correlation between steps in hominid evolution over the last 4.5 million years and climate change in Africa. From a vast forested tropical plain, it rapidly evolved to a mosaic of many small ecosystems through the effects of the East African Rift and associated crustal doming. Rift initiation correlates well with the molecular evidence for divergence of the western chimp line (that continued to live in a vast forest biome) from the hominid one to the east. Primate survival meant entering the dominant savannahs surrounding remnant upland forest, which demanded upright gait. Arboreal apes were well prepared anatomically and in brain power for that shift and their teeth provide decisive evidence for dietary change.

2 The first tools coincide roughly with the first continental glaciation in the Northern Hemisphere at 2.6 Ma. High latitude cooling related to tropical drying, and this must have been a time of intense dietary stress. There can be no doubt that stone tools enabled a decisive switch to a high-protein, high-energy meat-rich diet. As well as reducing the need for powerful jaw muscles and robust crania on which to seat them, thereby freeing a trend to lighter, expandable skulls, a rich diet allowed body resources to transfer from gut to brain. It also freed time to explore the potential of new habitats. At a single bound, newly human humans freed themselves from imprisonment in a single ecological niche. Wherever there was meat and water they could go. Thereafter we see no link between climatic change, no matter how dramatic, and steps in human evolution. We moved according to what was increasingly our own agenda.

3 Adopting an upright gait forced limiting changes on primate anatomy, descended exclusively from quadripedal forbears, as well as freeing the hands from which tool use and the growth of conscious-ness spring directly. Most important it limited further development of the pelvis on which all two-legged locomotion depends for maintaining uprightness. Since the mammalian birth canal must pass through the pelvis, that cradle for the upper body confers an upper limit on the size of a foetal head relative to its mother's stature. Resolving the dilemma between increasing consciousness and brain size, that has a strong positive pressure for selection, and pelvic limitation on foetal size seems to have been achieved by birth at an increasingly early stage in foetal development. From this stems a host of important consequences. The most important involve a longer period of infant development than in other primates, when the brain grows from an unusually small size in relation to that of maturity, to a final size that is larger in proportion to total body mass. The protracted helplessness of infants demands longer, more intensive care, and so feeds positively to reinforce both semipermanent pair bonding and social co-operation; again hallmarks of humanity.

4 The various waves of human migration out of Africa do correlate with environmental change. The pressure for shifts was undoubtedly climatic, and the paths followed routes permitted by climate, geography and, in the case of the opening of the Straits of Bab el Mandab, by tectonics. In coastal areas and some island chains, global fall in sea-level opened routes during full glacial episodes, permitting colonisation of islands and then long interglacial isolation when sea-level rose. Except for the colonisation of Australasia, which demanded boats or rafts, every continental area was ultimately accessible on foot from the African heartland.

5 The rise of humanity, particularly fully modern humans also marks enormous people-induced environmental change. Whereas the pace of faunal extinction in Africa is about as expected by natural causes in a period of rapid environmental change, for the other continents the period since 100 ka marks a mass of extinctions. It has the pace and is on the scale of those in the past that signalled division of post-Cambrian geological time into

Eras, Periods and Epochs. However, it is mainly limited to large mammals, both predators and prey. The close coincidences in time between colonisation and extinctions, for the Americas and Australasia especially, seems to point directly to human causes — we seem to have eaten our way through three continents. The contrast between faunal devastation on other continents and little such sign in Africa may mark our ancestors' meeting with faunas with no contact with human hunting. Compared with the animals of the African plains, those elsewhere were naive and easy prey. Carnivores in such circumstances were outcompeted to extinction. Humans also seem to have modified vegetation. Both Africa and Australia have large tracts dominated by fire-tolerant flora, the first dating back to about 1 to 2 Ma ago, the latter to the period following human colonisation about 50 ka ago. Holocene environments became human-distorted relics of a once natural world.

FURTHER READING

Coppens, Y. 1994. East side story: the origin of human kind. *Scientific American*, May, 62–9. Brief account of human evolution of East Africa.

deMenocal, P.B. 1995. Plio-Pleistocene African climate. *Science*, 270: 53–9. Relationship between climatic change in Africa and human evolution.

Jones, S., Martin, R.D., and Pilbeam, D.R. 1992. *The Cambridge Encyclopedia of Human Evolution*. Cambridge University Press, Cambridge. Monumental compendium of biological, anthropological and archaeological aspects of human evolution.

Kingdon, J. 1993. *Self-made Man and His Undoing*. Simon & Schuster, London. A social–Anthropological view of human origins – evolving human ecology.

Leakey, R.E. and Lewin, R. 1992. *Origins Reconsidered: In Search of What Makes Us Human*. Little, Brown & Co., London. A personal view of the search of human origins, and the interpretation of the fossil evidence.

Pinker, S. 1995. *The Language Instinct*. Penguin, London. An account of the genetic basis of language and its origins from the standpoint of a psycholinguist.

Stringer, C. and Gamble, C. 1993. *In Search of the Neanderthals: Solving the Puzzle of Human Origins*. Thames & Hudson, London.

Stringer, C. and McKie, R., 1996. *African Exodus: The Origins of Modern Humanity*. Jonathan Cape, London. Exposition of the 'Out of Africa' hypothesis for modern human origins.

Tattersall, I. 1997. Out of Africa Again and Again? *Scientific American*, April: 46–53. Review of the evidence for human origins to lie in Africa, and argues for a comparatively recent origin for *Homo sapiens* in that continent.

Wood, B. and Turner, A. 1995. Out of Africa and into Asia. *Nature*, 378: 239–40. Evidence for the earliest migration out of Africa by *Homo erectus*.

Wood, B. and Turner, A. 1999. We are what we ate. *Nature*, 400, 219–20. Review of current ideas on hominid evolutionary trends.

See also references to figure sources for Chapters 9 and 10 (pp. 258–9).

11

HUMAN IMPACT DURING THE HOLOCENE

ENTER ECONOMICS: THE EMERGENCE OF AGRICULTURE

The swift emergence from the last Ice Age 10 thousand years ago meant that grasses of various types spread beyond the tropics, as they had in earlier interglacial periods. None was a new species, and the seeds of each were edible for humans, but no means of gathering them efficiently had previously been possible (there is evidence for gathering of wild cereals from sites around the Sea of Galilee dated at 19 ka). The widespread and new use of small flakes from stone tool making – so called microliths – signifies that grass seeds did enter the human diet in volume in the Neolithic. By making composite tools, such as sickles of microliths embedded in wood or bone, humans began to adopt eating habits previously confined to paranthropoids. No human in full possession of their senses or a care for their teeth would contemplate chewing uncooked grain for sustenance. Cooking grain or grinding it and then baking makes grain an easy and abundant source of carbohydrate and protein. By 6 ka agriculture based on selective planting of the most productive grains and, through that, unconscious breeding, arose in the Middle East and the Indus valley (wheat, barley and oats), Africa (sorghum and millet), east Asia (rice) and central America (maize). Figure 11.1 shows when and where the world's dominant agricultural crops and domesticated animals originated.

Plant characteristics selected during domestication included increased seed size and number, self-pollination (this retains selected characteristics through cloning), better harvesting properties, more rapid growth and maturation, reduced nutrient and water needs, and improved flavour. The history of wheats provides a good example of the complex genetic history of organisms adopted and bred for use by humans (Box 11.1). Techniques that retain selected properties by propagation (cuttings and grafting) did not arise until Roman times. Surprisingly, many soft fruits that had been gathered from the wild for millennia were not cultivated until the Middle Ages. Recent trends focus increasingly on the profitability of mechanised agriculture to suit global markets – uniformity of size, synchronous ripening, adaptation to machine harvesting, lengthy storage life and many others. Now aided by genetic engineering, agriculture tends towards monoculture, where single fields up to hundreds of hectares lie under crops with a single genetic make-up. This uniformity brings advantages but also the problems of controlling disease and pests; if one plant is susceptible to attack then so are all the others.

At roughly the same time as wild plants began to be deliberately cultivated there began the domestication of naturally herding or gregarious animals (many of them ruminant grass-eaters). Selective breeding (conscious or not) favoured a variety of characteristics, including docility, reproductive success, rapid muscle and thus meat development, softer wools, higher milk yields and even aspects of body shape and stature for specialised tasks, most obviously among horses and dogs. Interestingly, early domestication of the formidable wild aurochs, the ancestral cow,

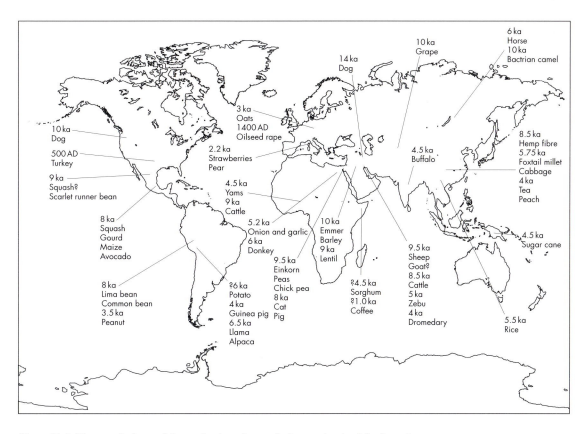

Figure 11.1 Times and places of domestication of several plant and animal food species.

favoured much smaller animals more easily managed and suited to smallholders' needs for milk and occasional meat. Figure 11.3 shows how this miniaturisation trend has reversed in the last 500 years as conscious breeding has sought to maximise milk and beef yields.

The continuing agricultural revolution increasingly demands pesticides, fertilisers and irrigation to meet economic demands. As well as blighting natural species diversity, repeatedly applying pesticides fosters evolution of resistant strains of pests. Numbered among these are so-called 'super-rats' that are immune to Warfarin, resistant weeds, fungi and insects, and perhaps most worrying, bacteria and other pathogens that have evolved resistance through the increasing use of antibiotics in livestock. Some, like the agents involved in BSE and CJD may even be

the direct products of routine use of mutagenic organo-phosphate insecticides, as well as disturbingly unnatural feeds.

The obvious fact that controlling a food supply is a great deal easier and more productive than perpetually looking for one, meant that more people than those needed to be involved in supplying food could survive. Figure 10.6 shows how world population may have grown following the domestication of food plants and animals. Agriculture by increasing the productivity of human labour does not merely allow population to grow, but by expanding generates a surplus either in store or on the hoof. Surplus production and the potential for at least some people to have time free from manual labour seems to have spurred the start of trading and thereby various forms of economy. Barter, based on exchange of

Box 11.1 Wheat

Modern wheat, *Triticum aestivum* has six sets of chromosomes in three pairs; it is *hexaploid*. Gene studies of wild wheats reveal some aspects of the origins of our farm wheat (Figure 11.2).

Wild diploid (Box 8.2) wheats have low yields and live on poor soils. Their ears are brittle and shatter when ripe to disperse the seed. This would make harvesting difficult, and it is obvious that early unconscious selection produced wheat with a less brittle head. Archaeologists have found evidence from about 10 ka for cultivation of semi-brittle wheats (*einkorn*) and an early tetraploid form (*emmer*) in the Near East. Emmer wheats were being cultivated in the British Isles by 6 ka.

Early cultivated wheat seeds, although fairly large, were tightly held in strong husks, that hinder removal of the clean seed during threshing. Selection, presumably unconscious, eventually produced ears with loosely held seed that is more easily threshed. Such tetraploid *durum* wheats are hard grained and grow well in dry regions. They are still a major crop in the Mediterranean, Australia and India. Low in gluten they produce flour suitable for pasta and unleavened bread, but do not produce the elastic doughs needed for breads with raising agents.

Evidence for hexaploid wheats first appeared in Turkey at about 9 ka, but the free-threshing mod-

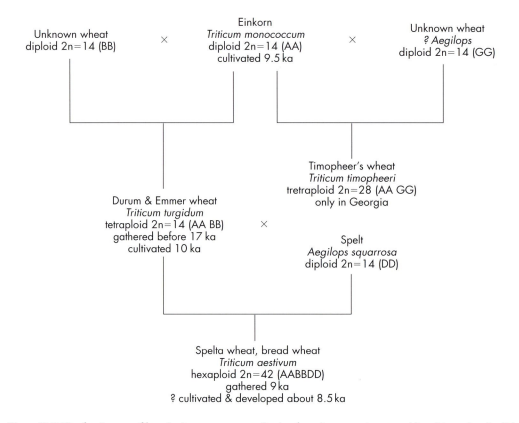

Figure 11.2 The family tree of bread wheat reconstructed using breeding experiments with cultivated and wild wheat forms.

ern wheat types (*T. aestivum*) appear in what is modern Syria and Iraq after 8 ka. Its cultivation was on irrigated plains and by about 7 ka the Egyptians used it on the Nile floodplain. Hexaploid wheats have a high gluten content making the dough sticky and elastic, so that it rises when mixed with yeast.

Wheats have dominated as a food crop in temperate continental climates. There are now over 25 000 different cultivars and the rather haphaz-

ard, unconscious or selective developments by early farmers has given way to multimillion pound breeding programmes. Yields have been increased by increasing grain size, number of grains in the ear and number of ears per plant. Crops are now more pest resistant, shorter stalked to eliminate collapse before harvesting and require highly fertilised soils. They are a far cry from the lowly diploid of rocky, nutrient poor soils of the Near East.

goods according primarily to their usefulness, has limits related to the volume of such trade. By having more than a certain amount of a particular product makes it increasingly difficult for a trader to obtain

suitable goods of other kinds by barter. A more convenient means of trade is using money that has a value that is exchangeable for any kind of product; a universal equivalent of value. For a host of reasons (among which are its inherent uselessness combined with the mystique of its rarity and attractive properties) gold achieved pre-eminence as a universal money before the rise of pharaonic Egypt. The universal buying power of money presents the opportunity of turning economy focused on the usefulness of produce on its head. Instead of remaining an intermediate convenience, it and the exchange value that it represents can be used to acquire produce only in order to resell it at a profit, thereby increasing money in the form of capital. Instead of various useful goods, the products of labour thereby become commodities whose intrinsic usefulness is secondary to their marketability. That is the underlying mechanism to modern capitalism, although such trade existed in embryo for millennia before it took on its present form.

Moving produce to market demands protection, for which a price must be paid. Individuals and groups freed from the need to work the land are able to trade and to protect transportation. By transferring surpluses to traders, those remaining to work the land become progressively enslaved to it, particularly when part of their surplus is demanded in taxes to assure protection. This is a crude explanation for the rise of feudalism and its inevitable suppression and control of both the tiller and those engaged in commerce. At its base is the peasant producer, tied to the land, and a deep conservatism founded on the

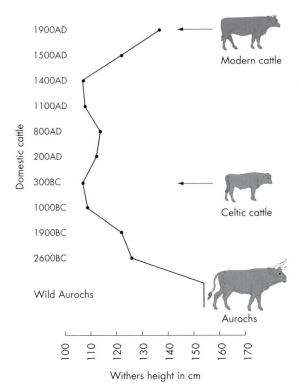

Figure 11.3 Size changes in British cattle since the domestication of aurochs. Many Celtic breeds, such as the surviving Dexter, were little taller than large dogs.

insecurity of primitive agriculture. To survive for millennia meant that ruling elites had to stifle any alternative forms of economy and means of production, either by force or through the fixed ideologies represented by the religions that evolved. Without exception, all sought to place people in god-given, fixed roles. Feudalism can be looked on as the lid of a pressure cooker. Beneath it contradictions grew between its repressive and rigid form, and inevitable discoveries about the world, which presented new opportunities for improving human life and profiting from the exploitation of such discoveries. Bursting from its feudal confines through the social revolutions in seventeenth century Europe allowed it to explode into a globalised economy through the Industrial Revolution. It escapes very few of us that modern relations between humans and the rest of the world centre on virtually every part of the environment either being a commodity or related in some way to commodity production and trade. Capital has become an environmental, indeed geological agent!

This partial side-step through agriculture into paranthropoid habits has in 400 generations moved humans from the Stone Age to using Pentium processors, into various forms of class society, nation states, global warfare, racism, massive pollution, dental decay and BSE. Whether or not pre-historic humans had a signal effect on their environment, and the chances are that they did, their consciousness extended only to their immediate surroundings and how they might ensure survival and well being. A mere 300 generations since Britain began to be converted to an entirely human environment, which it is without doubt today, human consciousness encompasses the globe and begins to grasp its interconnected workings. To some extent every human beyond tender years is aware of a deteriorating framework for their life. In every facet this has one root in a global economic system that is increasingly moribund, yet others that emanate from universal natural processes. In the next section we look at some specific environmental changes that stem directly from the Holocene episode of human social evolution, before examining likely generalised effects of modern humanity's actions.

DIRECT HUMAN IMPACT ON HOLOCENE ENVIRONMENTS

Despite the fact that millions of Europeans seek to protect their surroundings, particularly landscapes that have great cultural connotations, there is very little in modern Europe that is appropriate to the climatic conditions of an interglacial. Evidence from previous warm—wet periods shows that much of the continent was cloaked in dense forest where now there is only patchy tree cover. You saw evidence from the pollen records of sediment cores in Chapter 8 that large tracts in Britain had been cleared before 5 ka. Part of the clearance was to create agricultural land, but an increasing use of timber in dwellings and boats, for fuel and, after metals began to be used, as the essential reducing material in smelting further depleted the "wildwood" that previously blanketed most of lowland Britain. An area in SW England provides persuasive evidence for the pace at which deforestation took place in antiquity.

Deforestation

The Somerset Levels, between the Mendip and Quantock Hills is close to sea-level and the repository of sediments laid down in post-glacial swamps. Cores through the peat now exploited for horticultural uses show that it rests on a base of marine clays laid down before about 7 ka when rising sea-level outpaced the slow recovery of the land from glacial loading. Eventual emergence left a treacherous waste of swampland in which peat accumulated from the debris of mosses, reeds and sedges, until about AD 200 when the peat began to be cut for fuel. Modern mechanised stripping has unearthed evidence of how early people coped with this barrier to communication by building boardwalks. Figure 11.4(a) shows the earliest known trackway, dated using the ^{14}C method at about 5.7 ka. It consists of poles cut from mature hazel, ash, willow, holly and lime that support planks of oak and elm. Figure 11.4(b) shows a later track. It is very different from the first, being constructed of thin poles woven into hurdles. Growth rings show that the poles are 4 to 9 year-old regrowths. They are the product of a woodland management practice known as coppicing, where instead of felling an entire tree for timber, sufficient is left for the trunk to regrow a crest of thin poles. These can then be harvested for construction (as the base for mud in daub-and-wattle housing), for hurdles and for fuel. This later trackway is 5 thousand years old. Clearly within about one thousand years woodland in SW England had become so sparse as to warrant management measures.

Coppicing was widespread until recent times, preserving small, highly modified 'oases' of the pre-colonisation wildwood that cloaked the whole of lowland Britain in the early Holocene. Studies of the age of regrown poles in archaeological sites show a consistent management cycle of 4 to 9 years until about AD 1400, after which the cycle lengthened to 11 to 24 years before AD 1700. This coincides with the minor climate change known as the 'Little Ice Age', that slowed down regrowth. Increased demand for increased indigenous food production, following

(a)

(b)

Figure 11.4 Somerset trackways. (a) The Sweet Track dated at 5.7 ka, exposed in peat workings on the Somerset levels that extends for almost two kilometres. (b) The Walton hurdle trackway on the Somerset levels whose construction is entirely of hurdles made from coppiced hazel rods. It is dated at 5 ka.

the Second World War and Britain's loss of its Empire, saw between one third and a half of remaining ancient woodland destroyed between 1945 and 1975. A similar pattern emerges from studies throughout the rest of lowland western Europe.

In parts of the Scottish Highlands patches of natural open woodland dominated by conifers, show that the present climate can support thriving forest cover. That it did so in the early Holocene shows clearly in pollen records from northern sites, only to dwindle during the last few thousand years to be replaced by *Sphagnum* moss spores and heather pollen. Without any sign of dramatic climate shift upland Britain gradually became deforested. This perhaps started with the use of uplands for hunting and a source of constructional timber during the Mesolithic and was completed by the clearance of the Highland's human population in the eighteenth and nineteenth centuries for private hunting and runs for the newly bred Cheviot sheep that can survive Highland winters without shelter. What has replaced the upland equivalent of the wildwood is a uniform blanket of peat bog and heather moorland. Bogland is probably a human creation, though encouraged by natural processes. When wooded, soil subject to the intense rainfall of Europe's western rim maintains a low water table because the trees transpire vast amounts of water. Once deforested more water is retained until the soil becomes waterlogged. Trees will not regenerate from seeds as root systems rot and cannot sustain the symbiotic aerobic fungi and bacteria essential to their nutrient uptake. Only mosses and heathers can colonise the newly barren soil, further driving it into saturation by increased water retention. Rotting of dead moss in anaerobic conditions generates highly acid and reducing water chemistry. Excavation of blanket bogs to use peat as fuel reveals the human source of upland scenery in Britain, for at their base is generally a thicket of tree stumps, some showing clear signs of having been deliberately cut down.

Given the human origins of Britain's treasured landscapes the resistance to reforestation is full of paradox. It cannot happen naturally because of waterlogging stemming from earlier intervention. Instead bogs are deep ploughed with seedlings planted between the furrows that drain water away. The tragedy is that the transformed water chemistry is capable of mobilising a host of metals toxic to fish and their prey invertebrates, as well as being highly acid. While anthropogenic bog remains undrained, water slowly seeps to watercourses, precipitating its dissolved load and being neutralised *en route*. This also regulates the pace at which water enters drainages, while reforestation schemes accelerate run-off to give erratic water levels. Eventually a hydrological and chemical balance will develop as plantation trees reduce the water table, aerate the soil and establish root masses that slow run-off. Although commercial forestry, dominated by single, often alien species, is not ecologically very diverse, plantations of mixed native tree species can encourage wildlife diversity.

Deforestation at lower latitudes is a more recent phenomenon, partly because of different soil types, greater evaporation and faster regrowth of cut areas, and explosive population growth in previously disease-prone areas. It takes the form of clearance by slash and burn in newly colonised areas, but also through the devastating effect of increasing herds of livestock. Most damaging among these are goats, which unlike grass-eating sheep and cattle consume and digest anything green, despite shrubs and tree seedlings' often thorny protection. Increasing herd sizes, which are growing faster than human populations, is cultural in origin. Many societies reckon a family's wealth and standing from the number of its herd animals rather than their quality or usefulness. Goats wrecked the natural ecosystems around the Mediterranean by the time of the Greek civilisation, and have spread throughout sub-Saharan Africa. Adding to these factors are the use of wood for fuel and commercial logging for timber. Another agent, though not always directly human in origin is the increase in the number and area of wildfires, encouraged by episodes of drought. The blotting out in Malaysia and Indonesia of an area the size of Europe by smoke from wildfires in the Indonesian forests, and equally devastating fires in south-west USA and in Australia, signify huge losses in low-latitude forest cover.

Although the most obvious outcomes of tropical deforestation are loss of habitat and biodiversity, and short-lived, intense agricultural production before soil structure and sparse nutrients become irrecoverably exhausted, low-latitude forests have an important impact upon climate. They are the largest carbon repository at the Earth's surface, burning adding to the atmosphere's CO_2 content and the 'greenhouse' effect. Moreover, through transpiration they recycle water from soil to atmosphere, and by respiring fine particles of dust and resins they help water vapour to condense as clouds that shed rain. The consequences are clear from deforestation around the Panama Canal in Central America. Less rain now falls so that the reservoirs from which the lock system draws its water are at all-time low levels. Traffic from the Atlantic to Pacific oceans is already limited to avoid completely draining the reservoirs, and may halt should rainfall decline much further. Worrying as these trends are, loss of natural vegetation cover indirectly threatens human survival in many parts of the world.

Soil Erosion

Whether from trees or grass, vegetation roots serve to bind soil and the weathered rock on which it develops. Once this biological binding is lost soil is exposed directly to rainfall and flowing water. With a cover of leaves in tree and shrub canopies, falling rain drops are intercepted to flow more quietly to the soil surface where they infiltrate to recharge subsurface water. Leaf shade also slows evaporation, while rootlets and dead plant matter improve the permeability of the soil's surface. Heavy rain falling directly on to soil pounds it to a mud that dries to brick-like consistency, thereby preventing water from infiltrating. Instead it runs off the surface unhindered by any plant cover. Even on slopes as slight as a few degrees such sheet flooding develops rilles that begin rapidly to erode the soil (Figure 11.5). Both soil 'hardening' and erosion sterilise affected land from agricultural use. The great famines that accompany drought in sub-Saharan Africa are exacerbated by lost agricultural land that follows deforestation. In a vicious

circle, soil erosion and increased run-off feeds into attempts to harvest water as an insurance against drought. Reservoirs of all sizes rapidly fill with silts that ultimately render their construction useless. Sediment cores in natural lakes chart the onset of soil movement as agriculture spread and developed. Under mature forest cover virtually no soil enters rivers. In grassland about 2.5 tonnes per hectare is lost annually, while bare ground sheds up to 18 tonnes per year, depending on its slope. Much of the soil lost from the famine-prone and recently deforested Ethiopian Highlands ends up either in the arid wastes of the Red Sea lowlands or transported into the Blue Nile (which incidentally has been brown not blue for most of this century). Ironically, the fertile soils lost from Ethiopia once replenished the agricultural lands of the lower Nile valley and delta during annual floods. The building of the Aswan Dam in the 1950s to harness Nile water for hydro-power and regulate irrigation below it prevented this annual benefice. Lake Nasser is rapidly filling with silt.

The history of the Mayan civilisation in Central America reveals the devastating impact of agricultural society early in its history. Lake beds within the boundary of the Mayan Empire contain a layer of clay, dating from about 300 BC, that was deposited as soil erosion increased while the Classical Mayan culture cleared indigenous forest. This Maya Clay contains pollens of maize, the staple of Mayan economy, and weeds. Although maize cultivation continued, the Mayan culture collapsed around AD 850. Open forest developed and the Maya Clay ceased to be deposited, as soil erosion fell. Growing population and limited land in valley bottoms forced the Mayans to cultivate slopes and intensify their farming. Towards the end of their Classical period, skeletal remains reveal a pandemic deterioration in health and increased mortality among infants and children. Maize requires deep, fertile soil, and a combination of erosion and exhaustion of nutrients, possibly through leaching and over-intense cultivation, fostered harvest failures and thereby malnutrition and eventual population collapse.

Erosion is not accomplished by water alone. Where soils are fine-grained, as in the floodplains of large

Figure 11.5 Typical gullied surface of eroded soil on gently sloping land.

rivers or in areas of glacial tills deposited by Pleistocene ice sheets, they can be stripped by wind action once devegetated and dry. This is directly akin to the formation of loess by winds blowing across unvegetated till during the last glacial period. Intense, monocultural cereal cropping in the American midwest stripped the prairie soils (glacial tills and glacio-fluvial sediments) of their permanent grass cover. After repeated dry summers the bare soil began to blow away to create the infamous 'Dust Bowl' of the early 1930s. In one four–day wind storm alone, 300 million tonnes of soil blew away. Exactly the same fate threatens the world's largest irrigation scheme in the Ghedaref and Gezira areas of the Sudan that use water from both Blue and White Niles to develop the fertile silts of the hyper-arid land between them. As well as a vast network of carefully engineered canals, huge fields were mechanically levelled and put under cotton and cereals. Violent seasonal winds continually strip soil from this massive project, accounting for dust storms and even sand dunes in the middle of Sudan's capital Khartoum. Potentially capable of feeding the whole of sub-Saharan Africa, Sudan remains a net recipient of food aid.

Desertification

Irrigation seems a revolutionary solution to food supply, and schemes for making deserts bloom date back to the very earliest civilisations in the Tigris and Euphrates basin and that of the Indus. Observation of modern schemes highlight great dangers. Where evaporation rates are high irrigation water seeps into the soil, but a combination of surface evaporation and capillary rise draws it back to the surface. Although seemingly fresh, the water contains a dissolved load and also dissolves soluble salts from the soil itself. After initially high crop yields, evaporation gradually increases the salinity of the upper soil. Eventually it develops saline encrustations and becomes barren. Only annual flooding, leaching of salts and their transport into drainages prevents such salinisation, a natural process in the Nile valley until the building of the Aswan High Dam. The problem is made worse by over-watering as the water table rises and yet more moisture evaporates by capillary rise to leave its dissolved content in the upper soil layer. Archaeology records the fate of the Sumerian civilisation in Mesopotamia, which depended on irrigation from the Tigris and Euphrates. The Sumerians' written records show that yields of grain were over 2500 litres per hectare 4350 years ago. In 300 years this had fallen to 1500, and in another 400 years yields fell below 900 litres per hectare. Clearly this early agricultural society delayed its downfall by deliberately changing its crops. The early productive emmer wheat was replaced by more salt-tolerant einkorn, and finally wheat was abandoned for the even more tolerant barley. The efficiency of the Sumerians' water engineering rebounded on them to destroy the civilisation and power shifted to the less affected areas around Babylon to the north, higher in the drainage system. To this day the great plains of southern Iraq and south-west Iran have barren, saline soils.

Soil erosion by wind and water, together with salinisation, render land infertile. The affected areas become deserted, being essentially useless for human occupation. This is desertification proper and it is not reversible within human lifespans. The onset of aridity, and loss of vegetation at the interface between bare land and that with some vegetation cover does indeed see the spread of deserts in the popular sense of sand dunes and rocky wastes. However, this is as much the product of drought cycles as human intervention. The notorious Sahel belt in sub-Saharan Africa recovers and shrinks rapidly after a few seasons of high rainfall, in a smaller version of the long-term advance and retreat of aridity that follows the astronomical climate cycles. Albeit overgrazed, the arid front of human existence is not the victim of irreversible damage. The greatest dangers come from human intervention in the natural hydrological and nutrient cycles, and from grossly disturbing natural ecosystems, particularly in industrial agriculture that depends on irrigation, huge fields and single-species cropping.

Pollution

Until about 6 ka, human activity centred on the purely biological world, taking only common rocks for tools. The man found frozen in a glacier on the Austrian–Italian border in 1993 perished from exposure at around 5 ka, but among his possessions was a sign of another great change in human activity. He carried a well-crafted copper axe hafted with yew wood. Copper and other base metals show a sudden, though small rise in Greenland ice cores at depths equivalent to 4 thousand years ago. This may link to the rise of metal smelting in China during the Shang Dynasty. The levels rise to a peak in the period 500 BC to AD 300, at the acme of the Roman Empire, then decline in the European Dark Ages. Another peak marks the Chinese Sung Dynasty around AD 1000 (Figure 11.6). The reflection of this record in the remote Greenland ice cap signifies the global effect of even small, localised centres of metal production. In the 200 years since the Industrial Revolution production has risen by more than three orders of magnitude from those of the Middle Ages. Fortunately, modern smelting is more efficient than that of earlier times, when up to 20 per cent was lost compared with 0.2 per cent today. However the thousand-fold increase in production still means that emissions of metals are 10 times their level during Roman times.

Metals were first smelted from rich ores, but increasing efficiency and improved extraction has meant that the metal concentration needed in ore has dropped. Today it is only a few orders of magnitude above that found in common rock. The thousand-fold increase in metal production has been accompanied by the production of many million times the amount of waste. This does not merely leave unsightly holes and spoil heaps, but exposes rock with greater than normal metal contents to geological processes. Many of the ores contain sulphide minerals that rapidly oxidise in air and surface water, to form sulphuric acid. This not only acidifies surface water, but helps dissolve metals in streams. Smelting ores also releases sulphur from ore as the gas sulphur dioxide, and that adds to acid rainfall. The magnitude of this chemical mobilisation by mining has reached such a level as to match or exceed the natural movement of dissolved metals in all the world's rivers. Of course much is locked in useful and valuable products, but a visit to any town dump shows that even the most valuable commodities are eventually returned to the environment at a rapid pace. Entry of metals into water at unusually high concentration not surprisingly

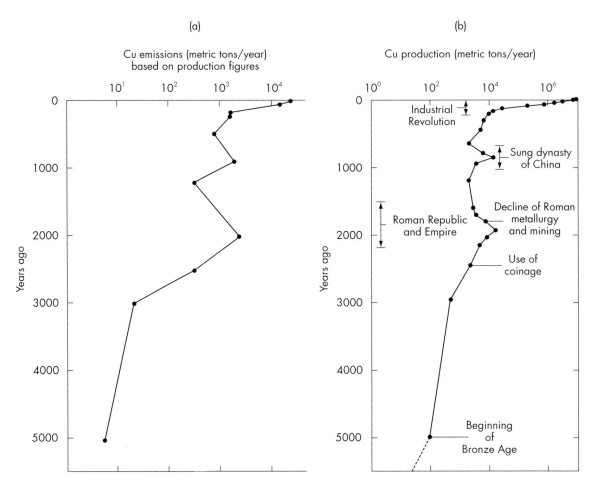

Figure 11.6 (a) Changing release of copper to the atmosphere, due almost entirely to smelting since the Bronze Age, calculated from (b) estimated world copper production.

damages aquatic life, and this extends to the open oceans as well as rivers in industrialised countries. The toxic effect of heavy metals lies in the inability of living cells to excrete them so they build up to damage cell structure. Such concentrations pass upwards through the food chain so that higher members, such as predatory fish contain higher metal levels. Humans sit at the very top, and horrifying outbreaks of mercury and cadmium poisoning among Japanese children in the 1960s, due to their high intake of locally caught shellfish, first spotlighted the feedback of human economics to people though natural processes.

Rainfall is naturally acid, through its content of dissolved CO_2. Such carbonic acid is responsible for the solution of limestone to form cave systems, and is the main agent (more properly the hydrogen ions released by acids) for rock weathering. It is however a weak acid, only lowering pH to about 5.6. Sulphur dioxide dissolves to form sulphuric acid which pound for pound releases far more hydrogen ions than does dissolved CO_2. Rainfall in large tracts of Europe, North America and parts of south Asia now has a pH as low as 4, due to sulphuric acid. Though released naturally by volcanoes, and in the form of stratospheric aerosols, SO_2 is implicated in climatic cooling. It is increasingly released by burning fossil fuels, as well as by smelting sulphide ores. Both coal and petroleum formed under conditions that not only encouraged the preservation of organic hydrocarbons but also the precipitation of iron sulphide from unoxygenated water. Falling as sulphuric acid in rain, this worsens the mobilisation of metals dissolved in stream water, and leaches soils of critical metals such as calcium while concentrating others, such as soluble aluminium compounds. Increasing use of fossil fuels, particularly in large thermal power stations whose emissions vent high in the atmosphere to travel large distances, is matched by declines in tree growth and the increased die-off of forests. Part of the damage seems to stem directly from acid and metal effects on leaves. But studies of recent tree rings reveal a decline in calcium content. The last might link to either leaching of soils by acid rain, or from the build-up of soluble aluminium compounds in soil, which inhibits calcium metabolism in root systems. Acid-rain damage to forests is distributed invariably downwind of major industrialised areas, often hundreds of miles distant.

The list of emissions is long, topped in terms of mass by carbon dioxide release through burning fossil fuels, of which more shortly. Close to the bottom of the list are the chloro-fluoro carbon compounds developed in the 1950s as propellants for aerosol sprays and used as much more efficient heat engines in refrigerators than earlier compounds, and in electricity transformers. Only tiny amounts emerge, but they are insoluble and almost indestructible. So they find their way into the stratosphere. There they wreak havoc out of all proportion to their abundance. Each carries a 'dangling' chlorine atom in its molecular structure, and this catalyses the breakdown of ozone (O_3) to ordinary oxygen. As upper-atmosphere ozone is the main protection from ultraviolet radiation for land animals and plants, patchy 'ozone holes' form a major health threat and may endanger crop productivity. Ironically, the inefficient burning of fossil fuels in internal combustion engines and the formation of photochemical smog from their emissions actually increases ozone low in the atmosphere. As it is an irri-

tant (indeed it is toxic) such anthropogenic ozone is implicated in the rapid increase in asthma and various allergies in urban environments. Being an even stronger oxidising agent than oxygen it interferes with the fundamental reducing reaction bound up with photosynthesis. Increased lower atmosphere ozone threatens also to reduce biological productivity.

Changing Oceans

In terms of volume, the oceans represent by far the largest habitat. So vast, and until recently so poorly known, that humanity has long taken the oceans for granted as both a lucky dip and a dumping ground. River pollution mentioned earlier comes to rest in the oceans, and increasingly everything from raw sewage to nuclear waste and redundant munitions is deliberately and cheaply disposed offshore. While any yachtsperson witnesses flotsam in even the most remote wastes, the worst scenarios are presented by enclosed seas, such as Europe's North Sea. Virtually every river that flows into the North Sea carries a toxic load so virulent that rumour has it that water from the Rhine can be used to develop black and white film! They also carry a rich cocktail of both nutrients from washed-off fertilisers and pathogens. This was demonstrated horrifically in the late 1980s. Satellite images showed the development of a huge patch of red algae at the outlet of the Baltic in May 1988. This was a 'bloom' triggered by the outpouring of fertilisers dissolved in the spring floods from the German and Baltic lowlands. Within four days the red algal bloom enveloped Denmark and tidal currents carried it westwards across the North Sea. Die off and decay of this red tide consumed oxygen from the sea water, to decimate near-shore fish stocks. Shortly afterwards Common and Atlantic Grey seals began to die in their thousands on the shores of both Britain and continental Europe. Their deaths were from a variant of canine distemper virus, undoubtedly emanating from rivers, since dogs are not well known for chasing fish. While impossible to prove, the coincidence may have reflected a common, though highly complex cause for both algal bloom and the seal die-off. The virus crossed species in an unprecedented way that seems likely to have involved some mutation. The event is reminiscent of the later discovery of trans-specific mutation of the BSE agent to pass from sheep to cattle to humans, which seems increasing likely to implicate the use of organo-phosphorus pesticides.

Probably from the time of *Homo erectus*, humans have exploited the sea for food. Only in the last two centuries has this cropping ventured beyond coastal waters. Increasing efficiency of netting seemed to promise a limitless harvest of high-quality protein, both for human and livestock consumption. This neglected what now is a seemingly obvious fact. Vast as they are, the oceans are biologically much less productive than the land surface. Exhaustion of fish stocks was inevitable, and so it has proved for every popular species in the North Atlantic. The teeming Newfoundland Grand Banks by 1993 were almost devoid of the great staple of the high street fish and chip shop, cod, and to eat herring in Britain is a luxury. The repercussions of overfishing large food species are unusual and far-reaching. All are low in the food chain, depending on either phytoplankton (marine plants) or zooplankton (small invertebrates and larvae). Population numbers of smaller species, particularly sand eels, exploded. Bird species favouring small fry, such as fulmars and puffins, multiplied rapidly too. One of the most startling transformations in the seaside scene is that fulmars are everywhere, whereas once they were restricted to the far north-west of Britain's shores. Despite their grace and excellent lines, fulmars are aggressive birds, and even quite sturdy herring gulls have been driven from their nesting sites, perhaps by the fulmars' early nesting habits, or because they vomit spectacularly when disturbed. Now commercial fishing seeks to harvest the sand eel, mainly as input to livestock feed, by literally sucking them out of the sea. The unfortunate puffin, renowned for its 'moustache' of sand eels held in its parrot-like beak, is now in decline, outcompeted by the voracious fulmar.

FORWARD TO THE PAST, QUICKLY?

The distortion of Holocene environments from what they 'should' have been in the normal course of an interglacial, to an increasingly humanised and indeed commodified condition of biological life is increasingly obvious to all. Each of the processes briefly discussed on pp. 240–8 threatens to rebound on human society as well as to transform the rest of the biosphere. On pp. 228–33 we examined the most recent mass extinction, for which the agency was neither a comet nor some geological or climatic catastrophe, but the competing and increasingly conscious role of the human species and its culture, its 'second nature' that allows us to live almost anywhere. The great question is this: are we to become victims too, or at least is our culture doomed because of itself? Many people with understandable concerns about human impact on the environment fear most a loss of the Earth's richness from which our culture ultimately stems. The foregoing does indeed pose such a threat of a dreary, depleted planet. Loss of agricultural land does threaten the lives of hundreds of millions, particularly at the fringes of arid lands and those most prone to soil loss by erosion. However, none seems likely to bring about extinction of either our species or its technological ability to cope with most of the problems brought on by an economy dominated by exchange-value over usefulness. Yet we live in environmentally turbulent times as a matter of natural course. The climatic record shows with little room for doubt that interglacials are about one tenth the length of cold–dry glacials.

The Holocene has lasted for 10 thousand years, and every previous one that we can resolve over the last 2.6 million years has never lasted much longer than that. Both the entry to and emergence from glacials have been abrupt and even chaotic especially the exit. The warming from the depth of the last Ice Age 18 thousand years ago was interrupted by the onset of a millennium of almost full glacial conditions during the Younger Dryas from about 11 to 10 ka. Its effects are recorded globally, and they signal cold and drought. It emerged within a matter of a few decades, maybe even a few short years. The most significant lesson from the discussion of glacial periods and shorter stadials in Part 1 is that vegetation, particularly forests, did shrink with decreased humidity in both hemispheres. What we regard as the most agriculturally productive climate–vegetation zones moved southwards and they contracted hugely, to be replaced by a much expanded zone of tundra at high latitudes and at lower elevations than now. Winds were more violent to set vast amounts of dust and sand in continual motion so that vegetation could not get a foothold in much larger sandy deserts, both cold and hot. This too forced back the boundaries of potentially fertile land.

Despite its outward expression of an industrial–technological culture, the 6 billion strong human population depends entirely on feeding itself through cultivation of a quite narrow range of staple crops, dominated by grasses, and an even narrower selection of livestock. It is immaterial whether the economy whereby this produce is distributed is dominated by exchange-value or usefulness. The simple fact is that the present distribution of the world's population and of the crops and agricultural methods on which its different parts depend has become established over 300 to 400 generations. The last five generations that populated the world have been witness to global conflicts of the most savage kind triggered only by disputes over the distribution of capital, not the real resources on which humanity depends. Should those dominantly food resources decline suddenly by a few per cent because of climatic shift to cool, dry, stadial conditions, the prospects of violent social upheaval on a global scale are difficult to avoid. Food production would be compressed to narrow belts at mid latitudes, separated from an equatorial humid belt by expanding arid zones. Populations would be compressed into these belts, abandoning the present main centres at high and tropical latitudes. Perhaps there would be some relief by an expansion of grasslands suitable for cereals and herds at low latitudes, but moving perhaps 3 billion people to places where 3 billion already live has apocalyptic overtones. A glimpse of the social outcomes of minor cooling in the past is outlined in Box 11.2.

The dominant fear among sections of politicians is

not such a cooling event, although it cannot be far off in the 'natural' course of events. Most warnings to which we are daily exposed concern global warming, the outcome of 'greenhouse' gases, mainly CO_2, building up in the atmosphere from burning of fossil fuels. Setting aside the estate agents' and insurance brokers' worries about drowned real estate following melting of ice caps and sea-level rise, what are the dangers from that? Surely, you might think, that would expand the fertile areas by an increase in humidity and a polewards expansion of frost-free winter conditions, earlier germination and later harvesting. And arid lands would retreat. Superficially, that looks like a rosy prospect. But the atmosphere transfers moisture evaporated from equatorial regions polewards in the Hadley circulation, and the trade winds add to that poleward transfer by picking up water vapour from the oceans that they cross. Precipitation would increase in areas already well-endowed with rain, and that includes much of northwest Europe and as far as the polar front that hovers around the latitude of Iceland. That however is only part of the likely scenario. Atmospheric circulation transfers heat polewards too, both in sensible form and locked in water vapour as latent heat. Both would increase. At this point you would do well to recall our conclusions about one of the dominant controls on the climate of the last 2.6 million years, that of the deep circulation of the Atlantic ocean that seems to explain why the Milankovich signal is biased towards its effects on Northern Hemisphere insolation (Chapter 4).

Studies of past climate show convincingly that the short-term advances and retreats of glacial conditions during the last ice age link not to any astronomical forcing, but to a less regular turning on and off of the deep-water conveyor. The most likely is an influx of freshwater from melting ice. When that happens, the flow of warmer tropical water, the surface Gulf Stream, stops. Global warming through economic activity presents entirely new conditions, hard to model by example from the past. However, we can speculate in the light of the known generalities. Global warming by itself should encourage two things; more melting of the Greenland ice cap and

Box 11.2 Greenland and Iceland

The Little Ice Age from the fifteenth to nineteenth centuries was preceded by a warmer period starting about AD 700 called the Little Optimum. This heralded bouts of North Atlantic exploration and colonisation. The Irish Scots found and lived on Iceland (which they called Thule) from about AD 770 to 870 when they appear to have left in a hurry on the arrival of the Vikings. The Norse people settled Iceland permanently and by AD 930 they had turned all the suitable land into farmland.

By AD 975 Iceland was overpopulated and famine became a real threat. Soon after this, Eric the Red, who had been exiled from Iceland for three years, set off to explore land further west. He landed there and named it Greenland. This sounds rather incongruous, as we now know it primarily for its ice sheet, but the coastal strips were a rich green turf, and Eric was trying to tempt other settlers there to join him, so he needed a favourable name!

The Greenland Norse population grew rapidly, and may have reached over 6000 people. Here, crops were impossible to grow, so they survived on fishing, their cattle and sheep, and hunting whales, seals and reindeer. Trade in walrus ivory, eider feathers and animal skins probably provided the other essentials.

All looked to be going well, but in the 1400s the colony dwindled and finally disappeared. The exact reason for this is still unknown, and indeed it was probably due to a series of problems. There is evidence from ice cores that the climate was deteriorating and temperatures were 1–2°C lower than during colonisation. This would have led to bad years when the hay harvest was insufficient to feed the farm animals through the winter and deteriorating weather could have tipped an already overgrazed ecosystem over the edge. Increasing sea ice made fishing and communication difficult and there seems to have been conflict with other inhabitants of the islands.

The Inuit arrived in Greenland from the northwest at about the same time as Eric the Red in the south. For 500 years the two groups lived in an uneasy relationship which seems to have included both trade and skirmishes. However, graves of the Norse show that they never learned the art of warm dressing in skins from the Inuit and continued to wear the latest Scandinavian fashion right to the end. The Inuit, who remain to the present showed that survival in a hunting culture is quite possible on Greenland, but importing a different lifestyle could lead to disaster.

---

less formation of sea-ice in winter. Both work to reduce the formation of cold brine. Warmer tropics encourage greater evaporation. Part of the load of atmospheric water vapour moves polewards to shed as rain and snow at high latitudes. This too reduces salinity of sea water at high latitudes. For this decisive part of the climate system these tendencies work to shut down the Gulf Stream and its warming influence on Western Europe. Even stabilising CO_2 emissions to present levels (the maximal agenda of the 'great and the good' at ecofests) merely delays the onset of cold conditions on the eastern flank of the

North Atlantic for global warming will not stop through that device. That would mean keeping to present rates of energy use, and is not welcome news to those who seek to control and expand capital. But a regional cooling is not the only issue.

More water vapour moved to colder high latitudes means that more will fall as snow to remain unmelted at lower elevations. For most of Scandinavia large glaciers could reform if the summer snow line descended by a mere few hundred metres (existing ones have expanded since the 1980s). By reflecting away solar heat in the summer, their bright surface would drive climate to lower temperatures still. This is a scenario for the return of glaciation on a regional scale; a transformation of warming into frigidity, perhaps even the start of another ice age. Such is one scenario that emerges from detailed study of past climatic change. It does not sit well with the human cultures that have emerged during the Holocene by the turn to agriculture, industrialisation and the surplus food that these can produce. Perhaps we should ask ourselves 'Is it possible that the dominant way of thinking honed through the easy millennia of the Holocene will itself be humanity's downfall?' It is to be hoped that, having read this book, you may now seek a different outlook! Human survival, evolution and expansion through the 2.6 million years of the Great Ice Age seems very likely to have left its mark at the genetic level. Our ancestors had to think fast and to act differently according to a huge variety of circumstances. Although the Holocene has lulled us into a false sense of security, 400 generations since facing the Younger Dryas is too little time to have fatally blunted our true heritage. What is emerging about our distant past gives every cause for optimism.

SUMMARY

1 The human population increased only very slowly until about 10 thousand years ago when a slight increase in growth rate occurred. 10 ka marks the onset of the Holocene climate improvement as well as the start of human domestication of plants and animals.

2 Before 10 ka humans gathered plant foods and hunted for meat. Their effect on the environment was comparatively small, but there are a number of extinctions of large herbivores associated with the appearance of humans in a region. However, many Pleistocene extinctions occurred before humans appeared in an area and the extreme deterioration and oscillation of the climate as the ice ages developed is thought to have caused the extinction of Tertiary browsers as well as triggered the evolution of a large number of new species.

Evidence for human hunting causing extinctions

is clearest in America, where humans arrived relatively late (perhaps after 18 ka) and their arrival coincides with the loss of over 40 large mammals. Some bones have been found with stone arrowheads in them. It is suggestive that most of the large herbivores surviving in North America are late arrivals from Eurasia where their ancestors evolved in the presence of early hominids. Areas where hominids and other mammals evolved together show earlier and less severe phases of extinctions. Large islands where humans arrived in the last 2 thousand years show a similar devastation to the wildlife. In our seas modern hunters (fishermen) have become super-efficient by using sonar to detect shoals and are overfishing northern waters.

3 From 10 ka humans increasingly utilised their surroundings, coppicing woodland for hurdle material and to attract animals, gathering and sowing wild crops for food. Deforestation of upland and areas on poor sandy soil may have triggered blanket bog and heath development. Both deforestation and soil degradation have continued to be serious effects of human farming practices during the Holocene.

4 As agriculture developed, domestication of plants and animals resulted in many breeds which are unable to thrive without careful management. Increasing selection and understanding of breeding has resulted in humans changing the course of evolution for many breeding lines and creating completely new forms. Recent intensive farming has demanded uniform crops which are easy to harvest mechanically; such crops usually need applications of fertilisers and pesticides to thrive creating completely unnatural vegetation and destroying much of the remaining wildlife.

5 With improved travel, many species have been introduced to new areas. Most of these are domestic or otherwise useful species, but in an ecosystem where they have not evolved their effects have been devastating: overgrazing by rabbits, woodland damage by pigs, hunting by cats and dogs. Both plants and animals can get out of hand without their natural diseases and predators.

6 Another group of organisms which have multiplied because of agricultural practices are weeds provided with rich open soils and frequent disturbance. Many were rare survivors from the glacial tundra and grasslands which re-expanded with the deforestation.

7 The industrial revolution is often considered to have begun about AD 1750 but use of fossil fuels and extensive metal smelting were common in Roman times. Atmospheric pollution by sulphur dioxide and toxic metals have thus occurred for over 2 thousand years, but only in the last hundred years have emissions reached really high levels. Atmospheric pollution is global in extent and has the potential to seriously damage many ecosystems. Acid depositions leach the soil of minerals and mobilise heavy metals and aluminium to toxic levels, low altitude ozone interferes with photosynthesis and lowers vegetation productivity, heavy metals accumulate in food chains and can damage health and development.

FURTHER READING

Goudie, A. and Viles, H. 1997. *The Earth Transformed: An Introduction to Human Impacts on the Environment.* Blackwell, Oxford.

Roberts, N. 1998. *The Holocene: An Environmental History, 2nd edition.* Blackwell, Oxford. A review of environmental change over the past 10 thousand years, including human impacts.

FIGURE SOURCES

THE GREAT ICE AGE

1.2 Modified from Figure 5.2 of Skelton, P. , Spicer, R.A. and Rees, A., 1997, *Evolving Life and the Earth*, The Open University, Milton Keynes.

1.3 Modified from Figure 4.7 in Whittaker, R.H., 1975, *Communities and ecosystems* (2nd edition), Macmillan, New York.

1.4 From Figure 4.7 in Cox, C.B., and Moore, P. D., 1993, *Biogeography: an ecological and evolutionary approach* (5th edition), Blackwell Scientific Publications, Oxford.

1.5 (a) R.C.L. Wilson; (b) Courtesy of Cynthia Burek.

1.6 Figure 1.4 in Wilson, C. (ed.), 1994, *Earth heritage conservation*, The Open University, Milton Keynes.

1.7a Cambridge Air Photograph Library.

1.7b British Geological Survey.

1.8 Figure 4.30 in McLeish, A., 1986, *Geological Science*, Thomas Nelson and Sons Ltd.

1.9 *Trafalgar Square*, Holiday Geology Guide, Earthwise/British Geological Survey, 1996.

1.10c Modified from Figure 4.9 of Van Andel, T.H., 1994, *New views on an old planet*, Cambridge University Press, Cambridge.

1.11b From Walcott, R.I. 1972, 'Past sea levels, eustasy and deformation of the Earth'. *Journal of Quaternary Research*, 2, 1–14.

1.12 Compiled from a number of sources including:

- Active sand dunes today from Goudie, A., 1983, *Environmental Change*, 2nd edition, Clarendon Press, Oxford; by permission, Oxford University Press.

- Active sand dune 18 000 years ago from Sarnthein, M., 1978, *Nature*, 272, 43–6.

- Ice sheets today and 18 000 years ago from Figures 3.1 and 3.2 of Williams, M.A.J., *et al.*, 1993, *Quaternary Environments*, Arnold, London.

- August sea surface temperatures today from Figure 10.3 of Williams *et al.*; op cit.

- August sea surface temperatures 18 000 years ago from CLIMAP, 1976, 'The surface of the ice-age Earth', *Science*, 191, 1131–7.

- Rainforest distribution today from Figure 1.4, this book.

- Rainforest distribution 18 000 years ago from Hamilton, A.C., 1976, 'The significance of patterns of distribution shown by forest plants and animals in tropical Africa for the reconstruction of upper Pleistocene palaeoenvironments: a review'. *Palaeoecology of Africa*, 9 , 63–97

1.13 Foram abundances from Ruddiman, W.F., Sancetta, C.D., and McIntyre, A., 1977 in *Philosophical Transactions of the Royal Society of London*, Series B. Foram electron photomicrographs provided by Professor Geoffrey Boulton FRS.

1.14 Figure 2.8a in Anderson, B.G., and Borns, H.W., 1994, *The Ice Age World*, Scandinavian University Press.

1.15 From Figure 10–12 in McIntyre, A., *et al.*, 1981, 'Glacial North Atlantic 18,000 years ago: a CLIMAP reconstruction'. *Geological Society of America Memoir* 145, 43–76.

1.16 Compiled from Figures 20–3 in Ruddiman, W.F., and McIntyre, 1981, 'The North Atlantic Ocean during the last deglaciation'. *Palaeogeography, Palaeoclimatology, Palaeoecology*, 35, 145–214.

1.17 From Glennie, K., Pugh, J.M., and Goodall, T.M., 1994, 'Late quaternary Arabian desert models of Permian Rotliegend reservoirs'. *Shell Exploration Bulletin*, 274, 1–19. Reproduced by permission of Abu Dhabi Onshore Operating Co. (ADCO).

1.18 From Figure 4 of N. Petit-Maire, 1991, 'Recent Quaternary climatic change and Man in the Sahara'. *Journal of African Earth Sciences*, 12, 125–32.

1.19 From Figure 3.6 of Van Andel, T. H., 1994, *New views on an old planet*, Cambridge University Press.

1.20 From Figure 10 (by Nicole Petit-Maire) in Brown, G.M., 1994, 'Understanding Earth's environment'. *Nature and Resources* (UNESCO), 30, 21–33.

1.21 From p. 46 of Philip's *Great World Atlas*, 1991, George Philip Ltd, London; © George Philip Ltd, Cartography by Philips.

UNDERSTANDING PRESENT AND PAST CLIMATES

2.2 Combined from figures in: W.D. Sellers (1965) *Physical Climatology*, University of Chicago Press and Van der Haar and V. Suomi (1971), *Journal of atmospheric science*, 28, 305–14, American Meteorological Society.

2.3 From Figure 1 in Woods, J.D., 1985, 'The world ocean circulation experiment'. *Nature*, 314, 501–11.

2.4 and 2.5a Figure 2.26 in Colling, Angela (ed.), 1997, *The Dynamic Earth*, The Open University.

2.5b From Figure 1 (top panel) in Gill, A.E. and Rasmussen, E.M., 1983, 'The 1982–83 climate anomaly in the equatonal Pacific'. *Nature*, 306, 229–34.

2.6a and b from Figures 1 (top) and 2 respectively in Broecker, W.S., 1997, 'Thermohaline circulation, the Achilles heel of our climate system: will man-made CO_2 upset the current balance?' *Science*, 278, 1582–8.

2.8 From Hansen, J.E., *et al.* 'Climate sensitivity: analysis of feedback mechanisms'. In: *Climate processes and climate sensitivity* (J.E. Hansen and T. Takahashi (eds). Geophys. Mono., 29, American Geophysical Union, 130–63.

2.10a Data from Hadley Centre for Climate Prediction and Research (b) and (c): adapted from Houghton, J.T., Jenkins, G.J. and Ephraums, J.J.

(eds), 1990, *Climate change: the IPCC Scientific assessment*. Cambridge University Press, ©IPCC.

2.11 From Figures 1.1 and 1.2 in Aitken, M.J., 1990, *Science-based dating in archaeology*, Longman, London.

UNDERSTANDING THE CRYOSPHERE

3.1 Modified from Figure 1.1: Benn, D.I., and Evans, D.A.J., 1998. *Glaciers and Glaciation*. Edward Arnold, London.

3.2 United States Department of Interior.

3.3 Modified from Figure 4.8 in Sugden, D.E., and John, B.S., 1976, *Glaciers and landscapes*, Edward Arnold, London.

3.4 From Figure 21.11 in Duff P. McL. (ed.) 1994, *Principles of physical geology*, Edward Arnold, London. Reproduced by permission of Stanley Thornes Publishers Ltd.

3.5 From Ahlmann, H.W., 1948, Royal Geographical Society Research Series No.1 (Figure 22 in Open University, 1980, S333 Palaeoclimatology Case Study).

3.6 From Figure 5.5 in Sugden, D.E., and John, B.S., 1976, *Glaciers and landscape*, Edward Arnold.

3.7 (a), (b), (c) from Figure 20.20 and (d) from Figure 20.23 in *Principles of physical geology*, Edward Arnold, London. Reproduced by permission of Stanley Thornes Publishers Ltd.

3.8 From Figure 21.18 in *Principles of physical geology*, Edward Arnold, London. Reproduced by permission of Stanley Thornes Publishers Ltd.

3.9 From Figure 21.19 in *Principles of physical geology*, Edward Arnold, London. Reproduced by permission of Stanley Thornes Publishers Ltd.

3.10 Figure 4.4 of Van Andel, T.A., 1994, *New views on an old planet*, Cambridge University Press.

3.11 From Figure 2 in Hughes, T., 1992, 'Abrupt climate change related to unstable ice-sheet dynamics: toward a new paradigm'. *Palaeogeography, Palaeoclimatology, Palaeoecology (Global Planetary Change Section)*, 97, 203–34.

3.12 From Figure 4 in Hay, W.W., DeConto, R.M. and Wold, Ch.N., 1997, 'Climate: is the part the key to the future?' *Geologisches Rundschau*, 86, 471–91.

3.15 Modified from Figure 2 in Imbrie, J., *et al.*, 1984, in Berger, A., *et al.* (eds), *Milankovitch and climate*, Reidel, Dordrecht, 269–305; with kind permission from Kluwer Academic Publishers.

THE DEEP SEA RECORD

4.1 From Anderson, R.N. 1986, *Marine geology*, John Wiley and Sons and Figure 1.4 in Open University, S330: *Ocean Chemistry and deep sea sediments*.

4.3 A. McIntyre, Lamont-Doherty Geological Observatory.

4.4 Figure 70 in *Palaeoclimatology case study*, 1980, Open University.

4.6 From Figure 9 in Shackleton, N.H., and Opdyke, N.D., 1973, *Journal of Quaternary Research*, 3, Part 1.

4.7b Modified from Figure 4.1 in Van Andel, T.H., 1994, *New views on an old planet*, Cambridge University Press.

4.8 and 4.9 From Figures 3 and 4 in Shackleton, N.H., and Opdyke, N.D., 1973, *Journal of Quaternary Research*, 3, Part 1.

4.10 From Figure 2 in Saltzman, B., and Maasch, K.A., 1990, 'A first-order global model of late Cenozoic climatic change'. *Transactions of the Royal Society of Edinburgh: Earth Sciences*, 81, 315–25.

4.11 Courtesy of Dr David Dunkerley, School of Geography and Environmental Science, Monash University, Victoria, Australia.

4.12 From Figure 11b in Aharon, P., and Chappell, J., 1986, 'Oxygen isotopes sea level changes and the temperature history of a coral reef environment in New Guinea over the last 10^5 years'. *Palaeogeography, Palaeoclimatology, Palaeoecology*, 56, 337–9.

4.14 Courtesy of Nat Rutter and Liu Tungsheng.

4.15 Simplified from Figure 4 in Abreu, V.S., and J.B. Anderson, 1998, 'Glacial eustasy during the Cenozoic: sequence stratigraphic implications'. *American Association of Petroleum Geologists Bulletin*, 82, 1385–1400.

REVEALING THE MILANKOVICH PACEMAKER

5.1, 5.4 and 5.6 From Figures 3 (top and bottom) and 1 respectively in Imbrie, J., *et al.*, 1993 on the structure and origin of major glaciation cycles 2. The 100 000 year cycle. *Palaeoceanography*, 8, 699–735.

5.2 Figure 3.13 from Warr, K., and Smith, S., 1993, *Changing Climate*, Open University.

5.3 see source for Figure 3.15

5.5 and 5.15b Based on part of Figure 2 in Berger, A., and Pestiaux, P., 1984, 'Accuracy and frequency stability of the Quaternary terrestrial insolation'. In *Milankovitch and Climate* (eds: Berger, A. *et al.*), Reidel, Dordrecht, 83–111; by kind permission of Kluwer Academic Publishers.

5.7 From Figure 21.30 in Duff, P. McL. (ed.), *Principles of physical geology*, Edward Arnold, London. Reproduced by permission of Stanley Thornes Publishers Ltd

5.8 From Figure 3 in Winograd, I.J., *et al.*, 1992, 'Continuous 500 000–year climate record from vein calcite in Devil's Hole, Nevada'. *Science*, 258, 255–60.

5.9a From Figure 3.1 in Goudie, A., 1983, *Environmental Change*, 2nd Edition, Clarendon Press, Oxford.

5.9b From Figure 7.5 in M.A.J., Williams *et al.*, 1993, *Quaternary Environments*, Edward Arnold, London. Reproduced by permission of Stanley Thornes Publishers Ltd

5.10 Courtesy of Nat Rutter and Lui Tungsheng.

5.11a Part of Figure 2 from: Yan, Z., and Petit-Maire, N., 1994, 'The last 140 ka in the Afro-Asian arid/semi-arid transitional zone'. *Palaeogeography, Palaeoclimatology, Palaeoecology*, 110, 217–33.

5.11b From Figure 1 (top) in Petit-Maire, N., 1994, 'Natural variability of the Asian, Indian and African Monsoons over the last 130 ka'. NATO ASI Series, 126, 3–26, Springer-Verlag, Berlin, Heidelberg.

5.11c From Figure 1c in Petit-Maire, N., Fontugne, M., and Rowland, C., 1991, 'Atmospheric methane ratio and environmental changes in the Sahara and Sahel during the last 130 kyrs'. *Palaeogeography, Palaeoclimatology, Palaeoecology*, 86, 197–204.

5.11d and e From Figure 2 E and A respectively in

Prell, W.L., and Kutzbach, J.E., 1987, 'Monsoon variability over the past 150 000 years'. *Journal of Geophysical Research*, 92, D7, 8411–25.

5.13 Figure 2 from Jouzel, J., Lorius, C., Johnsen, S., and Grootes, P. , 1994, 'Climate instabilities: Greenland and Antarctic Records'. *Compte Rondue Academie des Sciences, Paris*, Serie II, 65–77.

5.14 From Figure 1 in Petit, J.R., *et al.*, 1997, 'The ice-core record: climate sensitivity and future greenhouse warming'. *Nature*, 347, 139–45.

5.15a and c From Figure 2e and d respectively in Petit, J.R., Mounier, L., Jouzel, J., Korotkevich, Y.S., Kotlyatior, V.I. and Lorius, C., 1990, 'Palaeoclimatological and chronological implications of the Vostok core dust record'. *Nature*. 343, 56–8.

5.16a, b and d: Figures 2a, 2b and 1a respectively in source for Figure 5.15a and c. c and e from Figures 3f and b respectively in Jouzel, J., *et al.*, 1993, 'Extending the Vostok ice-core record of palaeoclimate to the penultimate glacial period'. *Nature*. 364, 407–12.

5.17 From Figure 1 in Lorius, C., Jouzel, J. *et al.*, 1990. 'The ice-core record: climate sensitivity and future greenhouse warming'. *Nature*, 347, 139–45.

5.18 From Figure 3.19 in Colling, A. (ed.), 1997, *The Dynamic Earth*. The Open University, Milton Keynes.

5.20 From Figure 3 Petit-Maire, N., 1994, 'Natural variability of the Asian, India and African Monsoons over the last 130 ka'. NATO ASI Series, 126, 3–26, Springer-Verlag, Berlin, Heidelberg.

EVIDENCE FOR RAPID CLIMATE CHANGE

6.1a Data from Hadley Centre for Climate Prediction and Research (b) and (c): adapted from Houghton, J.T., Jenkins, G.J. and Ephraums, J.J. (eds), 1990, *Climate change: the IPCC Scientific assessment*, Cambridge University Press, ©IPCC.

6.1b-j from Figure 21.8 in Duff, P. McL. (ed.), 1994, *Principles of physical geology*, Edward Arnold, London.

6.2 Devised by Mark Maslin.

6.3 (a) Modified from Figure 3 in Mooers, H.D., and Lehr, J.D., 1997, 'Terrestrial record of Laurentide ice sheet reorganisation during Heinrich events'. *Geology*, 25, 987–90. (b) Modified from Figure 3 in Lowell, T.V., *et al.*, 1995, 'Interhemispheric correlation of Late Pleistocene glacial events'. *Science*, 269, 1541–9.

6.4 From right side of Figure 3 in Bond., G.C., and Lotti, R., 1995, 'Iceberg discharges into the North Atlantic on millenial timescales during the last glaciation'. *Science*, 267, 1005–10.

6.6a From Figure 5 in Hay, W.W., DeConto, R.M. and Wold, Ch.N., 1997, 'Climate: is the past the key to the future?' *Geogisches Rundschau*, 86, 471–91.

6.6b From Figure 3 in Fairbanks, R.G., 1989, 'A 17 000 year glacio-eustatic sea-level record: influence of glacial melting rates on the Younger Dryas event and deep-sea circulation'. *Nature*, 342, 637–42.

6.7a From part of Figure 3 in Bond, G.,*et al.*, 1993, 'Correlations between climate records from North Atlantic Sediments and Greenland Ice'.*Nature*,365,143–7.

6.7b From Figure 1 in Alley, R.B., 1998, 'Icing in the North Atlantic'. *Nature*, 392, 335–6.

6.8 From R.Z., 1994, 'Linking ice-core records to ocean circulation'. *Nature*, 371, 289.

6.9 From Figure 2 in Brook, E.J., Sowers, T., and Orchad, J., 1996, 'Rapid variations in atmospheric methane concentration during the past 110 000 years'. *Science*, 273, 1087–91.

6.10 From Figures 1 and 2 in Blunier, T., *et al.*, 1997, 'Timing of the Antarctic cold reversal and the atmospheric CO_2 increase with respect to the Younger Dryas event'. *Geophysical Research Letters*, 24, 2683–6.

6.11 From Figure 3 in Sowers, T., and Bender, M., 1995, 'Climate records covering the last glaciation'. *Science*. 269, 210–14.

6.12 From Figure 1 in Greenland Ice-Core Project (GRIP) Members, 1993, 'Climate instability during the last interglacial period recorded in the GRIP ice-core'. *Nature*, 364, 203–7.

6.13 From Figure 2 (bottom two panels) in Wolf., E.W., *et al.*, 1997, 'Climatic implications of background acidity and other chemistry derived from electrical studies of the Greenland Ice Core Project core'. *Journal of Geophysical Research*, 102, No. C12, 26, 325–6, 332.

6.13 From Figures 1a, 4b, 4c and 4a in Taylor, K.C., *et al.*, 1993, 'The "flickering switch" of late Pleistocene climate change'. *Nature*, 361, 432–6.

6.14 From Figure 3 in Portier, C., and Zhisheng, An., 1995, 'Correlation between climate events in the North Atlantic and China during the last glaciation'. *Nature*, 375, 305–8.

6.15a and b From Figures 1 and 13 respectively in Behl, R.J., 1995, 'Sedimentary facies and sedimentology of the Late Quaternary Santa Barbara Basin, Site 893'. In Kennett, J.P. , Baldauf, J.G., and Lyle, M. (eds), 1995, *Proceedings of the Ocean Drilling Program*, Scientific Results, 146 (pt 2).

6.15c and d From Figures 1 and 2 in Behl, R.J., and Kennett, J.P., 1996, 'Brief interstadial events in the Santa Barbara Basin, NE Pacific, during the past 60 kyr'. *Nature*, 379, 243–6.

6.16(b) from Figure 64 in Bell, A., 1995. Energy I: Fossil Fuels. S268 Course Book, Open University. (c) from Woodcock, N. H., 1994, *Geology and Environment in Britain and Ireland*, UCL Press Ltd.

EXPLANATIONS

7.1 From Figures 1 and 2: Rea, D.K., and Prueher, M., 1995, 'Volcanic activity and global change: Probable short term and possible long term linkages'. In Isaacs, C.M., and Thorp, V.L. (eds), *Proceedings of the Eleventh Annual Pacific Climate (PACLIM) Workshop, April 19–22, 1994. Inter-agency*. Ecological Program, Technical Report 40; California Department of Water Resources.

7.2 From Figures 1a, 2b, 7a and b in Wright, J.D., and Miller, K.G., 1996, 'Control of North Atlantic deep water circulation by the Greenland-Scotland Ridge'. *Palaeoceanography*, 11, 157–70.

7.3 From Covey, C., 1991, 'Credit the oceans?' *Nature*, 352, 196–7.

7.4 and Figure 7.6 From Figures 11.1 and 11.6 respectively in Van Andel, T.H., 1994, *New views on an old planet*, Cambridge University Press.

7.5 From Figure 19.7 in Eyles, N., 1993, 'Earth's glacial record and its tectonic setting'. *Earth Science Reviews*, 35, 1–248.

7.7 From Figure 6 in Rea, D.K., Snoeckx, H., and Joseph, L.H., 1988, 'Late cenozian eoloan deposition in the North Pacific: Asian drying, Tibetan uplift and

cooling of the Northern Hemisphere'. *Palaeoceanography*, 13, 215–24.

7.8 From Figures 3 (top panel) and 4 (top and bottom panels) in Maslin, M.A., *et al.*, 1998, 'The contribution of orbital forcing to the progressive intensification of Northern Hemisphere glaciation'. *Quaternary Science Reviews*, 17, 411–26.

7.9 From Figure 19 Broecker, W.S., and Denton, G.H., 1989, 'The role of ocean-atmosphere reorganisations in glacial cycles'. *Geochemica at cosmochemica Acta*, 53, 2465–501.

7.10 From Figure 3 in Clark, P.U., and Pollard, D., 1998, 'Origin of the middle Pleistocene transition by ice sheet erosion of regolith.' *Palaeoceanography*, 13, 1–9.

7.11 From Raymo, M., 1998, 'Glacial puzzles', *Science*, 281, 1467–8 and Raymo, M., 1997, 'The timing of major climate terminations', *Palaeoceanography*, 12, 577–85 (part of Figure 5).

7.12 From Figure 2a in Hughen, K.A., *et al.*, 1998, 'Deglacial changes in ocean circulation from an extended radiocarbon calibration'. *Nature*, 391, 65–8.

7.14 From Figure 2 in Broecker, W.S., 1998, 'The end of the present inter-glacial: how and when?' *Quaternary Science Reviews*, 17, 689–94.

CLIMATE CHANGE AND LIFE ON LAND

8.1 From Figure SG3 in Study Guide and Glossary for *The Holocene* (set book), Open University, 1997.

8.2 Modified from Figure 2.6 in Roberts, N., 1998, *The Holocene*, Blackwell Publishers Ltd, Oxford.

8.3 From Figure 7 in Bennett, K.D., 1983, 'Devensian Late-Glacial and Flandrian vegetational history at Hockham Mere, Norfolk, England, 1: pollen percentages and concentrations'. *New Phytologist*, 95, 457–87.

8.4 From Figure 3 in Birks, H.H. and Matthews, R.W., 1978. 'Studies in the late vegetational history of Scotland, V'. *New Phytologist*, 80, 455–84.

8.5 From Figure 4.5 in Moore, P.D., Challoner, B., and Stott, P., 1996, *Global environmental change*, Blackwell Science, Oxford.

8.6a Amended and simplified from Figure 1 in Woillard, G.M., and Mook, W.G., 1982, 'Carbon-14

dates at Grande Pile: correlation of land and sea chronologies'. *Science*, 215, 159–61.

8.6b From Figure 4 in Grimm, E.C., *et al.*, 1993, 'A 50 000 year record of climate oscillations from Florida and its temporal correlation with the Heinrich events'. *Science*, 261, 198–200.

8.7 From Figure 2.16 in Goudie, A., 1992, *Environmental change*, 3rd Edition, Clarendon Press, Oxford; by permission, Oxford University Press.

8.8 Courtesy of Derek Ratcliffe and *Plantlife*.

8.9 Simplified from Figure 3 in Atkinson, T.C., Briffa, K.R. and Cooper, G.R., 1987, 'Seasonal temperatures in Britain during the past 22 000 years reconstructed using beetle remains'. *Nature*, 325, 587–92.

8.10 From Figure 19.2 in Chapman, J.L. and Reiss, M.J., 1999, *Ecology: principles and applications* (2nd edition), Cambridge University Press.

8.11 Adapted from Figure 7.4, p. 182 in Brown Junior, K.S., 1987, 'Conclusions, synthesis and alternative hypotheses'. In Whitmore, T.C. and Prance, G.T. (eds) 1987, *Biogeography and Quaternary history in tropical America*, Oxford Scientific.

8.12 From Figure 2 in Hooghiemstra, H., and Ran, E.T.H., 1994, 'Late and middle Pleistocene climatic change and forest development in Colombia: pollen record Funza II (2–158 in core interval)'. *Palaeogeography, Palaeoclimatology, Palaeoecology*, 109, 211–46.

8.13 From Figure 2 in Hooghiemstra, H., and Melice, J.L., 1994, 'Pleistocene evolution of orbital periodicities in the high resolution pollen record Funza I, Eastern Cordillera, Colombia'. *Special Publication International Association of Sedimentologists*. 19, 117–26.

8.14 From Figure b in Colinveaux, P., 1979, 'The ice-age Amazon'. *Nature*, 399–400.

8.15 Adapted from Figure 93 p. 185 in Frenzel, B., 1973, *Climate fluctuations in the Ice Age*, The Press of Case Western Reserve University, Cleveland and London.

8.16 From Figure 7 in Van Andel, T.H., and Tz. Edakis, P.C., 1996, 'Palaeolithic landscapes of Europe and Environments, 150 000–25 000 years ago: an overview'. *Quaternary Science Reviews*, 15, 481–500.

8.17 Modified from Figure 10A.5, p. 582 in Löve, A. and Löve, D., 1974, 'Origin and evolution of the arctic and alpine floras'. Ch. 10, in Ives, J.D. and Barry, R.G. (eds), *Arctic and alpine environments*, Methuen and Co Ltd, London, 571–603.

THE RECORD OF HUMANITY

9.1 and 9.2a Adapted from figures on p. 241, p. 343 and 351 in Jones, S., Martin, R.D., and Pilbeam, D.R., 1992, *The Cambridge Encyclopaedia of Human Evolution*, Cambridge University Press, Cambridge.

9.2b Adapted from Figure on p. 36, Kingdon, J., 1993, *Self-made Man and His Undoing*, Simon and Schuster, London.

9.4 From figure on p. 345 in Pinker, S., 1995, *The Language Instinct*, Penguin, London.

9.5 From figure on p. 320 in Jones, S., Martin, R.D., and Pilbeam, D.R., 1992, *The Cambridge Encyclopaedia of Human Evolution*, Cambridge University Press, Cambridge.

9.6 From figure on p. 29 in Kingdon, J., 1993, *Self-made Man and His Undoing*, Simon and Schuster, London.

9.7a, b From figure on p. 237 in Jones, S., Martin, R.D., and Pilbeam, D.R., 1992, *The Cambridge Encyclopaedia of Human Evolution*, Cambridge University Press, Cambridge, (c) From figure on p. 30 in Kingdon, J., 1993, *Self-made Man and His Undoing*, Simon and Schuster, London..

9.8a From figure on p. 325 in Jones, S., Martin, R.D., and Pilbeam, D.R., 1992, *The Cambridge Encyclopaedia of Human Evolution*, Cambridge University Press, Cambridge. (b) From figure on p. 28 in Kingdon, J., 1993, *Self-made Man and His Undoing*, Simon and Schuster, London.

9.9 From Lewin, R., 1995, 'Bones of contention'. *New Scientist*, 4 November, 14–15.

9.10a From figure on p. 241 in Jones, S., Martin, R.D., and Pilbeam, D.R., 1992, *The Cambridge Encyclopaedia of Human Evolution*, Cambridge University Press, Cambridge. (b) From figure on p. 43 in Kingdon, J., 1993, *Self-made Man and His Undoing*, Simon and Schuster, London (c) From Krantz, G., 1973, 'Cranial hair and brow ridges'. *Mankind*, v9, 109–11.

9.11 From figure on p. 343 in Jones, S., Martin,

R.D., and Pilbeam, D.R., 1992, *The Cambridge Encyclopaedia of Human Evolution*, Cambridge University Press, Cambridge.

9.12a and b From figure on p. 44 in Stringer, C., and McKie, R., 1996, *African Exodus: The Origins of Modern Humanity*, Jonathan Cape, London.

9.14a and b From figure on p. 204 in Leakey, R.E., and Lewin, R., 1992, *Origins Reconsidered: In Search of What Makes Us Human*, Little, Brown and Co., London. (c) From figure on p. 44 in Kingdon, J., 1993, *Self-made Man and His Undoing*, Simon and Schuster, London.

9.15a From *New Scientist*, 13 July 1996, p. 32; b From figure on p. 5 in Kingdon, J., 1993, *Self-made Man and His Undoing*, Simon and Schuster, London.

9.16 From figure on p. 23 in McKie, R., 1992, 'Genes rock the fossil record'. *Geographical Magazine*, (November), p. 2–25.

CHANGING ENVIRONMENTS AND CHANGING HABITS

10.2 From p. 83 in Stringer, C., and McKie, R., 1996, *African Exodus: The Origins of Modern Humanity*, Jonathan Cape, London.

10.3 From figure on p. 169 in Stringer, C., and McKie, R., 1996, *African Exodus: The Origins of Modern Humanity*, Jonathan Cape, London.

10.5 From figure on p. 94 in Kates, R.W., 1994, 'Sustaining life on Earth'. *Scientific American*, 271/10, 92–9.

10.6 From Figure 10.2A in Gingerich, P.D., 1984, 'Pleistocene extinctions in the context of origination-extinction equilibria in Cenozoic mammals'. In Martin, P.S., and Klein, R.G. (eds) *Quaternary Extinctions: a prehistoric revolution*, The University of Arizona Press, Tucson.

10.7 From Figure 3.7 in Roberts, N., 1998, *The Holocene*, Blackwell Publishers Ltd, Oxford.

10.8 From figure on p. 161 in Stringer, C., and McKie, R., 1996, *African Exodus: The Origins of Modern Humanity*, Jonathan Cape, London.

10.9 From Figure 6 in deMenocal, P. B., 1995, 'Plio-Pleistocene African Climate'. *Science*, 270, 53–9.

HUMAN IMPACT DURING THE HOLOCENE

11.1 Data from Zohary, D. and Hopf, M., 1993, *Domestication of Plants in the Old World* (2nd edition), Oxford Scientific; and figure 4.1 in Simmons, I.G., 1989, *Changing the Face of the Earth: culture, environment, history*, Basil Blackwell, Oxford.

11.2 Modified from Table 4, p. 29 in Zohany, D. and Hopf, M., 1993, *Domestication of plants in the old world* (2nd edition), Clarendon Press.

11.3 From Figures 6.7 and 8.7 in Davis, S.J.M., 1987, *The archaeology of animals*, BT Batsford Ltd.

11.4a and b From Plates 19 and XI respectively in Coles, B., and Coles, J., 1986, *Sweet Track to Glastonbury*, Thames and Hudson, London.

11.5 Steve Drury.

11.6a and b From Figure 2 and Table 1 respectively in Hong, S. Candelone, J.P., Patterson, C.C., Boutron, C.F. 1996, 'History of ancient copper smelting pollution during Roman and Medieval times recorded in Greenland Ice'. *Science*, 272, 246–9.

INDEX

Note: Page numbers in *italics* refer to figures.